The Animal Kingdom

Volume 6:
The Class Aves 1

Georges Cuvier
Edited and translated by
Edward Griffith

CAMBRIDGE UNIVERSITY PRESS

Cambridge, New York, Melbourne, Madrid, Cape Town,
Singapore, São Paolo, Delhi, Mexico City

Published in the United States of America by Cambridge University Press, New York

www.cambridge.org
Information on this title: www.cambridge.org/9781108049597

This edition first published 1829
This digitally printed version 2012

ISBN 978-1-108-04959-7 Paperback

CAMBRIDGE LIBRARY COLLECTION

Books of enduring scholarly value

Life Sciences

Until the nineteenth century, the various subjects now known as the life sciences were regarded either as arcane studies which had little impact on ordinary daily life, or as a genteel hobby for the leisured classes. The increasing academic rigour and systematisation brought to the study of botany, zoology and other disciplines, and their adoption in university curricula, are reflected in the books reissued in this series.

The Animal Kingdom

Georges Cuvier (1769–1832), made a peer of France in 1819 in recognition of his work, was perhaps the most important European scientist of his day. His most famous work, Le Règne Animal, was published in French in 1817; Edward Griffith (1790–1858), a solicitor and amateur naturalist, embarked in 1824, with a team of colleagues, on an English version which resulted in this illustrated sixteen-volume edition with additional material, published between 1827 and 1835. Cuvier was the first biologist to compare the anatomy of fossil animals with living species, and he named the now familiar 'mastodon' and 'megatherium'. However, his studies convinced him that the evolutionary theories of Lamarck and St Hilaire were wrong, and his influence on the scientific world was such that the possibility of evolution was widely discounted by many scholars both before and after Darwin. Volume 6 is the first of three books on birds.

Cambridge University Press has long been a pioneer in the reissuing of out-of-print titles from its own backlist, producing digital reprints of books that are still sought after by scholars and students but could not be reprinted economically using traditional technology. The Cambridge Library Collection extends this activity to a wider range of books which are still of importance to researchers and professionals, either for the source material they contain, or as landmarks in the history of their academic discipline.

Drawing from the world-renowned collections in the Cambridge University Library and other partner libraries, and guided by the advice of experts in each subject area, Cambridge University Press is using state-of-the-art scanning machines in its own Printing House to capture the content of each book selected for inclusion. The files are processed to give a consistently clear, crisp image, and the books finished to the high quality standard for which the Press is recognised around the world. The latest print-on-demand technology ensures that the books will remain available indefinitely, and that orders for single or multiple copies can quickly be supplied.

The Cambridge Library Collection brings back to life books of enduring scholarly value (including out-of-copyright works originally issued by other publishers) across a wide range of disciplines in the humanities and social sciences and in science and technology.

THE

ANIMAL KINGDOM

ARRANGED IN CONFORMITY WITH ITS ORGANIZATION,

BY THE BARON CUVIER,

MEMBER OF THE INSTITUTE OF FRANCE, &c. &c. &c.

WITH

ADDITIONAL DESCRIPTIONS

OF

ALL THE SPECIES HITHERTO NAMED, AND OF MANY NOT BEFORE NOTICED,

BY

EDWARD GRIFFITH, F.L.S., A.S.,

CORRESPONDING MEMBER OF THE ACADEMY OF NATURAL SCIENCES OF PHILADELPHIA, &c.

AND OTHERS.

VOLUME THE SIXTH.

LONDON:

PRINTED FOR WHITTAKER, TREACHER, AND CO.

AVE-MARIA-LANE.

MDCCCXXIX.

LONDON:
SHACKELL AND BAYLIS, JOHNSON'S-COURT, FLEET-STREET.

THE

CLASS AVES

ARRANGED BY THE

BARON CUVIER,

WITH

SPECIFIC DESCRIPTIONS

BY

EDWARD GRIFFITH, F.L.S., A.S., &c.

AND

EDWARD PIDGEON, Esq.

THE ADDITIONAL SPECIES INSERTED IN THE TEXT OF CUVIER

BY

JOHN EDWARD GRAY, Esq., F.G.S., &c.

VOLUME THE FIRST.

LONDON:

PRINTED FOR WHITTAKER, TREACHER, AND CO.

AVE-MARIA-LANE.

MDCCCXXIX.

LONDON:
SHACKELL AND BAYLIS, JOHNSON'S-COURT, FLEET-STREET.

LIST OF PLATES IN THE SIXTH VOLUME.

LIST OF PLATES IN THE SIXTH VOLUME.

Errata in the Plates.

General Character of Birds, Order Accipitres, for *Percnoptera*, read *Percnopterus.*

Wedged Tailed Eagle, for *fuscosa*, read *fuscosus.*

General Character of Birds, Order Passeres, Plate 1, for *Ramphocelinus*, read *Ramphocelina.*

Page 104, wood-cut of bird's wing, insert the word "primary."

On plate of structure of feathers, insert :—

Fig. 1. Portions of the shaft with the laminæ separated.

2. One of the laminæ much magnified.

3. A pair of the bristles greatly magnified.

Errata in Vol. VI.

Page	*Line*				
19	14	*for*	califorianus	*read*	californianus
21	20	..	laracaræ	..	caracaræ
24	11	..	feather	..	feathers
29	24	..	Holland	..	New Holland
30	26	..	Colvy	..	Cohy
46	10	..	vents	..	vent
59	9	..	fucosa	..	fuscosus
59	ib.	..	R.N.	..	R. A.
75	1	..	a more	..	is of greater
75	7	..	; a brown	..	or brown
77	3	..	grisseata	..	griseata
83	13	..	griscata	..	griseata
269	14 and 18	..	Lessron	..	Lesson
278	21	..	twenty-three	..	three
279	13	..	caudautus	..	caudacutus
279	22	..	alaspi	..	alapi
280	12	..	torque	..	breast
286	4 and 6	..	Prinops	..	Prionops
287		..	Aratamus	..	Artamus
290	21	..	Sitria	..	Tityra
291	5	..	Psarius	..	Psaris
293	10	..	Javanesis	..	Javanensis
296	20	..	Melanoti	..	Melaconoti
306	8	..	blanea	..	blanca
312	18	..	grapular	..	scapular
318	5	..	arremono	..	arremon
318	10	.	horny	..	hoary
322	8	..	; are unknown		
326	16	..	Rahia	..	Bahia
333	15	..	crisata	..	cristata
336	14	..	gray	..	Gray
341	14	..	Cochinsiensis	..	Cochinsinensis
347	6 and 7	..	*tail feathers long even sharp pointed*	..	tail feathers long even sharp pointed
358	18	..	insects, which are called *Peauhace*	..	insects, are called *Piauhau*
358	29	..	O. A.	..	O. d'Ind et d'Am.
361	11	..	Astro	..	Atro
368	1	..	flavigastor	..	flavigastra
371	4	..	lively	..	mobile
378	7	..	macronnus	..	macrourus
383	last	..	praursus	..	prasinus
395	7	..	vest-dere	..	vert-doré
396	17	..	Regilus	..	Fregilus
396	14	..	crinelus	..	crinitus
396	18	..	garrulus	..	graculus
396	28	..	head	..	lead
410	8	..	memloides	..	meruloïdes
414	9	..	æridotheres	..	acridotheres
416	20	..	turf yellow	..	tuft yellow
417	25	..	blue and green; varied nape;	..	blue and green varied; nape

Errata in Vol. VI.

Page	*Line*						
421	21	*for*	coracisa	..	..	*read*	coracia
425	21	..	Delophus	..	..	..	Dilophus
428	20	..	aurocapilla	..	..	..	atrocapilla
430	5	..	turfs ..	..	..	..	tufts
431	21	..	Pardanalotus	..	..	..	Pardalotus
435	13	..	whenchat	..	..	..	whinchat
440	12 and 16	..	scialis ..	..	..	..	sialis
448	18	..	Elgithina	..	..	..	Ægithina
449	1	..	Œrithina	..	..	..	Ægithina
460	4	..	Gm. ..	..	..	..	Swainson
464	25	..	Niniotello	..	..	..	Mniotilta
475	19	..	Bay ..	..	..	..	Ray
478	26	..	Trisch ..	..	..	..	Frisch
529	27	..	Cinclosonia	..	..	..	Cinclosoma

THE

ANIMAL KINGDOM.

CLASS, AVES *.

THE OVIPAROUS VERTEBRATED ANIMALS IN GENERAL.

ALTHOUGH the three classes of the vertebrated oviparous animals differ considerably from each other by the quantity of respiration, and by every thing relating thereto, as the power of movement and the energy of the senses; they display, nevertheless, many characters in common, when placed in opposition to the vertebrated vivipara.

The hemispheres of their brain are narrow, nor are they united by a *corpus callosum.* The tubercles *nates* are greatly developed, are penetrated by one ventricle, and not covered by the hemispheres, but visible below, or at the sides of the brain; the

* The reader will observe that in the present division of our work, instead of forming a tabular synopsis for the additional species, we have subjoined them, in the proper places, to the text of the Baron, and printed them, and any observations of our own, in a small type and inner margin.

crura cerebelli do not form that protuberance called the *pons Varolii;* their nostrils are less complicated; their ears have by no means so many little bones, and in many species have, indeed, none; the *cochlea,* when it exists, is much more simple, &c. The lower jaw, composed always of several pieces, is attached by a concave facet to a saliant portion of the temporal bone, but which is separated from the petrous portion. The bones of the cranium are more subdivided, or continue so longer, although they occupy the same relative places, and fulfil the same functions; thus the frontal has five or six pieces, &c. The orbits are separated from the sphenoïd only by a laminous bone. When these animals have anterior extremities besides the clavicle, which is often united with that of the opposite side, and takes the name of the *os furcatum,* or merry-thought, the omoplate is supported moreover on the sternum by a very long and large coracoïd apophysis. The larynx is more simple, and is without an epiglottis; the lungs are not separated from the abdomen by a complete diaphragm, &c. But, to speak of all these points, it would be necessary to enter into more anatomical details than can here be afforded. Suffice it, therefore, to have remarked the general analogy of the ovipara among themselves, greater, with reference to the plan on which they are constructed, than that of any of them with the mammalia.

Oviparous generation consists essentially in this,—that the young is not fixed by a placentum to the uterus, or oviduct, but remains separated by the most exterior of these envelopes. Its nourishment is prepared beforehand, and inclosed in a sac attached to its intestinal canal: it is this which is called the *vitellus*, or yolk of the egg, in which the young, at first imperceptible, is inserted and nourished, and augments by absorbing the liquor of the yolk. Oviparous animals which respire by lungs, have, moreover, in the egg, a membrane plentifully supplied with vessels, which seem to serve the purposes of respiration; they are attached to the bladder, and represent the allantoïs of the mammalia. It is not found in the fish, nor in the batracian reptiles, which, when young, respire like the fish, by gills.

Many of the cold-blooded oviparous animals bring forth their young developed, and without the shell or other membrane which separates them from the mother; these are called false vivipara.

THE SECOND CLASS OF VERTEBRATED ANIMALS.

THE BIRDS

Are oviparous, vertebrated animals, with double circulation and respiration, organized for flight. Their lungs are not divided, but fixed to the ribs, and are enveloped by a membrane pierced by great holes, which permit the air to pass into many cavities in the chest, lower belly, arm-pits, and even the interior of the bones; so that the exterior fluid not only bathes the surface of the pulmonary vessels, but also the surfaces of an infinity of vessels of the rest of the body. Thus birds respire, in some respects, by branches of their aorta, as well as by those of the pulmonary artery, and the energy of their irritability is in proportion to their quantity of respiration*. All their body is arranged to participate in this energy.

Their anterior extremities, destined to sustain them in flight, could neither serve the purposes of standing or holding: hence they are biped, and take things from the ground with the mouth: thus their body is inclined before the legs, the thighs carry them forward, and the toes are elongated to form a sufficient base; the pelvis much extended in length, to furnish attachment to the muscles which support the trunk on the thighs. There is, moreover, a set of muscles proceeding from the pelvis to the toes, and passing over the knee and the heel, so that the weight alone of the bird closes the toes, and thus they are enabled to sleep perched on one foot. The ischia, and especially the ossa pubis, are elongated behind, and widen, to leave sufficient place for the development of the egg.

The neck and the beak are elongated, to reach the ground, and the former has pliability enough to be bent backward when at rest. It has therefore many vertebræ. On the other hand, the trunk which supports the wings has very little motion; the ster-

* Two sparrows consume as much pure air as a guinea-pig.—Lavoisier, *Mémoires de Chimie*, i. 119.

num, especially, to which are attached the muscles which lower the wings in flight, is very much extended, and has its surface increased, moreover, by a laminous projection in the middle. It is formed of five pieces; one central, of which the laminous projection makes a part; two anterior lateral, for the attachment of the ribs; and two posterior lateral, for the extension of its surface. The degree of ossification of these last in each species denotes the proportion of vigour for flight.

The furca produced by the union of the two clavicles, and the two vigorous supports formed by the coracoïd apophyses, widen the shoulders; the wing, sustained by the humerus, by the fore-arm, and by the hand, which is long, and has one digit, and the vestiges of two others, carries along its whole length a range of elastic quills, which greatly extend the surface which resists the air. The quills adhering to the hand are called *primary*, and there are always ten; those belonging to the fore-arm are called *secondary*, and their number varies; feathers less strong, attached to the humerus, are called *scapular*; the bone which represents the thumb has also certain quills called *bastards*.

The bony tail is very short, but it has also a range of strong feathers, which, by spreading, continue to support the bird: their number is commonly twelve, but there are sometimes fourteen; in the gallinaceous birds there are eighteen.

The feet have a femur, a tibia, and a peroneum articulated to the femur by a spring, whose extension is maintained without effort on the part of the muscles.

The tarsus and metatarsus are represented by a single bone, terminated at the bottom by three pulleys.

There are generally three toes in front, and a thumb behind; the latter is sometimes wanting; and in the martins is directed forward. In the climbers, on the contrary, the external toe and the thumb are directed backward. The number of articulations increases in each toe, commencing with the thumb, which has two, and finishing with the external toe, which has five.

Birds are in general covered with feathers, a sort of tegument the best adapted to protect them from the effects of the rapid variations of temperature to which their movements expose them. The air cavities which occupy the interior of their body, and which even occupy the place of marrow in the bones, augment their specific lightness. The sternal portion of the ribs, like the vertebral, is ossified, to give more force to the dilatation of the chest.

The eyes of birds are so disposed, as to enable them to distinguish objects both far and near equally well; and a vascular and folding membrane placed at the bottom of the globe, at the edge of the cristalline, assists probably in displacing that lens. The anterior surface of the globe is moreover strengthened by a circle of bony pieces; and besides the two ordinary eyelids, there is always a third placed at the internal angle, and which, by means of a remarkable muscular apparatus, is able to cover the front of the eye like a curtain. The cornea is very convex, but the cristalline is flat, and the vitreous humour small.

The ear of birds has but one little bone between the tympanum and the oval aperture. Their cochlea is a cone scarcely bent; but their semicircular canals are large, and lodged in a part of the skull, where they are surrounded on all sides with air cavities which communicate with the arca. Night-birds alone have a large external ear, which, nevertheless, is not so prominent as that of quadrupeds; this opening is generally covered with barbed feathers, more fringed than the others.

The organ of smell, hidden in the base of the beak, has commonly only three cartilaginous cornets, which vary as to their complication; it is very sensible, although it has no sinus dug into the skull. The size of the osseous openings of the nostrils governs the form of the beak; and the cartilages, membranes, feathers, and other teguments, which straiten these openings, have an influence on the strength of the smell, and on the sort of nourishment.

The tongue has little muscular substance, and is sustained by a production of the hyoïd bone: it has but little delicacy in the majority of birds.

The feathers, as well as quills, which differ from them only in size, are composed of a stem, hollow at the base, and of barbs, each having others much smaller; their tissue, their brightness, their strength, and general form, vary infinitely. Touch must be weak in all the parts capable of it; and as the beak is almost always corneous, and possessed of little sensibility; and the toes are covered with scales on the upper side, and with a callous skin underneath; this sense must be but little efficacious in birds.

The feathers fall sometimes twice a year. In some species, the winter plumage differs from that of summer. In general, the female differs from the male, by colours less bright, and the young of both sexes resemble the female. When the adult male and female are of the same colour, the young have a dress peculiar to themselves.

The brain of birds has the same character as that of other vertebrated oviparous animals; but is distinguished by a size in proportion very considerable, often exceeding that of the same organ in the mammalia. It is principally to tubercles, analogous to the *corpora striata*, that the volume is referrible, and not to the hemispheres, which are very narrow, and without circumvolutions. The cerebellum is large, almost destitute of lateral lobes; and almost entirely formed by the vermiform process.

The trachea of birds has its annulations entire; at its bifurcation is a glottis, generally furnished with peculiar muscles, and named the lower larynx: it is there that is formed the voice of birds; the enormous volume of air contained in the air-vessels contributes to the force of their voice, and the trachea, by its various form and movements, to the modification of the voice. The upper larynx, very simple, has but little to do with this.

The face, or upper beak of birds, formed principally by the intermaxillaries, is prolonged backwards into two arcades, the internal of which is composed of the palatine bones, and the external of the maxillaries and jugals, and which are supported on a moveable tympanic bone; and on the upper part, this same face is articulated, or united to the skull by elastic laminæ: this mode of union leaves them, at all times, some degree of mobility.

The horn, which invests the two mandibles, serves the place of teeth, and is sometimes prickled, so as to represent them. Its form, as well as that of the mandibles which sustain it, varies infinitely, according to the nature of the food which each species takes.

The digestion of birds is proportioned to the activity of their life, and the force of their respiration. The stomach is composed of three parts; the crop, which is a folding of the œsophagus; the succentorial ventricle, a membranous stomach, furnished in the thickness of its surface with a multitude of glands, the secretion of which imbibes the food; and finally, the gizzard, armed with two powerful muscles, which two radiated tendons unite, and lined within with a cartilaginous coating. The food is ground there the more easily, by the bird swallowing little stones to augment the force of the trituration.

In the majority of species which live only on flesh, or on fish, the muscles and the surface of the gizzard are reduced to an extreme weakness; and it has the appearance of making only a single bag with the succentorial ventricle.

The dilatation of the crop is also sometimes altogether wanting.

The liver turns the bile into the intestines by two conduits, which alternate with the two or three by which the pancreatic fluid passes. The pancreas of birds is large, but their spleen is small; they have no epiploon, the uses of which are in part supplied by the partitions of the air cavities. Two appendages are placed toward the origin of the rectum, and a short distance from the anus; these are more or less long according to the food of the species. The herons have them very short; other genera, as the pici, are without them altogether.

The cloaca is a bag in which the rectum, the ureters, and the spermatic canals, or, in the female, the oviductus, terminate; it is open externally by the anus. Properly speaking, birds do not urinate, but their urine is mixed with the solid excrement. The ostriches only have the cloaca sufficiently dilated to admit of any accumulation of urine.

In most of the genera, copulation is effected simply by the juxtaposition of the anus. The ostriches, and many of the web-footed birds, nevertheless have a penis, which has a sort of gutta, or furrow, by which the semen is conducted. The testicles are situate in the interior, above the kidney, and near the lungs; there is only an ovary and an oviductus.

The egg, detached from the ovary, where nothing is to be seen of it but the yellow, imbibes, at the top of the oviductus, that exterior liquid called the white, and is furnished with the shell at the bottom of the same canal. Here incubation developes the young, unless when the heat of the climate is sufficient, as is the case with the ostriches. The young has at the tip of the beak a horny point, which serves to break the egg, and which falls off a few days after birth.

Every one knows the varied industry employed by birds in constructing their nests, and the tender care they take of their eggs and of their young: this is the principal part of their instinct. For the rest of their intellectual qualities, their rapid passage through the different regions of the air, and the lively and continued action of this element upon them, enable them to anticipate the variations of the atmosphere in a manner of which we can have no idea, and from which has been attributed to them, from all antiquity, by superstition, the power of announcing future events. They are not without memory, or imagination, for they dream; and every one knows with what facility they may be tamed, may be made to perform different operations, and retain airs and words.

DIVISION OF THE CLASS AVES INTO ORDERS.

Of all the classes of animals, that of birds is the most strongly marked, and that in which the species have the greatest resemblance, and which is separated from all the others by a wider interval. This fact, however, renders it more difficult to subdivide them.

These subdivisions are grounded, as in the mammalia, on the organs of food, and of prehension, that is, the beak and toes.

One is struck first with the *palmated* feet, that is, when the toes are united by membranes, a character which distinguishes all the *swimming birds*. The position of these feet behind; the length of the sternum, the neck often longer than the legs, to reach downward, the plumage close, shining, impermeable to water, agree with the feet in constituting the web-footed fowls good swimmers.

In other birds, which also have frequently some small webs to the feet, at least between the external toes, we observe elevated tarsi, legs denuded of feathers toward the base, a tall stature; in one word, all arrangements necessary for fording in shallow water, for the purpose of seeking their food. Such, indeed, is the regimen of the greater number of these; and although some of them live on dry land, they are named *Waders*, or *Grallæ*.

Amongst the truly terrestrial birds, the *gallinacea* have, like our domestic cock, a heavy carriage, a short flight, the beak moderate, with the upper mandible vaulted, the nostrils swelling out, and partly covered by a soft scale, and almost always the edges of the toes indented with short membranes between the bases of those before. They live principally on grain.

The *birds of prey* have the beak crooked, with the point sharp, and bent toward the base; and the nostrils pierced in a membrane, which invests all the base of the beak; the feet are armed with strong nails. They live on flesh, and pursue other birds; hence they have generally a powerful flight. The greater number have moreover a small web between the external toes.

The *passerine birds* include many more species than all the other families; but their organization is so analogous that they cannot be separated, although they vary greatly in size and strength. Their two external toes are united at the base, and sometimes some way up their length.

Finally, I have named (*Grimpeurs*) *climbing birds*, such as have the external toe behind like the thumb, because the majority of them are formed for a vertical position, to climb up the trunks of trees*.

Each of these orders subdivides into families and genera, principally by the conformation of the beak.

* Since my first elementary table, I have thought proper to suppress the order picæ of Linnæus, as it has no determined character. M. Illiger has adopted this suppression.

THE FIRST ORDER OF BIRDS.

Birds of Prey, (Accipitres, Lin.)

Are known by their bent beak and crooked talons, very powerful arms, by means of which they pursue other birds, and even weak quadrupeds and reptiles. They are among the birds, what the carnivora are among the quadrupeds. The muscles of their thighs and legs indicate the strength of their talons; their tarsi are rarely elongated; they all have four toes; the thumb nail and that of the internal toe are the strongest.

They form two families, the diurnal and the nocturnal.

The diurnal birds of prey have the eyes directed sideways; a membrane called the *cera*, which covers the base of the beak, in which are pierced the nostrils; three toes before, one behind without feathers, the two external toes almost always united at their base by a short membrane; the plumage is close; the feathers are strong, and the flight powerful; their stomach is almost entirely membranous, their intestines are but little extended, their cæcum is very short, the sternum large, and completely ossified, in order to give to the muscles of the wing more extension; and their *furca* is semicircular and very wide, the better to resist the violent falls of the humerus requisite to a rapid flight.

Linnæus made only two genera, which are two natural divisions, that is, the Vultures and Falcons.

The VULTURES, (VULTUR, Lin.)

Have the eyes close to the head, the tarsi reticulated, that is, covered with small scales; the beak long, bent only at the end; and a part, more or less, of the head, or even the neck, denuded of feathers. The power of their talons does not correspond with their size, and they rather make use of the beak. Their wings are so long that they hold them half extended when they walk. They are cowardly birds, and live more commonly on carrion than on a living prey; after eating, their crop forms a large protuberance, under the *furca;* a fetid secretion runs from the nose, and they are reduced almost to a state of stupidity.

The Vultures, properly so called, have a large and strong beak, the nostrils crosswise on the base, the head and neck without feathers, and a collar of long feathers under the neck. They have been seen only in the old world.

The *Fulvous Vulture, (V. fulvus,* Gmel.) *V. trencalos,* Bechstein. *Le Percnoptère,* Buff. Enl. 426, and *Le Grand Vautour,* Id. Hist. des Ois.i. in 4to., pl. v. * *The Vulture,* Albin. iii. t. 1. *Le Chassefiente,* Vail. Afr. *The Indian Vulture,* Latham and Sonnerat.

Of a gray or brown colour, approaching to fawn colour; the down of the head and neck cinereous, the collar

* The history of the great Vulture is that of the following species, but the figure belongs to this.

white, sometimes mixed with brown; the quill-feathers and the tail brown, the beak and feet lead colour. This is the most extended species, and is found on all the mountains of the whole ancient world. The body equals, and even exceeds, that of the swan.

It forms the genus *Gyps* of Savigny, having fourteen tail feathers; is found in Europe, Asia, and North Africa.

Kolb's Vulture, *V. Kolbii*, Lath. Vaill. O. A. t. 10. Sonnerat. Ind. y. t. 105,

Differs from the former by the feathers of the neck being long; found in Africa, India, and Java.

The *Indian Vulture*, Lath. *V. Indicus*, Lath. *Vail.* O. A. t. 11. pl. col. t. 26,

Has been established by Temminck as a distinct species, peculiar to India.

The *V. Chincou*, Lath. Vail. O. A. t. 12. is perhaps the young. The feathers round the neck are short.

The *Cinereous Vulture*, (*V. cinereus*, and *V. monachus*, Gm. Enl. 425.) The *Crested Black Vulture*, Edw. 290. The *Chincou* of China, Vail. Afr. *Arrian* of La Pérouse. *Black Vulture*, *Ashy Vulture*, *&c.*

Of a blackish brown, the collar remounting obliquely towards the occiput, which has itself a tuft of feathers: the feet and membrane of the base of the beak are of a blueish violet. It is not less extended than the last, and is still larger. It frequently attacks living animals.

The *Sociable Vulture*, Lath. (*V. auricularis*, Daud.) Vail. Afr. t. 9. Probably the *Vulture of Pondicherry*, of Sonnerat. Daudin, Ann. du Mus. ii. pl. 20.

Blackish, with a longitudinal fleshy crest on each side of the neck under the ears. Of Africa and the East Indies.

The *Arabian Vulture*, Lath. (*V. monachus*, Lin. Edw. t. 290) Vail. O. A. t. 12. pl. col. t. 426);

Has been established as a distinct species from the *Brown Vulture* of Europe and India. *V. cinereus*, Lin., and *V. Arrianus*, Picot, pl. Enl. t. 425, of which the Bengal Vulture of Lath. t. 1, is the young, and the *V. niger*, *V. cristatus* of Brisson, are varieties.

The *Pondicherry Vulture*, *V. Pondicerianus*, Lath. from Sonn. Ind. t. 104. pl. col. t. 2.

Is now proved to be a distinct species: it is black, with a fleshy caruncle on the side of the head: is perhaps the *Chocolate Vulture* of Latham, found in India, Java, and Sumatra.

The *Angola Vulture*, Penn. (*Falco Angolensis*, Gm.) Tour in Wales, 1. t. 19.

White scapulars; orbits naked, reddish; quills and base of tail black. Angola. Size of a goose: in British Museum.

The *Chincou*, Vail. *Vultur Chincou*, Daud. Vail. O. A. t. 12.

Brown crown with a loose downy crest, head, cheeks, and throat, with a fine black down, neck with a ruff of slender feathers, bill bluish-white. *China*. *V. Gingianus*, Gmel.?

The *Egyptian Vulture, V. Ægyptius*, Savigny, Egypt, pl. col. t. 407. the adult from Egypt, not the *V. niger*, of Brisson. *V. Galericulatus*, Temm. *V. Monachus*, pl. col. t. 13.

Found in East and North Africa.

Here may be perhaps added the *Madagascar Falcon*, Lath. *Falco Madagascariensis*, Daud. Sonn. Voy. Ind. ii. t. 103. Pale gray; beneath white, crown white, larger wing-coverts black tipt, quills white, dusky, barred, and black tipt.

America produces Vultures remarkable by the caruncles which surmount the membrane at the base of the beak. This is as large as the last, but the nostrils are oval and longitudinal. These are the *Sarcoramphus* of Duméril,

Gypagus of Vieillot, and a part of the *Cathartes* of Illiger and Temminck.

The *King of the Vultures*, (*Vult. Papa*, Lin.) Enl. 428.

As big as a goose, blackish when young, afterwards varied with black and yellow, and with the mantle yellow, and the quills and collar black when old. The naked parts of the head and neck are bright, and the wattle is indented like the crest of a cock. It is found in the plains and other hot parts of South America.

The young is the *Painted Vulture* of Lath. *V. sacra* of Bartram.

The *Condor*, or *Great Vulture of the Andes*. (*Vult. Gryphus*, Lin.) Humb. Obs. Zool. pl. viii. and pl. col. t. 103.

Blackish, with a spot on the wing, and the collar white. The upper wattle, moreover, is large, and not indented. The male has one on the beak, like a cock; the female has none. When young, this bird is of a yellow-brown colour, and without collar. It is the species rendered famous by the exaggerated account given of its size; but M. Humboldt states it to be about as big as our Bearded Vulture, (*V. barbatus*,) to which the Condor is assimilated in manners. It inhabits the highest mountains of the chain of the Andes in South America.

Mr. Vigors has placed *V. Califorianus*, Shaw, in this genus; but we have observed a fine specimen recently imported, which is without any wattle.

The PERCNOPTERA, (*Cuvier.*) GYPAETOS, *Bechstein.* NEOPHRON, *Savigny.* CATHARTES, *Illiger.*

Have the beak thin, long, swelling beyond its crook, the nostrils oval, longitudinal, and the head only, but not the neck, denuded. They are moderate-sized birds, and not at all equal to the Vultures, properly so called, in strength; hence they are more addicted to carrion and all sorts of filth, which attracts them from far; they do not even disdain to feed on excrement.

The name of *Neophron* has been restricted to the species found in the old continent, which have the front of the head only naked.

The *Percnopterus of Egypt, (Vult. Percnopterus,* Linn.) *Vult. leucocephalus* et *Vult. fuscus,* Gmel.) Enl. 427 and 249. *Vult. de Gingi,* Sonn. et Daud. *Origourap,* Vail. Afr. *Rachamah,* Bruce. *Pharaoh's Bird* in Egypt, and *Gingi Vulture,* Lath. Hist. t. 5.

As big as a crow, the adult male white, with the quill-feathers black; the young and females brown. These birds are spread throughout the old continent, and are particularly common in hot countries, which they purify of dead carcases. They follow the caravans in large flocks, to devour everything that may die. The ancient Egyptians respected them for the services done to their country, and even now they are never injured in that country. There are, indeed, some devout Mussulmans who bequeath property for the support of a certain number of these birds.

Monk Percnopterus, (*Cathartes Monachus,* Temm. pl. col. t. 222.)

Blackish-brown, quills black. Africa. spec. in Brit. Mus. from Exeter-Change.

The American species has been set apart under the names of *Cathartes,* by Illiger, and *Catharista,* by Vieillot. They have the head entirely naked.

The *Carrion Vulture. (Vult. aura,* Lin.) Enl. 187.

As big as the last, (the Percnopterus of Egypt,) with the beak a little shorter, and the body entirely blackish; common in all the hot and temperate parts of America, where it renders the same service as its congener in the old world.

The *Black Vulture*, (*V. atratus*, Bartram. *V. urubu*, Vieil. Wilson, Amer. Orn. F. 75, f. 2.)

Iridescent, with black neck; more feathers above than below; wings shortish; tail slightly notched; nostrils linear. This species has been much confounded with the *V. aura*, Vieil. Amer. Orn. t. 2. *Cathartes aura*, Tem.; but it has the feathers of the neck square all round; the wings do not reach beyond the tail, which is rounded, and the nostrils are oval.

The *Californian Vulture*, (*V. Californianus*, Shaw. *V. Vulturinus*, Tem. Nat. Mis. x. t. 301. pl. col. t. 31.)

Blackish; three feet long. Feathers of the collar and breast lanceolate; wings extending beyond the tail. There is a specimen in the British Museum, and another in possession of Mr. Leadbeater, both which have no wattle.

The *Tawny Vulture* is a *Gypaetos*. The *Cheriway* and the *Plaintive Vulture* are *Laracaræ*; and the New Holland Vultures are referred by Dr. Latham to the gallinaceous birds, and are said to be probably Falcons by Mr. Vigors, (Lin. Trans. xii.) and by M. Temminck.

THE GRIFFINS, (GYPAETOS, *Storr*. PHENE, *Savigny*.)

Were arranged by Gmelin in the genus *Falco*, but are more nearly allied to the Vultures by their manners and make; like them they have the eyes even with the head, the cera comparatively weak; the wings half spread when at rest; the crop, when full, bulging at the bottom of the neck; but their head is entirely covered with feathers. Their generic characters consist in a very strong beak, straight, bent at the end, convex at the bend; in nostrils covered with stiff hairs, directed forward, and in a pencil of similar hairs on the beak. Their tarsi are very short, and feathered to the toes; their wings very long, and the third quill is the longest of all.

The *Bearded Vulture*, Lath., or *Vulture Eagle*. (*Vult. barbatus*, Lin.) (*Falco barbatus*, Gm.) Edw. 106. *Nisser*, Bruce. *Gypaëte* of the Alps, Daud. ii. pl. 10.

The largest of the birds of prey of the old world, of which it inhabits, but in small numbers, all the high chains of mountains. It builds in steep rocks; attacks lambs, goats, the chamois, and even, as it is said, man while sleeping; and it is pretended that it has carried off children. It does not, however, refuse dead flesh. About four feet long, and nine or ten feet (French) in expanse of wings. Its back is blackish, with a white line down the middle of each feather; the neck and upper part of the body is bright yellow; a black band surrounds the head. There are specimens with the neck and chest more or less brown, but these appear to be young,

When it is *F. niger* of Gmelin. It is found in India.

The *Golden Vulture* of Willoughby, *V. aureus*, Bris. *Falco magus*, Gm., of Persia,

May probably be distinct. Savigny has indicated a species under the name of *Phene gigantea*.

The *Vulturine Eagle,* Lath. (*Falco Vulturinus,* Shaw. Vail. O. A. t. 6.)

Is referred to this division by Temminck, but it is placed with the Fishing Eagle, by other ornithologists. The wings are black, and much longer than the tail; legs dirty yellow. Size of the Golden Eagle.

Tawny Vulture, Lath. (*Falco ambustus,* Gm.) Brown, Illust. Zool. t. i., from Falkland Islands; appears also to belong to this genus.

THE FALCONS, (FALCO, *Lin.*)

Form the second, and much the most numerous, division of the diurnal birds of prey. They have the head and neck covered with feathers; their eyebrows are so prominent as to give the eyes the appearance of being sunk in the head, and to the whole physiognomy a character very different from that of the vultures. The majority of them feed on living prey; but they differ greatly among themselves in the courage displayed in the pursuit of it. Their early plumage is often differently coloured from that of adult age, and they do not assume the adult dress until three or four years old, a circumstance which has induced an improper multiplication of the species. The female is in general one-third larger than the male, which is, hence, sometimes called the *tercel.*

This genus should be first divided into two large sections.

THE FALCONS PROPERLY SO CALLED, (FALCO, Bechstein) commonly called the *noble birds of prey,*

Form the first. They are, for their relative size, the most courageous of the whole; their offensive arms, and the power of their wings, are proportioned to their courage. Their beak, bending from its base, has a sharp tooth on each side, at the point. The second quill feather is the longest; but the first is nearly as long, rendering the entire wing longer and more pointed. From these premises result peculiar habits; the length of the quill feather weakens their efforts at vertical flight, and renders it in a still air very oblique forward, and obliges them, when they wish to rise directly, to fly against the wind. They are very tractable birds, and are the most used in falconry, being taught to pursue game, and to return when called. All of them have the wings as long and longer than the tail.

The *Common* or *Peregrine Falcon,* (*Falco Communis,* Gm.*)

As big as a fowl, is always known by a sort of triangular black spot on the cheek; for the rest it varies in colour nearly as follows: the young has the upper part brown, and the feathers edged with red-

* We must not admit the pretended variety of *F. communis*, collected by Gmelin: thus the var. α Frisch, 74, is a buzzard, δ idem, 75, is a rough-footed buzzard, ε id. 80. The bird of St. Martin, ϑ id. 76, is a buzzard rather paler than common; κ Aldrov., a distinct species, &c. The *F. Islandicus*, *barbarus* et *peregrinus*, may, indeed, be no other than the common hawk in different states of moulting.

dish; the under part whitish, with oval longitudinal brown spots. As they advance in age, the spots of the belly and thighs become transverse blackish lines, and the white increases at the throat and bottom of the neck; the plumage of the back becomes at the same time more uniform, and is radiated brown, with blackish ash stripes; the tail is brown above, with pairs of reddish spots; and underneath are pale bands, which diminish in size with age. The feet and the arc of the beak are sometimes blue, and sometimes yellowish.

Found also in New Holland.

These differences may be observed, Enl. 470, the young; the Yearling Falcon, *F. Herotinus,* Bris.; 421, the old female; 430, the old male. *Frisch* gives but one young Falcon, pl. lxxxiii. Edwards has the old female, pl. iii.; the young, pl. iv.

Those called in the Pl. Enl. *Faucons pelerins,* (*Falco stellaris, F. peregrinus,* Gmel.) seem to be young, rather blacker than usual.

This is the celebrated species which has given its name to falconry. It inhabits all the north of the globe, and builds in the steepest rocks. Its flight is so rapid, that there is scarcely any part of the world it does not visit. It pounces on its prey vertically, as if it fell from the clouds. The male is used against magpies and other small birds, and the female against pheasants, and even hares.

The *Barbary Falcon,* Lath. *F. Barbarus,* Gmel. Alb. iii. t. 2. is a variety of this.

Europe produces five species of inferior size, *viz.*:

The *Hobby*, (*Falco subbuteo*, Lin.) pl. Enl. 432.

Brown above; whitish, spotted with brown, underneath; the thighs and bottom of the belly red, a brown mark on the cheek.

The *Orange-legged Hobby*, (*F. rufipes*, Bescht, the female. *F. vespertinus*, Gm.) Enl. 431.

Brown above, deep ash underneath; thighs and bottom of the belly red. The female has the head red, and all the other part barred ashy and black.

The *Merlin*, or *Emérillon*. (*F. œsalon*, Lin.) Enl. 468.

Brown above, whitish underneath, spotted with brown, even to the thighs; the smallest of our birds of prey. The *F. lithofalco* of Linnæus, Enl. 447; ashy above; reddish-white, spotted with brown underneath, is the old male. It builds in rocks.

The *Kestrill*, or *Cresserelle*. (*F. tinnunculus*, Lin.) Enl. 401 and 471.

Red, spotted with black above; white, spotted with pale brown underneath; the head and tail of the male ashy. Takes its name from its sharp cry. Builds in old towers, &c.

The *Lesser Kestril*, Lath. (*F. tinnunculoïdes*, Natter, Storr degl. Ucc. i. t. 25. ♂.)

Wings to the end of the tail; back and quills of the male without any spots; claws pure white. Inhabits eastern and southern Europe. Eleven inches long.

Severe Falcon, Lath. *F. severus*, Horsf. *F. Aldrovandi*, Reinw. pl. col. t. 128.

Above, and the two middle tail feathers, blackish-blue; quills black, lower part spotted with red; beneath reddish; bill bluish; cera and feet yellow. Length ten inches. Inhabits Java.

Banded-throat Falcon. F. monogamelus, Tem. pl. col. t. 314.

Ashy throat; tips of secondaries, tail covers, and belly, white; central longitudinal band on throat, quills, and many cross bands on belly, black; tail black, with a white central band; cera and feet red Length 13—14 inches. Of Senegal.

Double-bearded Falcon. F. biarmicus, Temm. pl. col. t. 324.

Above, dark ash; inner web of quills white, spotted; tail, many narrow white bands. Beneath reddish-white, with longitudinal streaks; back of neck reddish, throat whitish; a band from back of eye and angle of bill black. Central Africa. Length fifteen inches.

Uniform Falcon. F. concolor, Temm. pl. col. t. 330.

Bluish-gray; shaft of feathers, quills, and bill, black; tail obscurely banded. Senegal. Length 13—14 inches, in B. Mus.

White-throated Falcon. F. deiroleucus, Temm. pl. col. t. 348.

Black, throat white; spots on side of neck, breast, and thighs, red-brown; belly yellow, with broad black bands. Tail with five or six interrupted bands. Brazil. Length eighteen inches.

Orange-breasted Hobby, Lath. *F. aurantius*, Lath. Spix. t.

Bill and feet lead coloured, body blackish; back and base of tail white, interrupted bands on chest fulvous, thigh ferruginous. Surinam. Length fifteen inches.

The *Chicquera Falcon*, Lath. *F. chicquera*, Shaw. Vail. O. A. t. 30.

Above bluish; top of the head and nape reddish; beneath white, banded with ash colour; end of the tail red, with a black band; feet and bill yellow. Inhabits Africa. Length ten inches.

The *Crested Indian Falcon*. (*F. frontalis*, Daud. Vail. O. A. t. 28.)

Crested; slate-colour crest; nape, patch under each jaw, quills, brown; belly dirty white, black banded; tail long, with seven or eight brown bands. Of India.

The *Black-thighed Falcon*. (*F. tibialis*, Daud. Vail. O. A. t. 29.)

Above gray-brown, centre of feathers dark; throat white; beneath pale rufous, with dark brown streaks on the thigh, black; quill and tail dark; legs yellow; bill lead coloured. Size of a pigeon. Inhabits the Cape of Good Hope.

Rufous-backed Kestril, Lath. (*F. rupicolus*, Dand. *F. rupicola*, Licht. *F. capensis*, Shaw. Vail. O. A. t. 35.)

Above reddish-brown, spotted with black; head reddish-brown; wings black, tail red; beneath ash-coloured, rayed with black; throat white, bill black; feet yellow. Length fifteen inches. Of Africa.

The *Spotted Falcon*. (*F. punctatus*, Cuv. pl. col. t. 45.)

Rufous; beneath white, spotted with black; back and neck longitudinally lined with black; head and wings spotted with black; tail even, with seven black bands. Isle of France. Length ten inches.

Red Femoral Hawk. (*F. femoralis*, Temm. pl. col. t. 121. t. 343. ♂.)

Cinereous brown, beneath red; band above and behind the eye, black; thighs red. Inhabits the Brazils. Length twelve inches when full grown. *F. aurantius*, var. Minor, Lath. Var. *Major*, Licht. Azara, n. 39. *F. thoracicus*, Illig. Length fourteen inches; bill and feet much stronger.

Nankin Hawk. *F. Cenchroides*, Vigors. Mus. Lin. Soc.

Above red, beneath white; quills and tail feathers edged with black; tail pale gray, with a broad black band and white tips. Found in New Holland. Length twelve inches.

Orange-speckled Hawk. *F. Berigora*, Vigors.

Reddish-brown; throat and neck pale orange; quills and coverts brown, speckled with red; tail gray-brown, banded, with rufous tips. Found in Holland. Length ten inches.

The *Lanner*, Lath. *F. laniarius*. Lath. *F. stellaris*, Gmel.

Wings two-thirds as long as the tail; middle toe shorter than the tarsus; mustaches very narrow; feet bluish; two first quills notched at the end. Inhabits north of Europe. Length one foot and a half.

In some species of Falcons the tarsi are shielded, and not reticulated, and the wings are short. *Tinnunculus*, Vieillot.

American Sparrow Hawk. (*F. Sparverius*, Lin. ♀ *F. Domminicensis*, Lath. pl. Enl. t. 444. ♂. 465 ♀ Wils., O. t. 32. f. 2. t. 16. f. 1.)

Rufous beneath, pale, spotted with black; seven round spots about the head. *F. æsalon*, var. Lath., is the young.

Pigeon Hawk. (*F. columbarius*, Wilson, t. 17. f. 3.)

Dusky; beneath whitish, with blackish stripes; tail with four narrow white bands. Of Hudson's Bay

In others the edge of the beak is deeply bidenticulated. Tarsi scutulated, and the wings short, second quill longest. These form the genus *Hierax* of Vigors.

Bengal Falcon. (*F. cœrulescens*, Lin. Edw. t. 108. pl. col. t. 97.) Gal. Ois. t. 18.

Back bluish-black; temples inclosed in a white line; cera, eyebrow, feet, and lower part of body, yellow. The smallest of the order; length six inches. Bengal. The *Falco fringillarius* of some ornithologists.

In others the bill is two-toothed; the tarsi scaly; head crested, and the wings long.

Colvy Falcon, Lath. Hist. t. 10. *F. Lathami*, n.

Head crested, black; scapulars, rump, and beneath, white; breast and scapulars bay banded, latter black tipt. India.

The genus *Harpagus* of Vigors, and the *Bidens* of Spix, have the bill and tarsi of *Hierax;* but the third and fourth quills are the longest, as in the Sparrow Hawks.

Notched Falcon, Lath. (*F. bidentatus*, Lath. pl. col. 38. jun. 228.)

Above gray, brighter on the head and cheeks; beneath reddish, rayed with white; throat and lower tail-coverts white; bill ash gray; feet and cera yellow. Length fourteen inches. South America, when young. Pl. col. 228, and Spix, t. 7: it is the *Bidens albiventer* of Spix; and *B. rufiventer* of Spix, t. 6, is, perhaps, a variety.

Two-toothed Falcon. (*F. diodon.* Temm. pl. col. t. 198.)

Above blackish; back of head, cheeks, and side of the neck deep grey; below pale grey; throat white; thighs reddish; wings and tail rayed with black: length eleven inches. Brazil. It is *Bidens cinerascens*, and *B. femoralis* of Spix, t. 8.

The GERFALCONS, (HIEROFALCO, *Cuv.*)

Have the quills of the wings as in other noble birds, to which they are assimilated, except that the beak has only a festoon, as in the ignoble birds. The tail, long and displayed, exceeds the wings, although the latter are very long. Their tarsi, short and reticulated, are feathered to the upper third. Only one species is well known.

The *Gerfalcon (F. candicans, F. cinereus,* and *F. sacer.)* Gm. Enl. 210, 462, and Hist. des Oiseaux, i. pl. xiv. Edw. 55.

One-fourth larger than the falcon, is the most esteemed of all birds for falconry. It is brought principally from the north. Its common plumage is brown above, with a border of paler point to each feather, and transverse lines on the covertures, and quill feathers: whitish underneath, with long brown spots, which change with age on the thighs into transverse lines: the tail is radiated brown and grayish, but it varies so much in the prevalence of brown or white, that there are some with the body all white, with only a brown spot on each quill of the mantle: the feet and membrane of the beak are sometimes yellow and sometimes blue.

This genus has not been adopted generally, as the character is not constant, and only found in the adult specimens.

The second section of the great genus Falco is that of birds of prey called ignoble, because they cannot be easily employed in falconry; a tribe much more numerous than that of the noble, and which moreover it is necessary to subdivide considerably. Their longest wing feather is almost always the fourth, and the first is very short, which has the same effect as if the wing had been cut obliquely at the end; hence, *cæteris paribus,* their flight is weaker: their beak is also much less armed because it has not the lateral teeth near the point, but only a slight festoon in the middle of its length.

The EAGLES, (AQUILA, Brisson,)

Which form the first family of these, have a very strong beak, straight at its base, and bent only toward the point. Amongst these are found the largest species of the genus, and the most powerful of all birds of prey.

The *Eagles*, properly so called, (Cuv.)

Have the tarsi feathered, even to the base of the toes. They live in mountains, and hunt birds and quadrupeds. Their wings are as long as the tail, their flight both high and quick, and their courage exceeds that of other birds.

The *common Eagle*, (*F. fulvus*, Enl. 409, *F. melanaëtos*, when molting, *F. niger*, difference of age, *F. Mogilnik*, molting, Gm., *F. Canadensis*, Gm., when molting as in *F. melanaëtos.*

More or less brown, with the occiput yellow, the upper half of the tail white, and the rest black. This species is the most spread over mountainous districts of Europe and America.

The *Falco Cygneus*, Lath. is an albino variety.

The *Ring-Tail Eagle*, (*F. chrysaëtos*,) Enl. 410,

Differs from the last only in having a blackish tail marked with irregular ash-coloured bands. It is of it, nevertheless, that we are told the exaggerated stories of the ancients, touching the strength, courage, and magnanimity of their golden or royal eagle.

M. Temminck considers the *Common and Ring-tailed Eagles* as mere varieties of age, the latter being full-grown.

The *Spotted and Rough-footed Eagle, (F. nævius et F. maculatus*, Gm.) Savigny, Ois. d'Egypte, t. 2. f. 1. Adult. t. 2, f. 1. Jun. *F. melanaëtos*, Sav.

A third smaller than the other two; brown; tail black, with the tip whitish; some pale yellow spots form a band over the small coverts, one at the end of the great feathers, which mounts on the scapulary feathers, and one at the end of the secondary. The top of the wing has little yellow dots; the under part of the body is paler than the back, and the tarsi are thinner and less furnished than in the great eagles.

This species is common in the Apennines and other mountains of Southern Europe, but is rare in the north. It attacks only very weak animals. It has been found docile enough to be used in falconry, but it is said that it flies from and submits to the sparrow-hawk.

Imperial Eagle. F. Mogilkin, Gm., *F. Imperialis*, Beehst, *A. Heliaca*, Savigny, Ois. d'Egypte, t. 12. pl. col. t. 151, 154.

Wings longer than the square tail; five scales on the last joint of the middle toe; gape very long; one or more scapulars white. Egypt and Hungary.

Booted Falcon, Lath. (*F. pennatus*, Gm. pl. col. 83)

Feet feathered to the toes; some white lunules at the insertion of the wings; tail beneath brown.

Martial Eagle, Lath , *F. armiger*, Shaw, (*F. bellicosa*, Daud. Vail. O. A. t. 1.)

Brown, feathers pale edged, beneath whitish, quills black; tail even, one-fourth longer than the wings; legs pale, feathers to the toes. Size of an eagle. Africa.

Reinwardt's Eagle (F. Malayensis, Reinw. pl. col. t. 117.)

Brown black; tail feathers with whitish lunules. Indian Islands.

Crowned Eagle. F. coronatus, Azara. (not Buf.) pl. col. t. 234.

Blue ash; beneath paler; tail-coverts white tipt; quills and tail black; tail with two white bands and tips; crest long, erectile; neck whitish, with black streaks; tarsi naked. Brazil. Length twenty-eight inches.

Bonelli's Eagle. F. Bonelli, Temm. *A. intermedia*, Bonelli. pl. col. t. 288.

Tail square; tarsi feathered; blackish brown; cheeks and beneath reddish, marbled with white and chesnut; shaft and streaks black; tail, base ash, end black, and tips white. Length twenty-six inches. Found at Turin.

The *Black and white Eagle*, Azara. *F. Aguia*, Temm. pl. col. t. 301.

Above bluish ash; tail white tipt; side of the neck and breast bluish ash, marbled with white; lower wing and tail coverts white, with fine bluish rays; beneath pure white; tarsi beneath naked, yellow. Brazil and Paraguay.

Crown Eagle, (Edw. t. 224.) *F. coronatus*, Shaw, vii. t. 16.

Brown; feathers pale edged; forehead and orbits whitish; beneath white, black spotted; breast rufous; sides black banded; tail grey, with four black bands. Guinea.

New Holland produces eagles of the same form as to the tail, which is wedge-shaped, as

The *Wedge-tail Eagle*. *F. fucosa*, Cuv. R. A. t. 3. f. 1. pl. col. t. 32. *Milvus sphenura*, Vieil. Gal. t. 15.

Fulvous brown, varied with rufous. Length thirty inches.

The FISHER EAGLES, Cuv. (HALIÆTUS, Savigny,)

Have the same wings as the last, but the tarsi are feathered only on the upper half, and the other half shielded. They inhabit the banks of rivers and the sea-shore, and live principally on fish.

The *Sea-Eagle*, *Osprey*, or *Pygargus*, (*F. ossifragus*, *F. albicella*, and *F. albicaudus*, Gm.)

Form but one species, which, when young, has the beak black; the tail blackish, spotted with whitish; and the plumage brown, with a deep brown streak on the middle of the feather, (Enl. 112 and 415,) which with age becomes of an uniform grey brown, paler on the head and neck, with a white tail, and pale yellow beak (Frisch. lxx.) These changes have been verified in the menagerie of the French Museum. The *F. albicaudus* is the male of the great *F. albicella*.

This species attacks fish at all times, and is found all over the northern parts of the globe.

The *Bald Eagle*, Lath. (*F. leucocephalus*, Lin. Enl. 411. Wilson, iv. t. 36.)

Uniformly deep brown, with white head and tail, and yellowish beak, nearly as large as our common eagle. Lives in South America, and preys on fish. It seems that it comes sometimes into Northern Europe. When young, it has the body and head ashy brown, but it ought not to be confounded with the old sea-eagle with a whitish head.

Caley's Hawk, (*H. Calei*, Vigors. Lin. Trans. xv.)

Reddish brown, variegated with black; quills ash-coloured, black banded; pale tipt. New Holland. Length twenty-three inches.

Whistling Hawk. *H. canorus*, Vigors. l. c. xv.

Above ferruginous brown: wing coverts and quills fuscous brown; beneath white, varied with ferruginous. Length twenty-one inches. New Holland; perhaps, the young of *F. Novæ Zelandiæ*, Vigors.

Piscivorous Eagle, Lath. F. *vocifer*, Daud. Vail. O. A. t. 4.

Rusty brown, streaked with black; head, neck, breast and scapulars white, brown edged; tail white; quills black, outer web brown banded; belly and thighs rufous. Size of the osprey. Africa.

Crying Eagle. *F. axillaris* and *F. vociferus*.

Ash-grey; beneath white; smaller and larger wing coverts white; feet yellow. India and Africa. Size of a wood-pigeon.

Mace's Eagle, (*F. Macei*, Cuv. pl. col. t. 8, and jun. 223)

Reddish brown; head, nape, and upper part of the back red; eyebrows, cheeks, throat, and neck, in front, whitish; tail with a white band. Length twenty-six inches. India.

White-bellied Eagle, Lath. (*F. leucogaster*, Lath. pl. col t. 49.)

White; back, wing and tail dingy brown; tail tipt with white; bill and feet yellow. Length thirty-three inches. Pacific Islands.

Cuvier has here placed *F. Vulturinus*, Daud. Vail. Ois. Afr. t. 6; but Temminck and others have considered it as a species of *Gypætos*.

Marine Eagle, Lath. *F. Ichthyætus*, Horsf.

Brownish; vent, rump, tail, and thighs white; tail dusky at the tip. Length twenty-six inches. Java.

Fishing Eagle and African Pheasant, Lath. *Falco Piscator*, Gm. pl. Enl. 478.

Head long-crested, ferruginous brown; beneath white, brown streaked; wing coverts dove-coloured, with dark shafts; quills bluish brown, internally white spotted. Senegal.

Pondicherry Eagle, Lath., (*F. Pondicerianus*, Lath. Pl. Enl. t. 416.)

Chesnut; head, neck and chest white, varied with brown lines; six first quills black,ended. Length one foot and a half. India.

Blagre Eagle, Lath. (*F. Blagrus*, Daud. Vail. O. A. t. 5.)

Glossy white; head, nape, lesser wing coverts and tail pale gray brown; tail white-tipt; legs yellow, greater quills dusky black. Cape of Good Hope.

The Ospreys, or Bald Buzzards. (Pandion, Savigny.) *Triorchis* of Vieillot.

Have the beak and feet of the fisher eagles, but their nails are round underneath, while in other birds of prey they are bent and channeled; their tarsi are reticulated, and the second wing feather is the longest.

Only one species is known, which is spread over the fresh-water banks of nearly all the world, with little variations in plumage.

The *Osprey*, or *Balbuzard*, Fish Hawk of America. (*F. haliætus*, Lin.) Enl. 414, and Catesby, ii. Wilson, A. O. t. 5. f. 1.

One-third smaller than *F. ossifragus*; white, with a brown mantle, and a brown band descending from the angle of the beak toward the back; brown spots on the head and neck, and sometimes on the breast; the cere and feet sometimes yellow, sometimes blue.

Carolina Osprey. F. Carolinnensis, Lin. *H. Americanus*, Vieill. Gall. Ois. t. 11. *Aquila piscatrix*, Vieil. O. A. S. t. 4. are, perhaps, varieties.

The *Cayenne Osprey*, Lath. *F. Cayenensis*, Gmel. is, perhaps, a variety.

Some species differ in the tarsi being long, and the toes short and united at the base, which form the genus *Circætus* of Vieillot, which approaches the eagles.

Jean le Blanc, Lath. (*F. Gallicus*, Gmel. Pl. Enl. t. 413. *A. brachydactylus*, Meyer.)

Bill black; toes bluish; white, spotted with brown; back and wing coverts brown. Length two feet.

Gray French Eagle, (*C. cinereus*, Vieillot, Gal. Ois. t. 12.)

Dull ash; quills black; tail above brown; beneath white banded.

America produces fisher eagles, with long wings, like the preceding, in which a greater or less part of the sides of the head, and sometimes of the throat, is denuded. These are called *Caracara*, (see Azara, vol. ii. 30.)

And *Gymnops*, by Spix. Have been formed into the genera *Ibycter*, *Daptrius*, and *Polyborus*, by Vieillot, and *Milvago* by Spix.

The *common Caracara*, or *Brazilian Kite*, Lath. (*F. Brasiliensis*, Gm.) Gal. Ois. t. 17. Spix. t. 1. a. Jun.

As large as the balbuzzard, striped crosswise black and white, long and slender feathers, white at the throat, and a black crest a little elongated, in a tuft; the covering of the wings, thighs and end of the tail blackish. It is the most common predatory bird in Paraguay and Brazil. It is the Caracara of Margrave, but ill described; and the *F. cheriway*, Jacq. beyt. may be a variety of it;

The *Polyborus Vulgaris* of Vieillot, and the *Vultur Cheriway* of Jacquin, (Vog. t. 4,) and perhaps the

Falco Plancus of Miller, Cim. Phys. t. 17, and *Cook Voy.* ii. t. 32.

Red-throated Falcon, Lath. (*F. aquilinus*, Enl. 427. Gal. Ois. t. 16.)

Black, with the belly and lower covertures of the tail white; the throat naked and red.

Length eighteen inches. Is the *Ibycter leucogaster* of Vieillot: *F. formosus*, Lath.; and *F. nudicollis* of Daudin.

New Zealand Falcon, Lath. (*F. Novæ Zelandiæ*, Lath. Syn. t. 4 ♀ pl. col. t. 192 and 224, jun.)

Above gray brown, beneath paler, banded with red; tail yellowish gray; banded thighs ferruginous; bill bluish; cera and feet yellow. Length eighteen inches. New Zealand.

Negro Caracara, (*Daptrius ater*, Vieillot, *F. atterimus*, Temm. Gal. Ois. t. 5. pl. col. t. 37.)

Entirely black, except the white base of the tail; and yellow feet. South America. Length fifteen inches. When young, is *D. striatus*, Vieillot.

Yellow-headed Caracara, (*Milvago Ochrocephalus*, Spix t. 5. Jardines. Ill. Orn. t. 2.)

Dirty yellow white, with a black stripe from the eye to the ear; back wings and end of the tail black. Length twelve inches. Brazil. Brit. Mus.

Banded Caracara. *Gymnops fasciatus*, Spix. t. 4.

Black round the eyes; cheek and gullet naked; tail white, with five black bands. Brazil.

Streaked Caracara. Gymnops strigillatus, Spix. t. 4. a.

Brown; auricular spot blackish; side of the neck ferruginous; chest and belly ferruginous; crown streaked; wing and centre of tail dirty white, spotted and banded with black. Brazils.

Chimachima Falcon. F. degener, Illig. *F. crotophagus,* Pr. Max.

White; crown streaked with brown; back and wings black; tail base, with seven black bands, (beneath narrower,) end black; bill whitish; cera naked; feet lead-coloured. Length seventeen inches, tail eight inches, tarsi two inches one-third. When young dirty white; chest brown-spotted; back and wings sooty. *Chimachina* of Azara, ii. 6.

The Harpies, or Fisher Eagles with short wings. (Harpyia, Cuv.)

Are also proper to America, and have the tarsi very thick, strong, reticulated, and are one half only feathered, like the fisher eagles, properly so called, from which they differ only in the shortness of their wings; their beak and talons are even stronger than those of any other tribe.

The *Great Harpy of America,* or *Crested Vulture,* Lath.: *F. destructor* of Daudin; the *Grand Aigle de la Guiane* of Mauduit; probably, the *F. harpyia,* and the *F. cristatus* of Gmel.; certainly the *Yzguautzli* of Fernandes, who exaggerates its size in comparing it to a sheep; the *Vultur cristatus* of Jacquin; and, consequently, the *F. Jacquini* of Gm.; *F. harpyia* and *imperialis* of Shaw;

Is one of those birds which has the most terrible claws and beak. Its size is above that of the common eagle. Its plumage is ashy about the head and neck; blackish brown on the mantle and sides of the breast; whitish underneath, and with brown bars on the thighs; some elongated feathers form a black crest behind the head.

It is said to be so strong as to have sometimes cleft the skulls of men with a blow of the beak. The sloths form its common food, and it sometimes carries off young fawns.

Booted Harpy. H. Braccata, Spix. t. 3.

Black; tarsi densely feathered, dotted with white; rump white-spotted; tail long; four gray banded.

Vieillot also describes as a species *H. coronata;* and *H. ornata* of Spix is the *Crested Goshawk.*

The Eagle Hawks (Morphnus, Cuv.)
Spizaetus, Vieillot, and *Aquila,* Spix.

Have the wings shorter than the tail, like the last; but their long and spare tarsi and weak toes distinguish these from them. Some have the tarsi elevated, naked and shielded.

The *Tufted Eagle of Guiana; Aigle autour huppé (F. Guiannensis,* Daud.); *Petit Aigle de la Guiane,* Maud.

In the colours and the crest is extremely like the great fisher-eagle of the same country, but it is less; but its elevated, naked and shielded tarsi sufficiently distinguish it; the mantle is blackish, sometimes varied with deep grey; the belly white, with yellow claws, more or less marked; the head and neck sometimes gray, sometimes white, and the occipital tuft long and blackish.

The *Urubitinga, (F. urubitinga,* Lin.) pl. col. t. 55, Cuv. R. A. iv. t. 3. f. 2. Spix. t. 18.

Black; without a crest; with the rump and lower part of the tail white. This beautiful species seeks its prey in inundated places. (In Brit. Mus.)

One-banded Hawk, (F. unicinctus, Temm. pl. col. t. 313.)

Deep brown; forehead with two white spots; throat streaked with white; quills bandless; tail base, and end white; wing covers and thigh red, with dark spots. Brazil. Length nineteen inches.

Painted Hawk, (Aquila picta, Spix. t. 1. c.)

Blackish, spotted with red; tail longer than the wings, rather acute, blackish red, black banded, end fulvous, pencilled.

Snowy Falcon, F. niveus, Temm. pl. col. t. 127.

White; upper part of body, wings, tail brown; band on the quills, and tips brown. Length twenty-five inches. Java.

Cuvier proposed to place here *F. Novæ Zelandiæ,* Lath. t. 4.

Others have the tarsi elongated, and feathered the whole length; as

The genus *Spizaetus* of Vieillot, and the *Plumipeda* of Flemming.

The *Black-tufted Eagle of Africa (Huppart*, Vail. Afr. t. 2. Bruce, pl. xxxii., *F. occipitalis*, Daud.) *F. Senegalensis*, Daud.

As big as a crow; black; with a long tuft hanging from the occiput; the tarsi, the edge of the wings, and the band under the tail are whitish. Inhabits Africa.

The *Crested Goshawk*, Lath., *Urutaurana*, Margrave; *Autour huppé*, Vail. i. t. 26; *Aigle Moyen de la Guyane*, Maud.; *Epervier patu* d'Azara; *F. ornatus*, Daud.; *F. superbus* and *F. coronatus*, Shaw.

Crown of the head and tuft black; sides of the neck bright red; mantle black, varied with gray, waved with white; under parts white, with black bars on the flanks, thighs, and tarsi. It is a fine bird, of South America, which varies from black and white to deep brown.

It is the *Harpya ornata* of Spix, Vieil. Gal. t. 21.

Lake Falcon, T. limnæetus, Horsf. Java, t. pl. col. t. 134.

Brown; tail beneath, except at the tip, whitish ash-colour; tarsi feathered to the toe. Length twenty-four inches. Java.

Crested Falcon. F. cristatellus, Temm. pl. col. p. 282.

Tail long, square; tarsi quite feathered; crest of six or eight feathers, black, long, and narrow; reddish brown, beneath white; quills deep brown, obscurely banded internally; tail with seven or eight black-brown bands. Ceylon. Length twenty-four inches.

Noisy Eagle, Lath. (*F. albescens*, Daud. Vail. O. A. t. 3.)

White, spotted with black brown; tail black barred; hind head of male long crested; tail as long again as the wings; bill pale; legs yellow. Cape of Good Hope.

Spotted Eagle. *F. maculosa*, Vieillot, O. Am. Sept. t. 3*.

Black; throat and crest white, spotted with black; abdomen spotted with white; vents and thighs rust-coloured. Length twenty-four inches. Mexico.

Tyrant Eagle, Lath. *F. Tyrannus*, Pr. Max. pl. col. t. 73.

Tarsi short, feathered; crested; head, neck, and upper part of the back with white brown-tipt feathers; body brown. Brazil. Length twenty-six inches.

Black-headed Eagle. *F. Atricapillus*, Cuv. pl. col. t. 79.

Tarsi long, woolly and slightly feathered; white, with a spot on each side of the head between the beak and the eye; back of the head, back and wings black. Length sixteen inches. South America.

Chinese Eagle. *Falco Sinensis*, Lath. Syn. t. 3.

Reddish-brown; crown dusky; edge of feathers, quill, and base and middle of tail, and centre wing band, dark brown. China.

Finally, there are in America birds with beaks like all the last, and with short reticulated tarsi, half feathered in front; with wings shorter than the tail, and whose most distinctive character consists in the nostrils, which are nearly closed, and are like a mere cleft. Of these may be made a small tribe, under the name CYMINDIS, *Cuv.;* which is the Greek name of an undetermined bird of prey. Of these is

The *small Cayenne Eagle*, (*petit Autour de Cayenne*, Buff., *F. Cayenensis*, Gm.) Enl. 473. pl. col. 270.

Has moreover, as a character, a small tooth at the bend of the beak. The adult is white; the mantle bluish-black, with the head ashy, with four white bands on the tail; the young has the mantle varied with brown and red, with some black spots.

The *F. glaucopis*, Merrem. Beytr. ii. t. 7. is a common Buzzard. The *F. albus*, Shaw, in White's Journal, is an Hawk.

Hook-bill Eagle, *F. uncinatus*, Tem. p. col. 103. 104 ♀ 115 Jun.

Lead coloured, beneath paler; quills banded with brown ash; tail-base white, tips grayish; beak hooked. Brazil. Length 15—17 inches.

Crowned Eagle, Lath. *F. coronatus*, Azara. pl. col. 235.

Crested, black; head reddish-gray; belly white; thighs white, spotted with black. Grenada. Length thirty inches.

There are others which have similar beaks, nostrils, and wings, but their tarsi are short, and shielded, as

White-rumped Falcon, (*F. leucopygus*, Spix, t. 2.)

Blackish-gray; throat, abdomen and tail brownish; vent and base of tail, white. Amazon river, Brazil.

Long-beaked Eagle. *F. hamatus*, Illiger, pl. col. 61. 231 Jun.

Lead-coloured, quills black; base of tail, and lower tail coverts, white.

To this sub-genus may be added, as a section, the *Asturina* of Vieillot, peculiar for its lunate nostrils, short slender tarsi, and long claws.

Ashy Falcon. *Asturina cinerea*, Vieil. Gal. t. 20.

Ashy-blue, beneath white striped; tail with two black bands and white tips. Guiana. Length fifteen inches.

F. fuscus, Lath. Miller, Illust. t. 13. is perhaps of this section.

The Hawks, Autours, *Cuv.* (Astur, *Bechstein.* Dædalion, *Savig.*)

Which form the second division of ignoble birds, have, like the three last tribes of eagles, the wings shorter than the tail; but their beak bends from its base, as in all the following.

They are more particularly called Goshawks, which have the tarsi shielded, and rather short, the genus *Astur*, of Vigors.

The *Common Goshawk*, (*F. palumbarius*, Enl. 418, and 461, and the young *F. gallinarius*, Enl. 425, and Frisch, t. 62; probably also the *F. gyrfalco* and *F. gentilis* of Gm., so ill-determined are the species in modern works,)

Is the only species of this country. It is brown, with whitish eyelids; white underneath, barred across with brown in the adult; dotted when young; five browner bands on the tail. It equals the gerfalcon in size, but not in courage, falling always obliquely on its prey. It is nevertheless used in falconry for weaker game. It is common in all our hills and low mountains.

The *ash-coloured Hawk. F. atricapillus*, Wilson, is the very old specimen of this bird.

Ray's Hawk. Astur Raii, Vigors.

Above, ash-coloured; beneath white, varied with brown; tail pale gray, beneath whitish, banded with brown. New Holland. Mus. Lin. Soc. Length sixteen inches.

Banded Hawk. Astur fasciatus, Vigors.

Above fuscous brown; beneath white, with crowded brown bands; thighs red, banded. Length of the male seventeen, of female nineteen inches. New Holland. *A. approximans*, Vigors, is perhaps the young.

Broad-winged Hawk. F. Pennsylvanicus, Wilson, O. A. t. 54. f. 1. *F. latissimus*, Ord.

Dark brown; head streaked with whitish; beneath white, thickly spotted on the breast with brown arrow-heads; tail short, with two bars of white, and tipt with whitish; cera and feet yellow. North America. Rare.

Among the foreign Goshawks may be noticed that of *New Holland, White Eagle*, Lath. (*F. Novæ Hollandiæ*, Gmel.), and *F. albus* of Shaw, White's Jour. t. at p. 260, which is often altogether as white as snow; but it seems to be a variety of a bird of that country; ashy above, white underneath, with slight indications of gray in waves.

This is now proved, by many specimens, to be a distinct species, as there are many specimens in collections.

Short-toed Falcon. F. hemidactylus, Temm. pl. col. t. 3.

Ashy lead colour, beneath paler; tail beneath reddish, with two black bands; quills black, with a broad white band. Brazil. Length fifteen inches.

Slender Hawk. F. gracilis, Temm. pl. col. t. 91.

Ashy lead colour, beneath whitish, transversely streaked with cinereous lines; cheeks and throat white. Brazil. Length 18—19 inches.

Shining Hawk. F. nitidus, Lath. *F. striolatus,* Temm. pl. col. t. 87. 294. Jun.

Lead-coloured, beneath white, transversely waved with ash colour; tail black, with two narrow white bars; legs long, yellow. Brazil. Length 13—14 inches.

Yellow-throated Hawk. F. Xanthocorax, Temm. pl. col. t. 92.

Reddish-brown, beneath white, transversely striped with rufous; head, throat, and neck, cinnamon-red. Brazil. Length 12—13 inches.

Short-winged Falcon. F. brachipterus, Temm. pl. col. t. 141. and 116 Jun.

Dusky brown; beneath, and nuchal collar white, transversely striped with black; tail wedge-shaped, with three narrow white bands; eyelids white. Brazil. Length 18—20 inches.

White-necked Falcon. F. leuchauchen, Temm. pl. col. 306.

Brown above, beneath white; tail five narrow white bands; eyebrow, and spot on side of cheek, white, banded with black; front of cheek brown, beneath black; top of head, occiput, and half collar, black. Brazil. Length 12—14 inches.

Large-billed Hawk. F. magnirostris, Gm. pl. Enl. t. 46. pl. col. t. 86, Jun.

Ashy brown, neck and chest paler; quills bright red, black banded; tail gray, with four black bands; belly white, reddish-brown banded; thighs reddish, brown banded. Length fifteen inches. Brazil. Placed with the Sparrow Hawks by several authors.

Radiated Falcon, Lath. *F. radiatus,* Lath. Syn. t. 121. pl. col. t. 123.

Ferruginous, radiately spotted with black; wings and tail long, brown. New Holland.

Grey-breasted Hawk. F. poliogaster, Natterer, pl. col. 264, 295, Jun.

Slaty black, beneath ashy white; throat white; tail black, with three gray bands above, and four beneath. Hen reddish; wings and back dusky. Brazil. Length 16—17 inches.

Three-streaked Hawk. F. trivirgatus, Temm. pl. col. t. 103.

Brown; head and neck black; cheeks gray; tail with three dark bands; beneath white; throat with three black longitudinal lines: chest and legs with broad brown black-edged bands; sides of neck brown. Sumatra.

White-billed Hawk. *F. leucorhynchus,* Quoy and Gaimard, Freycinet, Voy. t. 13.

Blackish-brown; cera and feet yellow; rump white; tail cinereous, with three white bands. Brazil. Length thirteen inches.

One-banded Hawk. *F. unicinctus,* Temm. pl. col. t. 313.

Brown; scapulars, thighs, and edges of upper wing coverts red; throat feathers white edged; forehead with two white spots; quills pure brown, white tipt; tail white, with broad brown bands. Brazil. Length ten inches.

Vieillot refers *F. orientalis* and *F. Indicus,* Lath., to this genus, and describes, as new, *Sparvius cinereus,* and *S. monachus* of Brazil.

We may, moreover, associate with the Autours, or Gosshawks, some American species, with short wings and short but reticulated tarsi.

They are called *Physeta,* and since *Herpethotheres,* by Vieillot; and Mr. Vigors has restricted the genus *Dedalion* to them.

The *Laughing Falcon,* (*F. cachinnans,* Lin.) *Nacagua,* D'Az. Gall. Oist. t. 19. Spix, t. 3. a?

Named from its cry; white; the mantle, and a band from each eye, uniting at the neck, brown; the tail with brown and whitish bands. Of the marshes of South America, where it lives on reptiles and fish.

Streaked Falcon, Lath. *F. melanops,* Lath. pl. col. t. 105.

The size of a crow, black, spotted with white; beneath white; head and neck white, streaked with black; orbits black; tail black, with a white central band. Cayenne.

Surinam Falcon, Lath. *F. sufflator*, Lin.

Body whitish-brown; eyelid bony; cera and feet yellow. The genus *Physeta* of Vieillot.

We may call Sparrow Hawks (Nisus, *Cuv.*) those which have the tarsi shielded, and more elevated, the *Accipiter* of old authors.

The *Common Sparrow Hawk*, (*F. nisus*, Lin.) Enl. 412, and 467.

Has the same colours as the Goshawk, but its legs are higher, and its size about a third less. It is, nevertheless, employed in falconry. The young has the spots underneath arrow-shaped, and in longitudinal red dots; the feathers of the mantle are also edged with red.

There are some foreign species still smaller, as

Red-legged Falcon, Lath. (*F. gabar*, Shaw. Vail. O. A. t. 33. pl. col. 122. 140 Jun.)

Bill black; cera and legs red; above gray-brown; beneath bluish gray; upper and lower tail coverts white; quills dusky, beneath banded; tail even, banded; vent white, brown banded. Size of the Sparrow Hawk. Of Africa and Nubia.

Dwarf Falcon, Lath. (*F. minullus*, Shaw, Vail. O. A. t. 34.)

Brown, beneath white; throat brown spotted; belly and thighs brown banded; tail even and banded. Smaller than the Merlin.

Minute Falcon, Lath. (*F. minutus*, Lin.) Bris. i. t. 306. *F. Brissonianus*, Shaw.

Brown, rufous, variegated; crown variegated white; beneath white, with brown spots, and bands; tail with six darker bands. Of Malta.

Black Sparrow-hawk. *Sparvius niger*, Vieil., *F. Banksia*, Temm.—Gal. Ois. t. 22.

Black; upper neck-feather white based; tail white spotted; quills whitish gray, black spotted. Senegal. *Vieil.* New Holland, *Br. Mus.*

And there are also others much larger, as

The *Chaunting Falcon*, Lath. (*F. musicus*, Daud. *Faucon chanteur*, Vail. Afr. xxvii.)

Is as large as the Goshawk; ashy above; white, striped with brown underneath, and about the vent. It is found in Africa, where it hunts partridges and hares, and builds on trees. It is the only bird of prey known which sings well.

The *Collared Falcon*, *F. torquatus*, Cuv. pl. col. t. 43. 93. Jun.

Ash-coloured brown; neck reddish; beneath white, banded with red; quill, and tail feathers, banded with brown. Length twelve inches. Of New Holland. Mus. Lin. Soc.

Slate-coloured Hawk, *F. Pennsylvanicus*, Wils. A. O. t. 46. f. 1.

Slate-coloured; beneath white barred with ferruginous; tail with four broad black bands, tipped with white; cera dull green; irides and feet orange. When young this is the *Sharp-shinned Hawk, F. velox,* Wilson, A. O. t. 45. f. 1. pl. col. t. 67.

Streaked Hawk. F. virgatus, Reinw. pl. col. t. 109.

Ashy blue; front of neck, middle of breast, abdomen, and lower tail covers, white; lesser covers red, brown spotted; tail even, with three black bands. Java. Length ten inches.

Black-capped Hawk. F. pileatus, Pr. Max. pl. col. 205.

Cinereous; beneath whitish, with a brown longitudinal stripe on each feather; crown and wings blackish; thighs red. Brazil. Length thirteen inches.

Javan Sparrowhawk, F. Soloensis, Horsf. *F. cuculoides,* Temm. pl. col. t. 129, 110 Jun.

Cinereous blue; beneath dull iron grey; quills black; wing covers white at the base; tail, outer feathers excepted, banded with black, beneath whitish. Of Java.

Indian Sparrowhawk, F. Dussumieri, Temm. pl. col. t. 308 ♀ and 336 Jun.

Brown; neck reddish; beneath white, finely cross-banded with brown; quills and tail ash-brown, black banded, white tipt; central tail-feathers bandless. India. Fifteen inches.

The Insectivorous Sparrow Hawk, F. insectivorus, Spix, t. 8. a.

Ash-coloured brown; head, and chest, ashy, with large spots; abdomen whitish, red banded; vent whitish. Of South America.

Brown's Hawk, Lath. *F. badius*, Lath., Brown's Ill. t. 3. *F. Brownii*, Shaw.

Brown; beneath white, belly with yellow semicircular lines; wing coverts white edged; quills dusky, pale edged. Ceylon.

Long-tailed Falcon. F. macrourus, Lath.—Nov. Com. Petr. t. 89.

Cera and feet yellow; bill blackish; body above ashy, beneath white; neck ashy; quills white tipt. Russia.

The genus *Gampsonyx* of Vigors has the bill without notches, and short wings of the Hawk; but the second quill is the longest, and the tarsi are reticulated like the Falcons.

The *Falcon-like Hawk*. (*Gampsonyx Swainsoni*, Vigors, Zool. Jour. ii. 6.)

Ashy black, beneath white; forehead, cheeks, sides of abdomen, and thighs, orange; breast with a black spot on each side. Brazil. Length 9—10 inches.

The following indistinct species may probably belong to the Sparrow-hawks: *Sparius subniger*, South America. *S. cœrulescens*, North America. *S. semitorquatus*, Paraguay. *S. gilvicollis*, *S. magor*, Cayenne. *S. bicolor*, *S. guttatus*, Paraguay. *S. melanoleucus*, Paraguay. *S. cinereus*, Guyana. *S. tricolor*, South America. *S. superciliaris*, Paraguay. *S. cirrocephalus*, New Holland. *S. rufiventris*, (the *F. rufus* of Lath.) noticed by Vieillot.

The *Ictinia* of Vieillot differs from the Sparrow-hawk in the bill being short and slightly notched;

the tarsi short, weak and shielded, and the third quill is the longest. It has the habits of both the hawks and kites.

The *Spotted-tailed Hobby*, Lath. *Falco plumbea*, Lin. Lath. Hist. t. 12. Vieil. Gal. Ois. t. 17. Spix, Bras. t. 8. b. pl. col. t. 180.

Blackish ash; head, neck, and beneath paler; tail black; feet red. Is the *F. Mississipensis* of Wilson, A. O. t. 25. f. 1. and *Milvus Cenchris*, Vieil. America.

The Kites (Milvus, *Bechstein*). *Milvina*, Vigors.

Have short tarsi, with weak toes and nails, which, together with a beak equally ill proportioned to their size, render the species the most cowardly of all; but they are distinguished by their wings being excessively long, and by their forked tail, by which they have a most rapid and easy flight.

Some have the tarsi very short, reticulated, and half covered with feathers on the upper part like the last small tribe of eagles. The genus *Elanus* of Savigny.

Now divided into the true *Elanus*.

The *Blac*, Vail. Afr. t. 36, 37 (the *F. melanopterus*, Daud. Zool. Misc. iii. t. 122).

As large as a sparrowhawk, with the plumage soft and silky; the tail but little forked; ashy above, white underneath, with the small coverture of the wings blackish: the young is brown, varied with yellow. This bird is common from Egypt to the Cape. It hunts little else than insects. Also found in America, India, and New Holland.

The *Falco dispar* of Temmin. pl. col. t. 319, is the young. The *Elanus cæsius* of Savigny.

The *Nauclerus* of Vigors, and the *Elanoïdes* of Vieillot.

Riocour's Falcon. F. Riocourii, Vieill. pl. col. t. 85. Gal. Ois. t. 15.

White; upper part of head, neck, back, wings, and tail gray, with a line behind and before the eyes, and spot on the wing black. Length one foot and a quarter. Africa.

The *Carolina*, or *Swallow-tailed Kite*. (*F. furcatus*, Lin.) Catesby, t. 4; Wilson, a. A. O. 51. f. 2.

White, with the wings and tail black; the two exterior quill-feathers of the wing and tail very long: larger than the Blac. This attacks reptiles. Of South America.

The KITES, properly so called, have the tarsi shielded, and stronger.

The *Common Kite*. (*F. milvus*, Lin.) Enl. 422.

Fawn colour; the primaries of the wings black, and the tail red. Of all our birds this remains the longest and with most ease in the air. It attacks scarcely any thing but reptiles.

The *F. austriacus* of Gm. is the young of the common kite.

Black Kite. F. ater, Lin. Pl. Enl. 472. Jun. Vail. O. A. t. 22.

Head and throat banded lengthways black and white; above deep gray brown; beneath reddish brown, with long streaks on the centre of the feathers; thigh deep red; tail only slightly forked with nine or ten cross bands. Length one foot ten inches. South Europe and Africa. The *F. Egyptius* and *F. Forskahlii* of Gm., and the *F. parasiticus* of Shaw.

Vieillot describes a kite with a graduated tail, from New Holland; *Milvus sphenura*, Vieil. Gal. Ois. t. 15. which appears to be the *Wedge-tail Eagle*, *F. fucosa*, Cuv. R. N. t. 3. f. 1.

The HONEY BUZZARDS (PERNIS, CUV.) *Circus B.* Vieillot.

With the weak beak of the kites, these have a very peculiar character in the space between the eye and the beak, which in all the rest of the genus *Falco* is naked and furnished only with a few hairs, but in these is covered with feathers lying close and cut like scales; their tarsi are half feathered toward the top, and reticulated: for the rest they have the tail equal, the wings long, the beak bent from its base like all the following. We possess but one species.

The *Common Honey Buzzard*, (*F. apivorus*, Lin.) pl. Enl. 420.

Something less than the buzzard; brown above, variously undulated, with brown and whitish underneath: the head of the male ashy at a certain age. This bird feeds on insects, especially wasps and bees.

The *F. longipes* of Nilson, Orn. Suecica. i. t. is either this or a distinct species of buzzard.

There are some others in foreign countries.

The *Java Honey Buzzard; La Bondree huppée de Java,* Cuvier.

Altogether brown, with the head ashy like ours, but the tail black, with a whitish band over the middle, a brown crest on the occiput. Brought from Java by M. Leschenault.

The *F. Ptilorhynchus,* Temm. not Bechst. pl. col. t. 44.

The *Crested Buzzard. Buteo cristatus,* Vieil.

Crested; head white and brown; above feathers brown, red edged; beneath white; neck and crest with some brown spots; quills black; tail brown; beneath whitish; sides of neck and over eye a brown band. New Holland.

The BUZZARDS, properly so called, (BUTEO, Bechstein) *Circus A.* Vieillot.

Have long wings; the tail feathers of equal length; the beak bent from its base; the interval between it and the eyes featherless; the legs strong.

Some of them have the tarsi feathered to the toes. They are distinguished from the eagles by the beak curved from the base, and from the goshawks by the feathered tarsi and long wings. We have one species.

The *Rough-footed Falcon,* Penn. *(F. pennatus)* Frisch. lxxv. Vail. Afr. t. 18. is the F. lagopus, *Penn.* not the F. pennatus of Gm. See Temm. Man. 45.

Varied irregularly with brown more or less bright, and white more or less yellow; is one of the most extended species, being found almost everywhere.

It has been almost always considered a variety of some other bird. It is four times mentioned in Gmelin without ever being in its place. It is the *F. lagopus*, Brit. Zool. app. t. 1; *F. communis* and *leucephalus*, Frisch. 75; the *F. pennatus*, Brisson, app. t. 1; the *F. Sancti Johannis*, Arct. Zool. t. 9.

Black Hawk; F. Sancti Johannis, Gm.; *F. niger*, Wils. a. o. t. 53. f. 1, 2 Jun.

Black; above speckled with white; white round the eye; tail rounded, with narrow bands of pure white, and tipped with dull white. North America.

Winking Falcon, Lath. Supp. *F. connivens*, Lath.

Chocolate brown; beneath yellowish, brown spotted; back of neck and axillaries white spotted; quills and tail white banded; tarsi feathered. New Holland.

Black and White Buzzard. Buteo melanoleucus, Vieillot, Gal. Ois. t. 14.

Back, wings, and tail blackish brown; head, neck, beneath, and edge of secondaries white; tail with six black and pale bands. Brazils. Length eighteen inches.

But the *buzzards*, in general, have the tarsi naked and shielded. We have in Europe but one.

The *Common Buzzard, (F. buteo)* 1. Enl. 419.

Brown, more or less waved with white on the belly and throat. It is the most common and the most destructive bird of prey of Europe. It continues all the year in the forests; falls on its prey from the tops of trees, &c., and destroys much game. The *F. communis fuscus, F. variegatus, F. albidus, F. versicolor*, Gm., are all this bird in different states.

But we may notice among the foreign Honey Buzzards,

The *Bacha, F. Bacha*, Daud. Vail. O. A. t. 15.

As big as ours; brown, with small round spots, and white on the sides of the breast and belly; a black and white crest; and a large white band on the middle of the tail. It is a very cruel bird, proper to Africa, and makes its principal prey of the *Hyraces.*

It has been placed with *Cymindis*, found also in India and Java.

Red-tail Hawk, F. Borealis, Gm. Wilson, A. O. t. 52. f. 1.

Dusky; beneath whitish, with blackish hastate spots; tarsi partly feathered; tail ferruginous, with a black subterminal band. When young, the American Buzzard, *F. Leverianus*, Wilson, a. o. t. 52. f. 2. North America, the *Acc. ruficaudus*, Vieillot.

Tachard Falcon, Lath. *F. Tachardus*, Shaw; *Le Trachard*, Vail. O. A. t. 19.

Deep brown; feathers pale edged; beneath grayish yellow, blotched with brown; head grayish brown, white streaked; tail black banded; legs partly feathered, mottled. Africa.

Jackal Falcon, Lath. *F. Jackal*, Shaw, Vaill. O. A. t. 16.

Dusky brown; throat whitish; breast rufous; quills dusky, pale banded; tail short, deep rufous, and with a black spot. Size of the buzzard. Cape of Good Hope.

Desert Falcon, Lath. *F. desertorum*, Daud.; *Le Rougri*, Vaill. O. A. t. 17.

Rufous, beneath paler; throat and chin and vent whitish; quills black; tail beneath gray, obsoletely banded. Africa.

The Buzzaret. F. Busarellus, Shaw; Vaill. O. A. t. 20. *Le Buseray*.

Head and neck rufous white, varied with brown; back and neck rufous, spotted and streaked with dusky black; tail barred, base pale, end dusky; belly light rufous, with black brown bands; quill black, as long as the tail. Cayenne. Length nineteen inches.

Hobby Buzzard, Lath. Supp. *F. Buzon*, Daud. Vaill. O. A. t. 21.

Above varied rufous and black; head and neck and quills dusky; tail black, with the tips and central band white; beneath pale rufous, darker banded; quills one-half the length of tail. Cayenne. Length seventeen inches.

Speckled Sparrow-hawk, Lath. *F. Tachiro*, Shaw; Vail. O. A. t. 24.

Dull brown; beneath white, brown spotted; head and neck varied white and rufous, and brown spotted; quills white tipt; tail brown banded. Cape of Good Hope.

Banded-sided Hawk. F. Pterocles, Temm. pl. col. t. 59, 139 jun.

Slate-coloured; beneath white; sides of the belly and flanks transversely waved with rufous; tail white, with a black subterminal bar. Brazil. Length 16—17 inches.

Spotted Buzzard. F. poecilonotus, Cuv. pl. col. t. 9.

White; wings black, white spotted; tail with a black band, its base and tip white; bill black; legs yellow. Guiana.

Short-tailed Falcon. Falco ecaudatus, Lath.; *Le Batteleur,* Le Vail. O. A. t. 7, 8.

Head, neck, and beneath black; back and tail deep rufous; scapulars dusky, varied with gray; quills silver gray; tail very short. Cape of Good Hope. Larger than the osprey.

Whitish Buzzard, F. albidus, Cuv. pl. col. t. 29.

Crested; feathers deep brown, white spotted and tipt; tail three banded; head and lower parts white; head and back of neck spotted; breast and belly streaked, and thighs banded with brown. Pondicherry. Twenty six inches. Has some affinity to *Cymindis.*

Mantled Buzzard, F. palliatus, Marg. Cuv. pl. col. t. 204.

Feathers above dark brown, red edged; quill finely black banded; tail four black banded; head and lower part white, obscurely striated; occipital streak black; tarsi hid. Brazil. Nineteen inches.

Grey-cheeked Buzzard, F. Poliogenys, Temm. pl. col. t. 325.

Cheeks gray; throat white, with a longitudinal ashy band; above reddish brown; quills inner edge white, tips black; tail with four black bands; chest brown; belly and thighs white, with broad brown bands. Isle of Leçon. Length seventeen inches.

F. polyosoma, Quoy and Gaim. t. 14.

Cera and feet yellow; tail whitish, cross-lined with brown; tips black-edged; wings long. Malouin Islands.

F. desertorum, Vieil. O. Amer. Sept. t. 17, is most likely a variety of one of the other American species; as is also Buteo Americanus, t. 6.

The BUZZARDS, *Busards of Cuvier*, (CIRCUS, *Bechstein.*)

Differ from the last by having the tarsi more elevated, and by a sort of collar which the tips of the feathers covering the ears form on each side of the neck.

There are but two species in France which, by the variations in their plumage, have been multiplied by nomenclators.

The Buzzard, (F. pygargus,) Enl. 443 and 480.

Brown; above white, spotted with brown underneath, and the rump white. *L'Oiseau de Saint-Martin*, or *Hen Harrier*. (*F. cyaneus* and *F. albicans*,) Gmel. Enl. 459.

Ashy, with the quill feathers of the wings black, Appears to be no other than the old male Buzzard.

It is also the *F. communis, E. albus,* Frisch. t. 80; the *F. montanus,* B.; *F. griseus,* and the *F. Bohemicus,* Gm.

The *Grenouillard,* Vaill. O. A. t. 23, *F. ranivorous,* Shaw, is only the Buzzard; as is also the *Circus Hudsonius* of Viellot, American Birds, t. 9. *F. Hudsonius,* Lin. Edw. t. 107, is, perhaps, a variety of the common Bnzzard, not ascertained for certain; and when young, *F. uliginosus,* Gmel. Wils. t. f.

Colonel Montague first made this observation, and united them together under the name of *F. cyaneus,* adopted by Temminck. Found also in America. Called the *Marsh Hawk, F. uliginosus,* Wilson, t. 51, f. 1, Bonaparte, A. O. t 11, f. 1.

The *Harpy,* or *Moor Buzzard.* (*F. rufus,* Lin.) Enl. 460, (not 470.)

Brownish and red; the tail and the primary quills of the wings ashy. The Buzzard, (*F. æruginosus,* Gm.) Enl. 424; brown, with bright yellow on the head and breast; is the same bird at a year old. This bird generally resides near water, and preys on reptiles.

Montague Buzzard. F. cineraceus, Mont. Orn. Dict. t. ♂; Gal. Ois. t. 13; Naum. Voy. iv. t. 21, jun.

Confounded with the Hen Harrier, but the wings reach to the end of the tail, and the third quill is the longest.

The exotic species are

Winter Falcon, F. Hyemalis, Wils. A. O. t. 35, f. 1.

No collar round the face; wings, when closed, reaching but little beyond the middle of the tail; brown skirted, with ferruginous. When young, the Red-shouldered Hawk, *F. lineatus,* Wilson, t. 53. f. 3. North America.

Long-legged Falcon, Lath. *F. Acoli,* Shaw, Vaill. O. A. t. 33.

Breast with fine dusky linear stripes; legs very long, yellow; tail pale gray, long, end square; quills dusky black. Cape of Good Hope. Size of the Hen Harrier.

Salvador Falcon, Lath. *F. palustris,* Pr. Max. pl. col. t. 22.

Pale brown; beneath yellow red, with longitudinal brown stripes; throat deep brown; quills and tail cinereous gray, with brown cross stripes; eyebrows white. Brazil. Length 29.20 inches.

Golden-red Falcon. F. rutilans, Lechst. pl. col. t. 25; the *Aquila Buson,* Spix.

Golden red; beneath transversely striped with dusky; head streaked longitudinally; back and wings with cinereous brown spots. South America. Length 18—20 inches.

Black and white Falcon. F. leucomelas, Illiger, Azara, n. 28; the female *F. frenatus,* Illiger, Azara, n. 33. *Circus campestris,* Vieil.? From Brazil.

Quoys Buzzard, F. Historionicus, Quoy and Gaimard, Frey. Voy. t. 15, 16.

Above gray; beneath white, cross-barred with brown; cere and feet yellow. Malouine Islands..

Naked-cheeked Buzzard. F. gymnogenys, Temm. pl. col. t. 307; Son. Ind. t. 103.

Upper part and neck bluish gray; back and beneath finely banded black and white; wing coverts black spotted; quills and tail black, white tipt; tail with a white band. Madagascar. 21 to 25 inches.

Black and White Indian Falcon, Lath. *F. melanoleucus,* Lath. Indian. Zool. t. 2; *Le Tchong,* Vaill. O. A. t. 32; Sonnerat. ix. t. 182.

White; head, neck, back, axillæ, and quills black; feet yellow. India. Length sixteen inches.

The *Circus axillaris* of New Holland; *C. leucocephalus* and *C. rufulus, C. albicollis, C. malanopterus,* and *cinereus,* all from Paraguay, named by Vieillot, from Azzara, descriptions, and *C. variegatus* of South America; perhaps, belong here.

The Snake-eater, or Secretary (*Serpentarius,* Cuv. *Gypogeranus,* Ill.; *Gypogeranidæ,* Vigors; *Ophiotheres,* Vieillot);

Is a bird of prey of Africa, which has the tarsi at least as long again as the last, which caused it to be located by many naturalists with the grallæ; but these legs, entirely covered with feathers, the beak bent and cleft, the prominent eyelids, and all the details of its anatomy, place it in the present order.

The tarsi are shielded, the toes short in proportion, the region round the eyes denuded: there is a long rough crest on the occiput, and the two intermediate quill feathers of the tail greatly exceed the rest. It inhabits the dry and barren places in the environs of the Cape, where it pursues the reptiles; hence, it has the claws worn down by use. Its principal strength is in the leg. It is the *F. serpentarius* of Gm. Enl. 721.

The *Vultur Serpentarius* of Lath. and the *Secretarius reptilivorus* of Daud. figured; Miller Cym. Phys. t. 28.; Petiver Gaz. t. 12, f. 12; Phil. Trans. lxi. t. 2; Le Vaill. O. A. t. 25, copied by Shaw; and Lath. Hist. t. 7.

Nocturnal Birds of Prey *

Have the head large; very large eyes, directed forward, surrounded with a circle of slender feathers, the anterior of which cover the cera of the beak, and the posterior the opening of the ears. The enormous pupils of their eyes permit so much light to enter, that they are blind in open day. Their skull is thick, but of a light substance, with large cavities which communicate with the ears, and probably increase the sense of hearing; but their apparatus for flight is not very powerful; the *os furcatum* has no great resistance: their feathers, with soft barbs, and very downy, make

* Speaking of the divisions of this genus, an excellent ornithologist has observed "All these divisions are unsatisfactory as generic, not having, at least, *external* characters sufficiently distinct to constitute even sections."

no noise in flight. The external toe is capable of a forward or backward direction, at the will of the animal. These birds fly, generally, during twilight and moonshine. When attacked, or struck by any new object, in the daytime, they raise themselves up without flying, and assume ridiculous postures.

Their gizzard is muscular, although they subsist on animal matter, principally mice, little birds, and insects, but it is preceded by a large crop: their cœca are long and enlarged at the bottom. Some birds have a natural antipathy to these, and unite from all parts to assault them; hence, they are employed to draw birds to the net. There is but one genus made of them—

STRIX, *Lin.*

Which may be divided by their tufts of feathers usually called horns, the size of their ears, the extent of the circle of feathers which surrounds the eyes, and some other characters.

The species which have round the eyes a large complete disk of fringed feathers, surrounded itself by a circle or collar of scaly feathers, and between the two a large opening for the ear, are more removed in form and manners from the diurnal birds of prey than those whose ear is small, oval, and covered by fringed feathers, which extend only below the eye. Traces of this difference are distinguishable even in the skeleton.

Among the first species we shall name

The HORNED OWLS. OTUS, (Cuv.)

Such as have on the forehead two plumes of feathers, which are erected at pleasure, and whose ear conch extends in a half circle from the beak toward the summit of the head, and is furnished in front with membranaceous opercula. Their feet have feathers down to the talons. Of these there are in Europe,

The *Short-crested Owl.* (*St. ascalaphus*, Savig.) Brit. Zool. tab. b. iii. pl. col. t. 57.

One-fourth longer than the common species, and like it yellow, dotted with brown, and vermiculated on the wings and back, but the belly striped across with narrow lines, and the crests very short. Of Africa, but sometimes appears in Europe.

The *Common long-eared Owl.* (*St. Otus*, L.) Frisch. 89, Brit. Zool. t. 434, f. 1; Wilson, A. O. t. 51, f. 3.

Yellow, with longitudinal brown spots on the body, vermiculated with brown on the wings and back; crests half the length of the head; eight or nine bands on the tail. The *S. Mexicana et Americana* differs from this only in the spots being blacker and less diffused; but is considered distinct by the American ornithologists.

The *Short-eared Owl*, and *Brown Owl.* (*St. ulula* and *St. brachyotos*, Gm.) Enl. 438; Frisch. 100, Brit. Zool. t. b. iv. f. 2.

Nearly like the preceding as to colours; the back not reticulated, but narrow lines upon the belly, and four or five brown bands on the tail. The crests are only found in the male; they are so small and so seldom erected, that they have scarcely ever been remarked, or the species has been placed among those without crests, or has been divided. Also found in America.

This species has also been called *St. stridula*, *S. palustris*, *S. tripennis*, *S. arctica*, *S. accipitrina*, Pallas; and *S. tripennis* and *S. brachyura* by various authors.

Among the foreign species may be remarked.

The *Great American Horned Owl.* (*Str. bubo Magellanicus et St. Virginiana*, Gm.) Enl. 585. Edw. 70. Daud. ii. 13. *Jacurutu* of Marg. *Nacurutu* of d'Azara, (Wilson, O. A. t. 50, f. 1, and *B. pinicola*, Vieil. O. A. t. 19.)

Nearly as big as our great horned owl, striped across with brown underneath; brown, sprinkled with black, above. It is spread from one extremity of America to the other, and lives in the woods.

"Intermediate, between *surnia* and *ulula*," C. Bonaparte.

There is a species, a fourth smaller, at the Cape of Good Hope.

Spotted-eared Owl. St. maculosa, Vieil. Gal. Ois. t. 23; *St. Africana*, Temm. pl. col. t. 50.

Black; face and upper part of neck barred with brown, ash, and whitish; head and back spotted with white; quills banded brown and white; tail beneath brown, with five white bands; feet feathered. South Africa. Length 16—18 inches.

Oriental Eared Owl. St. orientalis, Horsf.; *St. strepitans,* Temm. pl. col. t. 174.

Brown, with ferruginous bands; shoulders, axillaries, belly, and shins white, banded with brown. Java and Sumatra. Length twenty.four inches.

Large-Billed Owl. St. Macrorhynchus, Temm. pl. col. t. 62.

Variegated brown, red, and whitish; beneath whitish, transversely banded with brown; breast white, dashed with brown; beak large. North America. Length nineteen inches.

White Horned Owl. St. lactea, Temm. pl. col. t. 4.

White, varied with brown, and striped with gray; beneath varied with brown; quills and tail yellow banded; wings with five large spots; tarsi white; toes naked. Senegal. Length twenty-four inches.

Long-Billed Owl. St. longirostris, Spix, N. A. t. 9, a.

Reddish above, and beneath streaked with brownish black; throat and below the eyes ferrugineous; bill long; legs long, hairy to the claws; wing shorter than the tail. Brazil. Length sixteen inches.

Noisy Owl. St. strepitans, Temm. pl. col. t. 174.

Dusky, waved with reddish; beneath whitish striped with brown; tail tips white; tarsi white, barred with brown. Length nineteen inches; toes yellow, naked. India.

We may keep the name of

HOWLERS, (ULULA, *Cuv.*)

For the species which have the beak and the ears of the last division, but not their crests. We have none of them in France, but they are found to the north in both continents; as, for example,

The Great gray Howler of Sweden. (*St. litturata*, Retzius.)

Nearly as large as a great-horned owl; mixed with gray and brown; above whitish, with longitudinal gray-brown spots beneath. It inhabits the mountains in the north of Sweden.

The *St. laponica*, Retz; not *St. litturata*, which is the Hawk Owl.

The Howling Owl of Canada. (*St. nebulosa*, Gm.) Wilson, A. O. t. 33, f. 2.

Rather less than the last; the neck and chest barred across brown and whitish; the back brown, with whitish spots; the belly whitish, with brown meshes; tail longer than the wings. Europe and North America.

STRIX, *Savigny*.

Have the ears as big as those of the eared owls, and provided with an opercule, which is still larger than that of those species; but their elongated beak bends only towards the end, while in all the other subgenera it is arched from the point. It is without crests; the tarsi are feathered, but they have nothing but hair on the toes. The mask formed by the fringed feathers which surround the eyes a more extent, and gives their physiognomy a more extraordinary appearance than in the other species.

The Common White or Barn Owl. (*St. flammea*, L.) Enl. 440; Frisch. 97; Wilson, O. A. t. 50, f. 2.

Appears to be spread all over the globe. Its back is clouded with yellow and ashy; a brown, prettily sprinkled with white dots, each dot inclosed between two black points; and the belly sometimes white, sometimes yellow, with or without brown sprinkling. It builds in towers and belfries; and it is this which the people consider especially as a bird of bad omen.

The *Strix Sylvestris*, *St. rufa*, *St. noctua*, *et St. alba* of Scopoli, and *St. Soloniensis* of Gmelin, and interlaced in his system, are too undetermined to be regarded but as varieties, and probably of this species.

St. Javanica, Gm. is the same; and, perhaps, the *Mouse Owl*, Lath. Hist. from New Holland.

The *Bay Owl.* *St. Badia*, Horsf. Zool. Java, t. pl. col. t. 318.

Bay, spotted with black; beneath pale; throat and chin white, with a brown collar; toes naked, rough, scaly. Length twelve inches. Java.

Tuidara Owl, *St. perlata*, Licht. not Vieil. *St. Tuidara*, n. *Tuidara*, Marcgr. *Effrayé*, Azzara, 46. Like *S. flammea*, but the legs are longer. Brazil.

The Syrnii. (Syrnium, *Savigny.*)

Have the disk of the fringed feathers and the little collar like the last; but the conch is reduced to an oval cavity, which does not occupy a half of the height of the cranium. They have no crests, and the feet are feathered to the nails.

The *Wood Owl of England.* (*St. aluco et stridula,* L.) Enl. 441, 437; Frisch. 94, 95, 96.

Is a little larger than the common or barn owl; covered all over with longitudinal brown spots marked on the sides with transverse indentations: there are some white spots on the skull and toward the anterior edge of the wing. The bottom of the plumage is grayish in the male, reddish in the female; whence the sexes have long been considered as two species. These birds build in the woods, or often lay in other birds' nests, and retreat into the old trunks of trees.

Brazilian Owl. *St. hylophila,* Temm. pl. col. t. 373.

Banded reddish brown and black; face pale brown, with four black bands; head and neck bay, with black crescents; chin white, black banded; belly white, with black edged bay crescents. Brazils. Thirteen inches.

We reserve the name of

DUCS, (BUBO, *Cuv.*)

For the species which have the conque as small, and the disk of feathers less remarkable, than the Syrnii. They have crests. That which is known by thick legs feathered to the nails, is

The *Great Horned Owl.* (*St. Bubo*), Enl. 434, Frisch. 94.

The largest of the night birds; yellow, with brown stippling on each feather: the brown prevails most above, the yellow underneath: the crests are nearly black.

Supercilious Owl, Lath. *St. grisseata*, Daud.; *St. superciliosa*, Shaw, (Vail. O. A. t. 43.)

Are other great-horned owls, with the crests or tufts wider from each, and placed more backward, and are erected with difficulty above the horizontal line. But one is known of Guiana, with a red or brown plumage, finely striped with blackish; the crests or tufts white at their internal edge, and some drops of clean white on the wings.

Tarsi hid by the leg feathers, clothed with a few fine hairs. Is it not rather a *Surnia?*

Others have all the appearance of the Ducs; but the tarsi and toes are quite naked, shielded in front and reticulated behind.

Hardwick's Naked-legged Owl. St. Hardwickii, n.

Pale brown; feathers of the upper part marked with a broad longitudinal band; beneath marked with a narrow longitudinal band, and some obscure cross ones; wings and tail banded with deep brown. Length twenty-two inches. India. Perhaps, the *Hutum Owl*, Lath. Hist. t. 13.

The FALCONINE OWLS. NOCTUA, (*Savigny.*)

Have neither crests nor wide or concave conchs to the ears, the opening of which is oval, and scarcely larger than in other birds. The disk of fringed feathers is smaller, and even less complete than in the Bubo.

Some are remarkable by a long, wedge-shaped tail. They have the toes very feathery, and are called

Hawk Owl, the *Surnia* (Dumeril). It seems that some species or varieties exist throughout the north. These are nearly allied, and badly distinguished under the names *St. funerea, Hudsonia, uralensis, accipitrina,* &c.

Hawk Owl, St. funerea, Lin. pl. Enl. 463,

Is the best known species of Siberia. Blackish brown above, with white spots in little drops on the head in transverse bars on the top of the head, and striped transversely white and brown underneath, with ten transverse white lines on the tail. This species hunts more by day than by night.

See Wilson, A. O. t. 50, f. 6.

This species is also *St. Hudsonia* and *St. ulula* of Gmel. and *St. Nisoria* of Meyer. Is found in North Europe and America. It is different from

Ural Owl, Lath. *St. Uralensis,* Pallas, Lepechin Voy. ii. t. 3. pl. col. t. 27.

Whitish, with large longitudinal spots; face whitish; tail greatly wedged, much longer than the wings. Arctic Regions. The *St. litturata,* Retz, not Cuvier; when young, *St. macroura,* Meyer.

The *Falconine Owl,* Lath. *Choucou,* Vail. O. A. t. 38; *St. Africana,* Shaw.

Of Africa. Entirely white underneath, with fourteen or fifteen lines on the tail; and, according to him, more nocturnal than the others.

Variegated Owl. St. Nisuella, Shaw, Vail. O. A. t. 39.

Brown, shaded, mixed with white; beneath barred with brown and white; tail banded dusky brown and rufous white, one-half longer than wings; eye-disks white, with dusky markings. Of Africa.

Coquimbo Owl, Lath. *St. cunicularia*, Molina? *St. grallaria*, Temm. pl. col. t. 146; Bonap. A. O. t. 7, f. 2.

Cinnamon gray, spotted with white; beneath white, spotted with brown; tail even, a little longer than the wings; feet with scattered bristles. North and South America. The *Urucurea*, Azara 47.

Others have the tail short, and the toes feathered. The largest, and at the same time the largest night bird without crests, is

The *Snowy Owl*, or *Harfang*, *(St. nyctea,)* Enl. 458;

Which nearly equals *St. bubo* in size. Its plumage, white as snow, is marked with transverse brown spots, which disappear as the bird gets old. It inhabits the north of both continents; builds on elevated rocks; hunts hares, moor-game, and ptarmigans.

The *White Owl* of Vaill. O. A. t. 45, is only an old Harfang, badly prepared.

In other parts of Europe there are much smaller species, as

The *Common Passerine Owl*. (*St. passerina et Tengmalmi*, Gm. *St. pygmæa*, Bech.) Enl. 439; *La Chevechette*, Vail. Ap. 46.

Scarcely larger than a blackbird; deep brown, with a white throat; brown round spots on the wings and breast; four white lines on the tail. There are several species nearly allied to this in America and in India, &c.

The *Red Passerine Owl* (*St. passerina*, Meyer and Wolf.)

Of a redder tint, both on the brown and on the white; a whitish half collar on the neck; some triangular red spots on the sides of the tail; the toes only covered. It is still less than the last, and in the head is almost altogether assimilated to the sparrow-hawk.

The history of the small Pàsserine Owl of Europe is not as yet clear. Almost every ornithologist has regarded the smallest species as the *St. Passerina*; whence has resulted the greatest confusion in the Synonyma.

Little Owl, Lath. *St. Passerina*, Lin. Edw. t. 228. pl. Enl. t. 439.

Size of a jay. Toes covered with a few white hairs; feathers of the head with a long pale line. Europe, Egypt, and Nubia. England, (*Edw.*) This is the *St. noctua*, Retz; *St. nudipes*, Nilson, not Daud.; the *Noctua* of the ancients, the emblem of Minerva.

Tengleman's Owl. St. Tenglmalmi, Gmel. Penn. B. Z. fol. t. B. 5. Gal. Ois. t. 23.

Size of a jay; toes and tarsi covered to the claws with a thick velvet; head feathers each with two rows of white dots. *S. darypus*, Bechst.; *St. noctua*, Tengm. and *St. fünerea*, Lin. Fauna Suec. Europe, the *St. Passerina* of Montague's collection.

Arcadian Owl, and *Dwarf Owl,* Lath. *St. Arcadia,* Gm. Vail. O. A. t. 46. Wils. A. O. t. 34, f. 2.

Size of a Blackbird. Tarsi and toes thickly downy, dark brown, spotted with white; beneath whitish, red spotted; tail as long as the wings. North Europe and America. So also *St. passerina,* Retz and Wilson, *St. pusilla,* Daud., and *St. pygmea,* of Bechst.

Dwarf Owl, St. pumilla, Illiger, pl. col. t. 39.

Red brown, spotted with white and black; beneath variegated red and white; tail dusky, with band formed of white spots. South America. Length five inches. *Carburé,* Azara, 49.

Ferruginous Owl, St. ferruginea, Br. Max. pl. col. t. 199. *St. phalænoides,* Vieil.

Red beneath, whitish striped with rufous; scapulars spotted whitish yellow; tail red; in young, brown barred. Brazil.

Chestnut-winged Owl, St. castanoptera, Horsf. pl. col. t. 98.

Transversely lined gray and dusky; scapulars and back chestnut; belly varied white and chestnut. Java. Length eight inches.

Pearl Owl, St. perlata, Vieil. (not. Licht.) Vail. O. A. t. 284.

Reddish brown, white spotted, and striped: cheeks, throat, and crop white, black shaded; crest red, varied with black; bill yellowish brown; toes hairy. Senegal.

Occipital Owl, n. *St. occipitalis*, Temm. pl. col. t. 34.

Varied brown and yellow, spotted with white; beneath whitish, striped with rufous; forehead and vertex rufous white, dotted; quills banded red and brown. Africa. Length seven inches. Toes and tarsi downy.

Sparrow-like Owl, *St. passerinoides*, Temm. pl. col. t. 344.

Gray-brown head, and white dotted; scapulars and wings white spotted, and banded; face, throat, and beneath white; sides splashed with brown; tail black, with four white bands. Brazil. Six inches.

Others have the tail short and the toes naked. Cayenne has several very fine species, especially the three following:

The *Cayenne Owl*, Lath. (*St. Cayennensis*, Gm.) Enl. 442.

Irregularly and finely striped with brown on a yellow ground.

The *Fasciated Owl*, Lath. *St. huhula*, Daud. *St. lineata*, Shaw. (Vail. Afr. 41.)

Striped white on a black ground; four white lines upon the tail. It avoids the light so little, that it is called the Day Passerine Owl. The size of these two is that of the *S. passerina.*

The *Downy Owl*, Lath. (*St. torquata*, Daud.) Vail. Afr. 42.

Brown above, whitish underneath; round the eyes brown, with a brown band on the breast; the throat and eyelids white. It is larger than the *S. aluco.* It is the *Nacurutu sans aigrettes* of D'Azara.

The *Spectacle Owl*, Lath. *St. perspicillata*, Daud. Lath. Hist. i. t. 15; and the *Masked Owl*, *St. larvata*, Shaw; *St. personata*, Daud. Vail. O. A. t. 44, are perhaps var. of age of the last species.

There are some in America which have the tarsi as well as the toes naked; such is

The *Bare-legged Owl*, Lath., *St. nudipes* of Daud. Vieil. Amer., t. 16, fulvous brown, neck and wings white, spotted beneath with long brown spots: legs brown.

See also *St. griscata*, Daud. Vail. O. A. t. 43.

The Scops, (Scops, *Savigny.*)

With the ears flush with the head, have the imperfect disks and the naked toes of the last.

The *Scops*, (*St. Scops,*) Enl. 436.

Scarcely as big as a blackbird. Plumage ashy, more or less clouded with yellow, prettily varied with small longitudinal narrow black streaks, and transverse vermiculated gray lines, with a suite of whitish spots on the scapular, and six or eight feathers to each crest. It is a very pretty little bird.

Red and Mottled Owl, *Str. asio*, Lin. *St. nævia*, Wilson, A. O. t. 19, f. 1.

Dark brown, mottled with black, pale brown, and

ash; wings spotted with white; beneath white, mottled with black and brown; tail even; feet covered with short feathers. North America. Length 8—10 inches.

The old birds, *St. nævia*, Gmel., and *St. alba*, perhaps *St. albifrons* of Latham.

Black-headed Owl, *St. atricapilla*, Natterer, pl. col. t. 145.

Yellowish, varied with black and brown; beneath white, with longitudinal stripes, spot, and zigzags of brown; head black, occipital band white, dotted with black; neck with a yellow spotted collar. Brazil. Length ten inches.

White-eared Owl, *St. leucotis*, Temm. pl. col. t. 26.

Brownish-white, beneath paler; feathers with the longitudinal shaft, and tips black, and reddish zigzags; face white; ears barred black; quills and tail ash coloured, waved with brown. Senegal. Length six inches.

Indian Owl, *St. Leschenaultii*, Temm. pl. col. t. 20.

Brown-red, striped with red; beneath reddish, transversely waved with brown; tarsi naked, blue; toes scaly.

Lempyi Owl, *St. Lempyi*, Horsf.? *St. noctula*, Temm. pl. col. 99.

Black or brownish, marbled with reddish; beneath reddish white, waved and spotted; neck with two collars, upper white, with brown spots, lower black, with reddish white spots. Java.

Crossed Owl, St. choliba, Vieil. *Strix decussata,* Licht. *Choliba,* Azara, n. 48.

Abdomen white, crossed by narrow brown lines. Length nine, tail three, tarsi one and a quarter inches. Bahia.

Sonnerat's Owl, St. Sonnerati, Temm. pl. col. t. 21.

Red-brown, beneath white, transversely barred with brown; head, and wing covers, white, spotted; eye-disks, face, and throat, reddish white; tarsi and toes red, downy. India. Length eleven inches.

Indian Owl. St. Brama, Temm. pl. col. t. 21.

Dusky brown, varied with white; beneath whitish, transversely spotted with brown; eyebrows and collar white, with ashy gray lunules; quills and tail with white bands. India.

Pagoda Owl. St. seloputo, Horsf. *St. pagodarum,* Temm. pl. col. t. 230.

Above rusty chestnut, with obsolete cross bands; beneath white, with deep rusty chestnut bands; throat white; face and eyebrows yellow-red. India and Java. Length eighteen inches.

Hairy Owl. St. hirsuta, Temm. pl. col. 289.

Brown, beneath whitish, brown spotted; forehead and ceres white; top of head and nape ashy brown; throat reddish; tail brown, with four ash bands, and white tip; toes marbled red and brown, edge naked, with yellow tubercles. Ceylon.

Mauge's Owl. St. maugei, Temm. pl. col. t. 46.

Ashy red, beneath rufous, spotted with white; scapulars and wing covers spotted with white; quills and tail feathers barred dusky and brown: throat ashy. West Indies.

Cross-bearing Owl, St. cricigera, Spix, Brazil. t. 9.

Above gray-brown, beneath dirty white, with brown longitudinal cross bands; thighs red, and tarsi short, rather woolly; feathers of the back white streaked.

White-edged Owl, St. albomarginata, Spix, Brazil. t. 10. a.

Brownish black, above and beneath purely white, waved; tarsi gray-black, woolly; tail black, with four narrow bands, and tips white*.

* There have been many other Accipitres described as separate species by some naturalists; but as they have not been figured, and are not, perhaps, otherwise sufficiently authenticated, I have thought proper to omit them. In forming my present list, I have chiefly depended on Cuvier's Notes, Temminck's Manual and Coloured Figures, and Prince Musignana's excellent examination of American birds.—J. Ed. Gray.

Generic Characters of Birds.

ORDER ACCIPITRES.

Fam. I. Diurnal

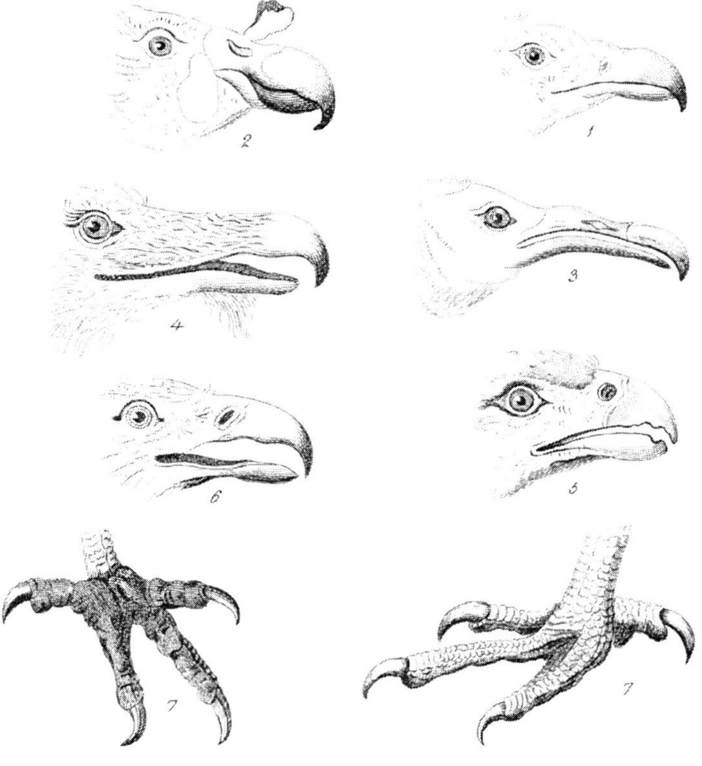

Fam. II. Nocturnal.

1 Vultur
2 Sarcoramphus
3 Percnoptera
7 Claw of Accipitres

4 Gypaetos
5 Falco nobilis
6 Aquila ignobilis
8. Strix

London Published by G. B. Whittaker, & Co. March. 1829

SUPPLEMENT ON BIRDS IN GENERAL.

Ornithology, the science of birds, includes two great divisions: 1st, All that relates to the physicalities of the class, and to their manners, habits, instincts, and intellectual qualities; and 2dly, The artificial classification of the species into orders, genera, and minor subdivisions, to assist us in the study of the interrelative peculiarities of the several species. As we shall dismiss the consideration of the second of these divisions in a very few words, it may be more convenient to enter upon it first.

Artificial ornithology, in common with the other branches of zoology, is attended with all the difficulties in which matters of indefinite excellence and human invention must ever be involved. True it is, that what is called a natural method has a point to arrive at in assimilating, or rather identifying, itself with the divisions of nature; but, as has been before observed, these divisions are, in fact, rather fancied than real; and however decided they may appear on a superficial view, close examination will soon detect the links which connect them with each other. Hence all systems, though founded on nature, must be to a great extent artificial, or the objects of their creation will be defeated; for, to adopt all the aberrations of nature would be to describe all the species.

Since his time, the system of Linnæus has very generally and very deservedly prevailed in the arrangement of this class. That of our author may, by no great latitude of expression, be said to be an improvement of his; and the *Règne Animal*, in this respect, may be considered an improved edition of the *Systema Naturæ*. Several other systems have also arisen; but the celebrity of the men, and the intrinsic merits of those of Linnæus and Cuvier, have fixed the public choice on them, and will in all probability consecrate their systems to general use when the others are neglected or forgotten. With a view, however, of condensing as much information as possible, we propose giving a brief view of such proposed arrangements of this class as deserve consideration, and proceeding immediately to the physicalities, &c. of the species.

Aristotle did not treat of birds in a very methodical manner. In the third chapter of his eighth book on animals, he notices the various modes in which birds subsist; observes that some are carnivorous, others granivorous, and others polyphagous; that some take their food on land, and others seek it in the waters. He speaks subsequently of birds which disappear in winter; and afterwards gives an enumeration of the species then known, under their names merely, and for the most part without description, so that it is impossible to recognise them. In one chapter, however, he treats of the eagles pretty largely, and especially of their habits.

Pliny, in treating of birds, notices a tolerable number of species, but neither describes nor classifies them. Belon (whom we have noticed in our preliminary sketch) is the first author in whom we find any thing like the elements of classification on this subject. His work, very remarkable for the period in which it was written, contains very just notions concerning the analogy of structure between the birds and mammalia, more especially in his comparisons of the skeletons. The divisions of chapters in his first book proved that he was well acquainted with those points of ornithology which must form the foundation of that science. In his second book he treats of birds of prey, diurnal and nocturnal; and the order in which he considers them, commencing with the vultures, &c. has not been changed by modern naturalists. He places the cuckow at the end of the diurnal birds of prey, and something in the form of the feet and colours of the plumage seemed to justify this approximation. But falling into the same error as the ancients, he places the bat among the nocturnal birds. The third book treats of palmipedes, such as ducks, cormorants, pelicans, &c. The fourth, of river-birds not palmipedes, as the crane, heron, ibis, curlew,

&c.; but among them are birds very different in organization and habits, such as the martin, fisher, &c. In the fifth book he treats of land-birds that construct their nests on the ground, as the ostrich, peacock, land-curlew, partridge, pheasant, quail, &c. To these, pretty exactly approximated together, Belon adds others that have but little analogy with them; for instance, the plover, the lark, and the woodcock. To this he was led by the peculiar habit which he selected as the characteristic of these birds, namely, the position of their nests. The sixth is occupied with birds of various habits and omnivorous diet, as crows, ravens, jays, pies, perroquets, pigeons, &c. The seventh and last describes birds that haunt the hedges, bushes, groves, &c. as the nightingale, linnet, tomtit, canary, sparrow, &c. &c.

Belon does not group the species into genera, but in general approximates together those which have naturally the greatest affinity. We may discover, however, in his work, indications of more general divisions of considerable value, and which may be termed *orders*. The second book, for example, evidently corresponds to the order denominated by modern naturalists accipitres, raptores, zoophagi, or birds of prey: the third, to the order palmipedes: the fourth, for the most part, to the grallæ. The first part of the fifth comprehends all the gallinæ; and the rest of it and the sixth contain those birds so difficult to characterise in a general way, and which constitute the order passeres.

Gesner's book, though full of erudition and very excellent remarks on the birds of Switzerland, is alphabetical in its arrangement.

Aldrovandus, though he gives no new descriptions, has yet classified all the species known in his time. He has not admitted genera, but he has established groups which may be compared to what we now term *families*. He was an indefatigable and indiscriminate compiler, and has swelled out his book to three folio volumes. His first volume contains twelve books, of which the following are the titles :—1. Of eagles in general: 2. Of eagles in particular; where there are many chapters on the chrysaëtos, haliætos, pygargus, morphnos, percnopterus, ossifragus, &c. of the ancients: 3. Of vultures in general, where many species of these birds are distinguished: 4. Of accipitres in general: 5. Of accipitres in particular; the sparrow-hawk, buzzard, the merlin, kite, cuckow, &c.: 6. Of falcons in general: 7. Of falcons in particular, in which the various species or races of falcons employed in hawking are described: 8. Of nocturnal birds of prey, as the great horn-owl, the owl, screech-owl, &c.: 9. *Birds of a middle nature, between birds properly so called and quadrupeds*, as the ostrich and bat: 10. Fabulous birds, griffins, harpies, &c.: 11. Of perroquets, in which many species of cockatoos, &c. are described: 12. Of ravens in general, and of some other birds which have a hard and powerful beak: here are noticed not only ravens, crows, and pies, but also the calao, birds of paradise, toucans, &c. &c.

The second volume contains six books: 13. Of wild gallinaceous birds, as the peacock, partridge, quail, &c.: 14. Of gallinaceous tame birds, as the domestic cock and all its varieties: 15. Of birds which, like the last, possess the triturating power, and yet seek water, as the different varieties of pigeons, turtles, and certain passeres, and inhabit the neighbourhood of streams: 16. Baccivorous birds, as thrushes, blackbirds, &c.: 17. Vermivorous, or more properly insectivorous birds, as the wren, swallow, &c.: 18. Singing birds, as the nightingale, &c. &c.

The third volume has but two chapters: 19. Palmipedes, swan, &c.: 20. Birds frequenting banks and shores, as cranes, herons, flamingo, woodcock, &c.

Johnston merely compiles from Gesner, Aldrovandus, and others. Gross errors are remarkable in his arrangement, such as placing the parrot and the ostrich among the birds of prey, and other such like inconsistencies. Nevertheless, his method, which is essentially that of Belon, does yet still form the basis of those which have been definitively adopted by modern naturalists, with this difference, that the latter are based on characters derived from external forms which were not used by Belon and Johnston.

The Ornithology of Willoughby, which appeared in 1678, is the origin of methods founded on external characters. The forms of the beak and feet are particularly adopted as the basis of his divisions; and like the naturalists before mentioned, he uses the habits and modes of subsistence as distinctive of the groups which he admits, and which are twenty in number. The first eighteen divisions are composed of terrestrial birds, and the two last of aquatic.

Ray, in his Synopsis, follows with very little variation the method of Willoughby. He uses, however, new characters derived more especially from the number of feathers in the tail, and the internal structure of the body.

Barrère, in 1741, instead of profiting by the judicious direction given to ornithology by the two last-mentioned writers, published a method totally artificial, in which the most different birds are ranged side by side, and those which approximate most in their organisation are separated by considerable distances.

The work of Klein is another artificial system just as unsatisfactory as that of Barrere. He has founded his first division on the number of toes, which has led him to class in one family birds totally different in all the rest of their organisation, and in their mode of living.

On the arrangement of Linnæus we shall not dilate, as we have already laid a tabular view of it before our readers in another place. We shall merely remark that his classification is one of the best that has ever been published in respect to the divisions and subdivisions of orders. Four of these orders are still generally retained; namely, the accipitres, grallæ, gallinæ, and anseres. Some genera, indeed, are not placed suitably to the characters of the division under which they are found:

for instance, motacilla is ranged in the section of passeres simplicirostres, whereas, from the character of the beak, it should be under emarginatirostres.

The method of Brisson is purely artificial. It is composed of twenty-six orders, and one hundred and fifteen genera. The birds are classed, 1st, according to the presence or absence of the membranes uniting the toes, and according to the greater or less perfection of such membrane where it does exist: 2d, according to the number and disposition of the toes: and 3rd, according to the form of the beak. The birds whose toes are without membranes compose the first seventeen orders. Those which have four toes and the legs covered with feathers to the heel are contained in the first fourteen. Those which have the four toes separated from their commencement are confined to the first thirteen. Those which have three anterior toes and one posterior are confined to the first twelve. The last nine orders are composed of birds whose toes are furnished with membranes in their entire length. We shall not trouble our readers with any minuter analysis of this system.

Schæffer, in 1774, published a methodical distribution of birds, in which he uses for the distinction of orders only the characters furnished by the feet.

The method of our countryman, Latham, is pretty nearly that of Linnæus, with the addition of two orders, the first of which comprehends only the pigeons, and the second the ostrich. A third order, borrowed from Schæffer, contains the pinnatipedes, or birds with a cleft instead of an entire membrane like the true palmipedes. This writer also added several new genera to those already established.

In this brief notice of systematic writers, we must not omit the name of Mr. Vigors, whose observations on the nomenclature of ornithology, and whose improvements in the classification of certain families of birds are of the utmost value. This gentleman, to the profoundest knowledge of the subject, unites the power of adorning it by the most elegant style of composition; and of illustrating it from the most varied and extensive resources of erudition. We trust, that he will continue his labours on this department of zoology, in the full conviction that a truly scientific and luminous system of nomenclature will be their result.

M. de Lacépède divided birds into two sub-classes, the first characterised by having the lower part of the leg furnished with feathers, and no toes completely united by a wide membrane. This sub-class is again separated into two divisions and four subdivisions. The first division is characterised by thick and strong toes, two in front and two behind: the second, by three toes in front and one or more behind. The first subdivision has the claws strong and very crooked; the second, claws not much crooked; external toes free or united only along the first phalanx; the third, external toes united almost through their entire length; the fourth, front toes united at their base by a membrane.

The second sub-class is characterised thus: Lower part of the leg free from feathers, or many toes united by a wide membrane. First division: Three toes before, one or none behind. 1st subdivision: Front toes entirely united by a membrane; 2nd. Four toes united by a membrane; 3rd. Three toes before, one or none behind. Second division: Two, three, or four very strong toes. 1st subdivision: Toes not united by a membrane at the base.

M. de Lacépède makes forty orders, all distinguished by some peculiarity of the beak.

M. Duméril, in his *Zoologie Analytique*, admits the same orders as M. Cuvier, and subdivides them into a great number of families.

We might very considerably extend this account of the systems of various ornithologists if our object were merely to augment our work without increasing its interest or utility. But as we have more respect for the time and patience of our readers, we shall avoid any further details on so dry a subject.

The information, in fact, to be derived in this way amounts to little else than multitudinous lists of synonymes which no human memory could possibly contain, or, if it could, would not be much advantaged by the acquisition. We have frequently taken occasion to observe, in the course of our labours on the Mammalia, the great detriment arising to science from this vain and troublesome pedantry. As we proceed downwards in our researches on animal existence, we find ourselves more and more impeded by it. Nor is ornithology the branch of natural history that suffers least from its pernicious influence. We have not always been able to avoid it ourselves, nor indeed can any writer do so whose business it is to give an account of what has been done by his predecessors in zoology. But we can assure our readers that it is by no means our inclination to indulge in this parade of pretended science, and that our principal object of condensing within moderate limits as much useful and interesting matter as we can, shall not be lost sight of in the subsequent portions of our work.

It is, however, but justice to remark that ornithology involves great difficulties of classification, and that this will in some measure account for its multiplication of systems and synonymes. Birds are not interdistinguished by such strong leading characters as the mammalia. Their internal organisation is not so varied, nor are even their higher subdivisions characterised by the same strongly marked differences. When we consider the different orders of the mammalia, we find each of them distinguished by some leading organ ; some traits of conformation prescribing the absolute necessity of certain habits and modes of existence. This is the case, more or less, from man down to the cetacea. What can be better or more naturally defined than the quadrumana, the carnivora, the rodentia, the ruminantia, the cetacea? If, in some instances, the grand division of the carnassiers, and the pachydermata, are less so, it must be attributed to the reluctance of some naturalists, more especially our author, to the precipitate multiplication of orders, and partly, in the case of the pachydermata, to those gaps left in that order by the destruction of so many ancient genera and species. Indeed, as to the division of the carnassiers, it can only be considered as a provisional one. There can be no doubt, but that a more perfect acquaintance with some of its tribes must induce some alterations of arrangement; that, at least, the cheiroptera and marsupialia must be separated from it. Similar observations are applicable to the edentata, from which some modern naturalists have seen the necessity of separating the echidna and the ornithorhynchus. The genus equus might also, perhaps, be removed with propriety from the pachydermata. Setting aside such exceptions, if they be so, there can be no hesitation in deciding that the leading distinctions of the mammalia are, in general, much more striking than those of the birds, and the generic and specific distinctions are not less so.

These are obvious reasons for the difficulties of classification, and the temptation to multiply systems. But where this is the case, the only alternative of the naturalist who desires to be useful, is accuracy of description. We would not, like Buffon, abandon system altogether; because it aids the memory, and, if not conducted in a manner altogether arbitrary, serves to show the actual inter-approximation of beings in nature itself. But, after the example of that great NATURAL HISTORIAN, we would lay much greater stress on facts than systems. We would consider the faithful description of an animal, of its disposition, and of its habits, as of infinitely greater import to the progress of real knowledge, than the most complete exposition of all the systems of nomenclators, which, while they enable pedantic vanity to shine in the coteries of scientific fashion and folly, materially impede the study of zoology.

As the grand divisions and races of mankind have appropriated distinct portions of the earth as their habitations, so the grand divisions of the animal world are, for the most part, located in their exclusive domains. Thus it has been allotted to the quadruped to live on the earth, to the fish to cleave the depths of ocean, to the bird to wing the wide regions of the air, and it is not a little remarkable that each of these beings bear, in their respective natures, no small analogy to the element which destiny has prescribed for their abode.

The fish, continually immersed in a cold and relaxing fluid, possesses a softer texture of conformation, a moist temperament, and a great flexibility of organs, in accordance with the natural inconstancy of the waters by which he is surrounded. The quadruped, situated on a terrestrial and stony soil, has contracted a solidity of organization, and a weight of limbs, which retain him attached to the earth; while the bird, continually traversing the subtler atmospheric medium, inhaling in expansive lungs, and through their appendages and prolongations, a considerable quantity of air, which penetrates his entire system, even to his bones and feathers, must, of necessity, acquire the peculiar lightness, buoyancy, and activity which distinguish him.

We may observe, indeed, this adaptation of which we are speaking in various proportions in animals, according to the nature of their more usual habitat. Do we not find that water-fowl, retaining in their bodies a great quantity of the humid principle, are much more gross and heavy, than the agile and exclusive tenants of the air? Have not the gallinæ, such as the turkey, partridge, hen, &c., constantly living on the earth, contracted a weight of body, to which the races habituated to live in the high atmospheric regions are strangers? It is thus we find the aquatic mammifera, such as the hippopotamus, the lamantin, and the seal, much more stupid and heavy than those which live on dry ground. Even among these last, how much more lively and delicate are the gazelle, the chamois, the wild goat, and other natives of the mountains, than the quadrupeds of the valley and the plain? Even in the fish, which prefer light and limpid streams with sandy bottom, we find a texture more solid, fibrous, and compact, than in the flabby, and indolent inhabitants of stagnant and muddy waters. Nay, even man himself is not exempted from these local influences. He becomes lax of fibre, corpulent, and dull on the marshy plain and in the humid valley; light, lively, muscular, and energetic in the bracing breezes of the highland and the mountain.

The air, then, must be the most influential element upon the birds, which are perpetually immersed in this vast atmospheric ocean which surrounds our globe. Their whole organization is penetrated by it, as a sponge imbibes water. They have immense lungs, adhering to the ribs, provided with aërial sacs, insinuating themselves into the abdomen. Their bones, cellular texture, feathers—in short, all parts of their system, admit more or less air into their interstices. The sanguine system being thus in perpetual contact with the air, it is easy to imagine that the oxygenation of the blood must be more powerful and complete in birds than in any other animal. The respiration of the bird must be a combustion more ardent and rapid than ours. In fact, it may be considered a sort of fever, analogous to that incident to phthisical subjects, with this difference, that, instead of consuming the body, it warms and animates it with redoubled energy. It constitutes the predominant function of the economy of the bird, which is altogether proportioned to this peculiar source of vital energy. A slight consideration of the constitution of birds will prove this. Their flesh is dry and fibrous, their muscles exceedingly contractile and robust, their disposition lively and impetuous. They are ardent in the sexual intercourse, furious in combat, wild, irritable, and in perpetual motion. They sleep little, and eat much. They seem to have received from nature stronger sensations, more vital force and activity than other animals, for they live a very long time, and are yet of a temperament extremely warm. Quadrupeds are of a colder and more moderately-tempered constitution. They have neither the activity, ardour, lasciviousness, nor vehemence of disposition discernible in all the actions of the winged tribes. They dwell, for the most part, peaceably upon the earth, and man either subdues them to obedience with facility, confines them to the desert waste, or strikes them with terror by his hostility.

But the bird, the untamed denizen of the air, easily evades the tyranny of man. Independent in the solitude of his native skies, he has little to fear from the chains of captivity, or the constraint of domestication. The eagle, the condor, the swallow, the bird of paradise, shooting through the air on rapid and energetic wing, seem almost to despise those heavy species whom their weight attaches to the earth, and subjects to the dominion of man. It is only the races mal-organized for flight, and, so to express ourselves, the most terrestrial, that man has been enabled to subdue, the gallinæ, a grovélling and gormandising tribe, or geese, ducks, and other clamorous and voracious species, which prefer the wretched boon with which we repay their servitude, to poverty with independence. Man, indeed, abuses his power and dexterity in imprisoning, from infancy, the enchanting musicians of the grove. He rather retains them as captives by violence, than as subjects by domestication; they are slaves, not friends, and if they sing in their captivity, it is less for the purpose of charming their masters, than of distracting their own ennui, and solacing their own cares: for birds are still greater lovers of liberty than quadrupeds, and the most untameable among them are also the best organized for flight, and the most generally agile. The more their wings are powerful and extended, the more the pectoral muscles that move them are robust, the less are the legs of these same birds adapted for walking. The ostrich, which runs so admirably, cannot fly; but the swallow, the martin, the sea-swallow, the gull, which fly so well, have feet so small that they can hardly make use of them. We might say that the one kind have wings at the expense of the feet, and that the others run at the expense of the capacity for flying; nature principally making more perfect the organs which are most exercised, and weakening those which are least employed. We may, thus, divine beforehand the habits of an animal, by observing the organs which are most developed. Thus we find the gallinaceous birds, which run remarkably well, fly with extreme heaviness, and the penguins, &c., which swim with such rapidity, have merely pinions incapable of sustaining them in the air; from this we see, that these animals are necessitated to adopt the mode of living which their organization has prescribed.

All birds provided with long legs, like the grallæ, must have a long neck, and many vertebræ, because they must seize their prey on the ground; but a long neck is not always accompanied by long legs, instance the swans and other palmipedes; for these aquatic species having only to plunge their heads to the bottom of marshy water, have need of nothing but short oars to swim with.

Birds with those long legs, or stilts, (from which circumstance they are called *échassiers* by Cuvier,) have no need of a tail so much extended as those with short feet, to serve as a helm in their flight. In fact, the grallæ turn their legs behind when they fly, and use them like a tail. On the contrary, those with short feet, as the promerops, aras, &c., have received from nature a tail remarkably long.

Notwithstanding that there are other species of animals capable of supporting themselves in the air, such as the vespertilio, the galeopithecus, the roussette, among the mammalia; the flying-dragon, among the reptiles, many species of flying-fish, and an infinite number of winged insects; and though the ostrich and some other birds cannot fly, still the capacity of flying is the principal faculty which distinguishes this class of animals. Their body is of an oval form, evidently conformed for the execution of this movement. The dorsal spine, ossified and inflexible, presents a basis of support for the violent action of the wing; a sternum, widened like a sort of breastplate, with a long longitudinal keel in the middle, presents powerful attachments to the motive muscles of the wing, and a considerable space for muscular play. The clavicles, or bones of the furca, joined in the form of a V, separate each shoulder in the opposite direction, and resist, with elasticity, the vigorous movements which the action of flight requires.

In the skeletons of birds, the vertebræ are found to vary con-iderably. Thus, in the sparrow, which has the fewest, there are nine cervical, and nine dorsal; while in the neck alone of the swan there are twenty-three. By the formation of a facette attached to each of the cervical vertebræ, the neck is preserved in a curve, as its natural unrestrained position, while the vertebræ of the back are either fixed to each other, or are so bound together by strong ligaments, as to render the whole series incapable of any inclination out of a straight line, an arrangement which evidently has reference to the faculty of flight, by affording a more effectual resistance to the muscular power employed by the wings, because, in such birds as do not fly, the spine is capable of a curve, or bend.

The number of vertebræ in the tail varies also, in proportion to the length of the organ in each genus.

The large square plate, called the sternum, convex in front, and concave behind, to which the muscles of the wing are attached, covers the thorax and the abdomen. In front of this is the laminar bone before mentioned, the size of which is always proportioned to the power of flight of the species, and in the ostrich, which does not fly, it is altogether wanting. On each side of the sternum are some long pieces, called sternal ribs, which connect it with the vertebral ribs, forming altogether a protection for the intestines.

The omoplate is small, forming a parabolic arch, and placed parallelly with the spine on the ribs. Its coracoïd apophysis forms a long and very strong bone, flatted from front to rear. The clavicles are united above the sternum, in front of the coracoïd apophyses, forming one distinct piece. This provision is evidently to afford the clavicle a greater elastic force, which tends to separate the two omoplates, when the bird puts its immense pectoral muscles into action, in lowering the wings in flight. The insertion of the coracoïd apophyses prevents the lowering of the shoulder-blade, by which the wing acts with greater effect upon the resisting air. In birds of powerful flight it is larger than the humerus, but in the gallinaceous birds these parts are of about equal length, and in the ostrich the humerus is longer than the radius or the cubitus.

Birds have three pectoral muscles, one of which weighs more than all the other muscles of the body put together. The middle pectoral muscle acts as a lever to the wing, and prevents the bird turning over in flight.

The extremity of the wing, analogous to the hand, or fore-feet of mammalia, has a range of carpal bones, a single metacarpal bone, and a bone called os styloïde, which represents the thumb and toe, with two phalanges, and another os styloïde smaller than the first. These bones have not, like ours, the movements of pronation and supination, but only those of extension and flexion. The muscles and tendons which move them with such vigour, will allow of no other; for the wing must be strong enough to resist the shock of the air, without turning, which would overthrow the bird.

Like quadrupeds, the birds possess the principal organs of life, as the intestinal tube, which no animals can want, a heart, with two ventricles and two auricles, a double and perfect circulation, lungs, brain, parts of generation, &c., all adapted to their peculiar nature of life.

But they are destitute of many parts which the quadrupeds possess. Thus, they have neither lips, teeth, oreillon, or fleshy tail. In the interior of the body, they are without the diaphragm, epiglottis, and urinary bladder. They pass some urine, however, into the cloaca of the excrements, through the ureters. Many parts are modified differently from their analogous ones in quadrupeds; thus, the female birds have but one ovary and *oviductus*, instead of the matrix of the vivipara. The males have no scrotum, but the testes are situated in the belly, near the reins and lungs.

The bird, using the anterior extremities for flight, and not for locomotion or prehension, is, like man, a biped. This posture elevates the head, and gives it a different air from that of quadrupeds.

The femur is always shorter than the tibia ; the peroneum is very slight, and never descends so far as the tibia; the single bone which represents the tarsus and the metatarsus, varies considerably in length, and on this depends the height of the bird on its legs. The toes have been sufficiently noticed.

The beak, already described, varies greatly in length and form, and, with the web, or interdigital membrane, will be found to form the groundwork of the most prevailing principles of artificial separation.

Sight is extremely perfect in birds, and they have the peculiar faculty of seeing objects near or distant equally well. The means by which this is effected are not satisfactorily explained, though a power of changing the convexity of the eye is probably the proximate cause. Like all other physical peculiarities, it is admirably adapted to the mode of existence of the class ; a quick and perfect sight of objects and perception of distances is necessary to the rapidity of movements and the securing of their prey to birds. All the genera, except the owls, see a single object but with one eye. The situation of these organs, however, enables them to take in a much larger field of view than animals whose eyes look straight before them.

Not to dwell with minuteness on some peculiarities which distinguish the eyes of birds, we shall pass to an additional word or two on the third eyelid, or nictitating membrane : this is folded in the angle of the eye next the nose, and is brought over the organ like a curtain, in a vertical direction, and not horizontally, or up and down, like the ordinary eyelids. This membrane is partially transparent, and one of its purposes seems to be, to prevent the access of too much light into the eye, when the bird is exposed to that inconvenience. With a few exceptions, the upper eyelid of birds is fixed, the lower one only moving.

The action of the nictitating membrane is highly mechanica and curious. Being partially pervious by light, it seems necessarily to be destitute of fleshy fibres, and could not, therefore, be attached in the ordinary way to a muscle. It is elastic, and lies, when unexcited, drawn back in the angle of the eye, but, when used, is put into action by two muscles attached to the posterior part of the globe of the eye, one of which is composed of fibres descending obliquely toward the optic nerve, and terminating in a tendon of a peculiar character, having no insertion or attachment, but forming a cylindrical canal, which bends round the optic nerve. The other muscle is attached above the eye, near the nose, and is composed of a little fibrous cord, which passes under the eye, to the lower edge of the nictitating membrane; the action of these two muscles draws the membrane across the eye.

Of the construction of the ear, what has been said in the text must suffice. The sense of hearing is very perfect in the class; smell, on the contrary, seems obtuse, except in the birds of prey, particularly the vultures, which seem led to their food very much by this sense. The apertures of the nostrils vary nevertheless in the different genera.

From the make of the tongue, covered with corneous papillæ, it does not seem probable that birds enjoy the sense of taste in a very high degree; and whether they are much influenced in the choice of the food proper to each by this sense, may be questioned.

The insensibility of the feathers, and callous character of the integuments in the parts without plumes, seem sufficiently to evince that the sense of touch also in this class is very imperfect.

A brief notice of the nature and construction of feathers, the common integuments of this class, may not be without interest.

Surrounded as we are on all sides with works of wonder and astonishment, some of such stupendous magnitude that the mind cannot embrace them, and others so infinitely minute, that it cannot seize them, it is perhaps but little surprising that we go through life, and hourly pass by many of the productions of nature, but highly deserving our attention, and alike calculated to produce admiration and astonishment. A common feather may be instanced as one of the unheeded, but curious, productions of creation.

The feathers of birds are of three kinds: First, the plume, or down; secondly, the coverts, or tectrices, and the scapulars; and thirdly, the remiges, or flag-feathers, including the primary, secondary, and tertial of the wings, and the rectrices, or those of the tail.

The wing and tail feathers are much used in dividing the class, and as they are frequently mentioned in all writers on ornithology, it may be useful to premise shortly, that the wing consists of seven bones: one in the brachium, two in the cubitus, two in the carpus, and two in the metacarpus, or spurious wing. The ten larger quill-feathers, called primores, spring from the carpus; from the cubitus, an indefinite number, called secondary, and from the brachium small feathers only. In the metacarpus are implanted three small stiff feathers, called the spurious wing, *ala spuria*, whose use is not apparent. The accompanying wood-cut may serve to illustrate this explanation.

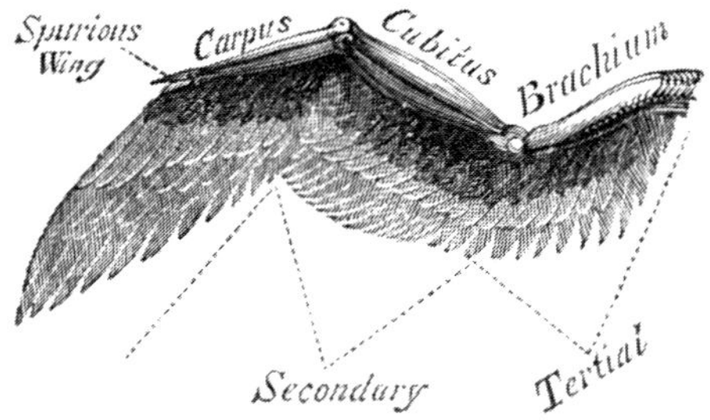

The feathers, which are instruments not merely of clothing, but of motion, are called remiges, flags, or quills. These, as every one knows, are composed of a shaft, hollow, cylindrical, and horny toward the bottom, which goes off into a subquadrangular, solid, but porous and light, substance, protected by

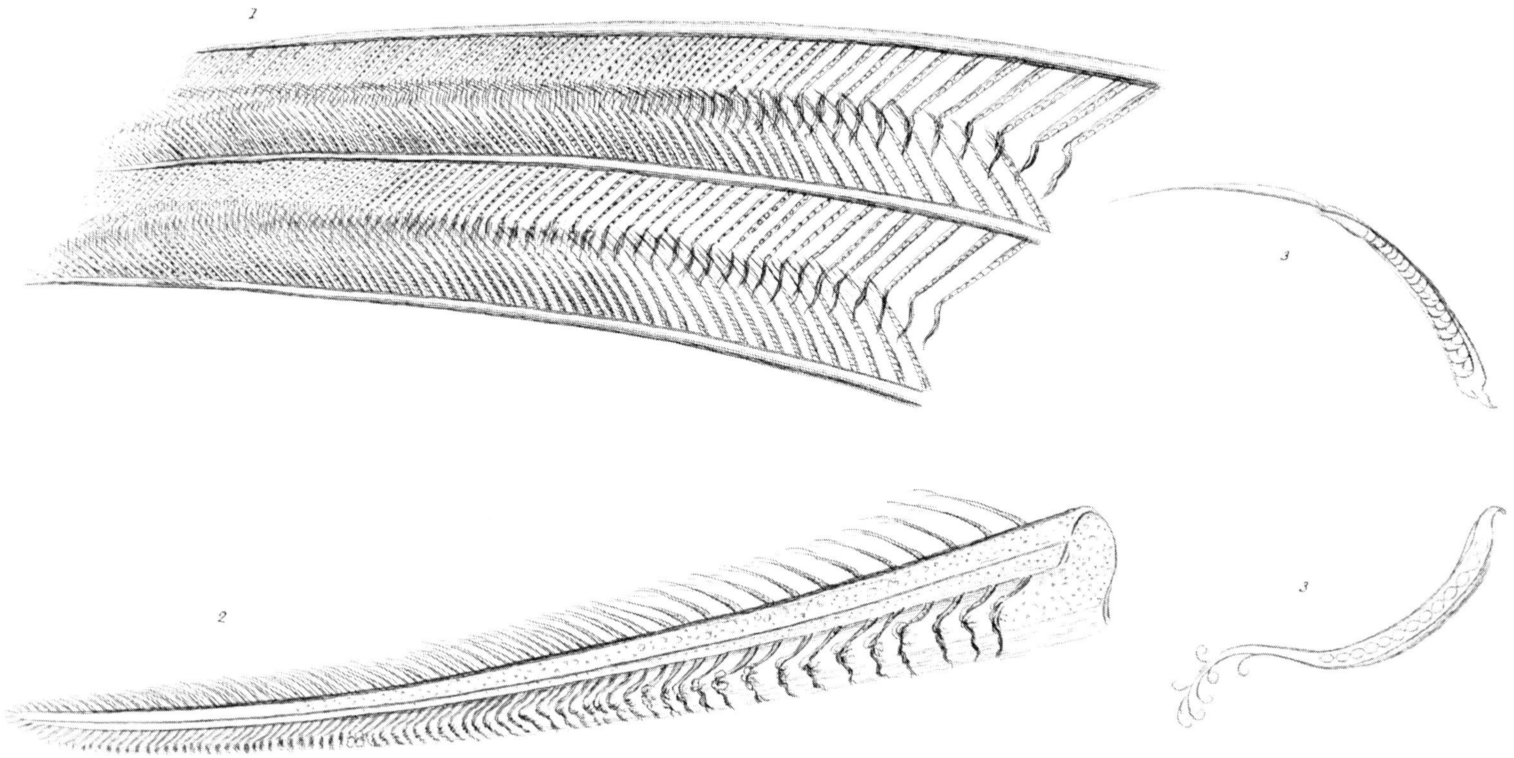

1

3

3

2

STRUCTURE OF FEATHERS.

London Published by G.B. Whittaker Nov. 1. 1827.

a horny exterior, and terminated in a point, from each side of which, above the cylinder, proceeds diagonally a vane, composed of proximate parallel laminæ ; and here common observation, with regard to a quill, terminates. To investigate its less obvious, but more curious incidents, we must make use of the microscope. By the aid of this instrument it appears, that these laminæ are not flat, as they appear to the unassisted eye, but are semitubular, having, on their outward edge, a series of bristles, set in pairs opposite one another, which clasp with the bristles of the approximate laminæ, and cause that adhesiveness observable between the several laminæ of the vane, and that readiness to reunite after they have been forcibly separated.

The bristles are not of the same form on each side of one lamina, the lower tier forming a simple and slight curve, while the upper terminate with three or four little hooks, which serve to catch the simple corresponding bristle of the next lamina.

This is the general plan by which the quill-feather of a bird, when opposed by its flat under-surface to the air, is made impervious to that subtle element, so as to support the sailing body on it alone. This general plan, however, is varied in its application to the several species. Thus, in the small and light species, the laminæ and the bristles are proportionably larger, compared with the feather, than they are in larger and heavier birds.

It is, as we have observed, only one edge of the semitubular laminæ, which is furnished with bristles; the opposite edge goes off in a little ledge, composed of longitudinal fibres, which little ledge is the only part of the feather found not to be of a cellular texture.

The prevalence of this texture, even to the minutest parts of the feather, is truly wonderful, for even in the smallest bristles a series of cells may be observed along their whole length, provided the magnifying power be sufficient; and thus the whole substance of the feather is rendered as light as possible, a quality of the first importance to its office.

Accident must frequently cause the separation of the laminæ, and though their natural elasticity, aided by the hooked bristles, will soon restore them to their contiguity necessary for flight, still the bird is enabled, by an oily secretion, with which it charges its bill, to anoint and adjust the delicate apparatus of the laminæ, by drawing the vanes of the feather gently through the bill. Birds of passage are generally observed to do this carefully, previous to starting on their protracted periodical flights.

This very short sketch of the feather of a bird destined to its locomotion, may be aided by the figures inserted of the structure of these feathers, as displayed to us by the microscope.

The feathers already shortly described are those calculated for the locomotion by flight of the bird. There are others fitted only for its clothing, which are of a very different construction, though not less curiously adapted to their intended purposes. These are set in a quincunx form over the whole body, thus, :·:. Immediately upon the skin is a covering, called down, composed of delicate plumes, of different sizes, and of extreme softness and pliability, flaccid, branching, and scattered. Instead of a shaft, beset with parallel laminæ, readily adhering or separating, the shaft of the down is furnished with rays which *will not* unite. Nor do they lie together, but are scattered in all directions, and furnished with knots, similar to those on a bamboo-cane, and applied closely to the skin, thus forming a general covering to the body, so essential to preserve and equalize the vital heat in all situations. To prevent the consequences of so light a substance as the down being blown about, a provision is made for confining it. This is done by the next tier of feathers, which is of a twofold structure, on the upper side partaking of the laminous formation of the flying feather, while the under side is lined with down, which, uniting with that immediately next the skin, confines it to its place, composing altogether a regularly spread under garment, thus braced, as it were, and wrapped round the body. Next to these compound feathers, which thus compose the second tier, lie the coverts, of different sizes and shapes, in an imbricated manner, each feather taking a curve adapted to the part of the body it covers, thus forming a sort of upper-garment, which, with the under one, is admirably calculated to preserve the heat of the body within, and to keep out the wet and cold from without, effectually protecting the animal from the various temperatures it must rapidly experience in passing through the air.

Thus conformed, and provided with this wonderful apparatus of wings and covering, the bird does not hesitate to shoot into the region of tempests, and proceed to most prodigious distances. Nothing is more wonderful to the contemplation of the natural philosopher, than this power of flight. Its mechanism is combined with such astonishing skill, and rests upon such powerful resources, that no machine, invented by the most able mechanician, has, as yet, been found capable of imparting such a faculty to man. All who, without the aid of a balloon, (which is not flying, but a sort of sailing,) have attempted to elevate themselves into the air, have shared the fate of Icarus.

We shall enrich our pages with a few of the reflections of the illustrious Buffon on this subject. "To give some idea of the duration and continuity of motion in birds, and likewise of the proportion of time and space which their courses occupy, we shall compare their swiftness with that of quadrupeds in their greatest progressions, whether natural or forced. The stag, the rein-deer, and the elk can go through forty leagues in a single day. The rein-deer, harnassed to a sledge, can make thirty, and continue this many days in succession. The camel can make three hundred leagues in eight days. The horse, educated for the race, and chosen from among the lightest and most vigorous, can perform a league in six or seven minutes; but his speed soon relaxes, and he would be incapable of supporting a longer career, with the spirit and celerity with which he commenced. We have cited the example of an Englishman who went seventy-two leagues in eleven hours and thirty-two minutes, having changed horses one-and-twenty times; thus the best horses can make no more than four leagues in an hour, nor more than thirty leagues a day. But the swiftness of birds is considerably greater. In less than three minutes we lose sight of a large bird; of a kite, for example, which proceeds horizontally, or an eagle, vertically, and the diameter of whose extent in flying is more than four feet. From this we may infer, that the bird traverses more than a space of four thousand five hundred feet in a minute, and that he can proceed twenty leagues in an hour. He may then easily proceed at the rate of two hundred leagues a day, flying for only ten hours. This supposes many intervals in the day, and the entire night for repose. Swallows, and other birds of passage, may thus proceed from our climate to the Line in less than seven or eight days. M. Adanson has seen and caught, on the coast of Senegal, swallows which arrived there the 9th of October, that is, eight or nine days after their departure from Europe. Pietro della Valle says that, in Persia, the carrier-pigeon makes greater way in one day than a man on foot can in six. The story of the falcon of Henry II. is well known, which, pursuing with eagerness a smaller bustard at Fontainebleau, was taken the following day at Malta, and recognised by the ring which she bore. A falcon from the Canary Islands, sent to the Duke of Lerma, returned from Andalusia to the Isle of Teneriffe in sixteen hours, which is a passage of two hundred and fifty leagues. Sir Hans Sloane assures us that, at Barbadoes, the sea-gulls proceed in flocks to a distance of more than two hundred miles, and return again the same day. A course like this, of more than one hundred and thirty leagues, sufficiently indicates the possibility of a voyage of two hundred; and I believe we may conclude, from the combination of all these facts, that a bird of elevated flight can traverse every day four or five times as much space as the most agile quadruped.

"Every thing contributes to this facility of motion in the bird. First, the feathers, whose substance is very light, whose surface is very extensive, and whose tubes are hollow; then the arrangement of these same feathers, the form of the wings, convex above and concave below, their firmness, their great extent, and the force of the muscles which move them; finally, the lightness of the body, the most massive parts of which, such as the bones, are much lighter than those of quadrupeds, for the cavities of the bones in birds are proportionally much greater than in quadrupeds, and the flat bones which have no cavities are much more slender, and less weighty. 'The skeleton of the onocrotalus,' say the anatomists of the academy, 'is extremely light. It weighs but three-and-twenty ounces, though remarkably large.' This lightness of the bones considerably diminishes the weight of the bird; and we shall find, in weighing the skeleton of a quadruped with that of a bird in the hydrostatic balance, that the first is specifically heavier than the other."

We have already observed on the strong and piercing sight of birds, which the extent, elevation, and rapidity of their flight necessarily presuppose.

"A hawk," says Buffon again, "sees from on high a lark upon a clod of earth at twenty times the distance at which a man or a dog can perceive it. A kite having soared to an elevation beyond our ken, can see the small lizards, field-mice, and birds, and select those upon which he chooses to pounce. This great extent of the visual power is accompanied with a precision equally great, for the organ being at once both extremely supple and extremely sensible, the eye grows round or flat, is covered or uncovered, contracts or dilates, and speedily and alternately assumes all the forms necessary to adapt itself to every degree of light or distance.

"Moreover, the sense of sight being the only one which produces the ideas of motion, the only one by which the degrees of space which are traversed can be compared, and the birds being of all animals the best adapted for motion, it is not surprising that they possess, in the highest degree of certainty and perfection, that sense which should be their principal guide. They are able to traverse a great space in a very little time: they, therefore, must be enabled to discern its extent and limits. Had nature, in bestowing on them such rapidity of flight, rendered them at the same time short-sighted, these two qualities would have been contrary, and the bird would not have dared to make use of his lightness, nor attempted a rapid flight: he would only have hovered slowly along under the dread of unforeseen shocks and resistances. The swiftness with which a bird can fly may indicate the extent of his reach of vision; not, however, absolutely, but relatively. A bird whose flight is quick, direct and sustained, certainly sees farther than another of the same form, which moves more slowly and obliquely; and, had nature ever produced birds with short sight and rapid wing, such species must have speedily perished from this contrariety of qualities, one of which not only hinders the exercise of the other, but exposes the individual to an infinite number of risks. From all this we may presume that the birds whose flight is shortest and slowest are also those whose power of vision is the least extended. Just as among the quadrupeds we find the unāu and the aï, which move but slowly, have the eyes almost hidden, and the sight but faint."

We have already spoken of the nictitating membrane. Birds have also in their eyes a large quantity of aqueous humour, especially birds of elevated flight, that the light may be so much the more refracted as the air in which they rise becomes more rarefied. The reverse is the case with the fishes, for the light is sufficiently refracted through the watery medium in which they are immersed, and which is so much denser than the air.

The power, which, however it may be explained, birds do certainly possess of altering the convexity of the eye, of rendering the sight more or less distant, according to the wants of the animal, by correcting the divergence of the visual rays, is the reason why many birds, as well as the owl family, are nocturnal. A considerable number are also partial to twilight, as, for instance, the majority of the grallæ.

With means like these, the bird is enabled to travel in the air. Its specific lightness; the vigour of its wings; the nimbleness of its motions; the directions of its tail, which serves as a rudder; permit it to ascend, to descend, to turn, to flutter in all directions, to cut in a right line, to shave the surface of the earth or water, to hide itself in the clouds, and, in a word, to sport at its pleasure in the immense field of the atmosphere. Sometimes it will descend to gather the seeds in the fields, and sometimes, elevating itself above the clouds, respire the pure and serene air under the azure sky, while terrestrial animals are battered by the tempest and menaced by the lightning. Birds of high flight, enveloped in a warm, thick, and downy plumage, fear nothing of the piercing cold of the loftiest regions of our atmosphere. It is remarkable that birds employed in falconry, which their trainers are desirous of preventing from flying to too great an elevation, never mount but to a moderate height when deprived of the feathers of the belly and sides, because they are then afraid of the effects of cold. The water-fowl, provided with a thick down and an oiled plumage, which do not suffer the moisture to penetrate, plough the surface of the seas and lakes with perfect safety. Nature, moreover, has provided all birds with a certain gland, which distils over the crupper an oily humour, with which they anoint their plumes, passing them between their beaks. But this humour is peculiarly abundant in aquatic birds. Their skin even imbibes it, and thence acquires a rancid flavour; and it insinuates itself through the entire plumage. From this it occurs that these birds, though perpetually immersed in the water, are never washed by it, the liquid rolling over them without moistening their plumage, even though they seem desirous of it—

"Et studio incassum videas gestire lavando."

Fishes, which by a figure of speech may be considered in some sort birds of the water, as birds might be called fishes of the air, are also provided with an oily gland to anoint their scales: but it is placed in front, so that the simple action of swimming suffices to spread this fatty substance over their scales, and thus defend them from the relaxing influence of the water. Such is the admirable foresight and ineffable contrivance of the Author of Nature!

The constant habit of living in the air, of experiencing its full influence, and of being exposed to all its variations, imparts to birds a knowledge of all the meteoric changes which take place in the atmosphere, of winds, of seasons, and of bad weather. The kite, says the prophet Jeremiah, knows his time in the sky. The turtle-dove, the stork, and the swallow, know the period of their returns. We find, indeed, that all animated beings, not distracted by other cares, can presage the changes of temperature. This is even the case with man, and especially with those whose nerves, from nature or indisposition, have received any peculiar sensibility.

It is well known to sailors, that when the divers and sea-gulls retire to the rocks on rapid wing, and make the shores re-echo with their clamours, as if to warn their companions; when water-fowl parade the strand with apparent anxiousness; when the cranes, quitting their marshes, soar above the clouds, and the swallows fly in circles over the surface of the water; the prudent navigator should lower his sails, and anticipate the storm. We see, again, black legions of ravens beating the air with their wings, and the rooks clamouring in the fields at the approach of rain. On such occasions, the heifer in the pasture snuffs in the air with elevated head; the frogs croak in the marshes, the ants bring back their chrysalides to the nest; and fishes come to the surface of the water to respire. All animals appear to presage the tempest; and it is thus that shepherds and labourers, constantly exposed to the atmosphere, divine all its variations by a sort of instinctive observation. But on the return of fine weather, we see a total change of all those symptoms in animated nature. The birds which inhabit strands and shores no longer come to dry their plumes in the sun; the screech-owl no longer utters his funereal cries in the evening; the hawk, on the contrary, circles in the pure azure sky; the smaller birds sport among the newly-budding leaves; the raven testifies his joy by sonorous croaking; and the cattle bound on the plains. One might even be led to imagine that birds possessed some knowledge of the future, and were gifted with a foresight superior to that of other animals. It was doubtless from this idea that the ancient augurs, destitute of our barometers, observed them with so much care, and drew presages from their movements. We are not yet, perhaps, acquainted with the fullest extent in which the modifications of the atmosphere, the weight, density or rarefaction, the moisture, dryness, or electric state of the air, can influence the organization and sensibility of animals, and even the character of men.

> "Verum, ubi tempestas et cœli mobilis humor
> Mutavêre vias, et Jupiter uvidus austris
> Denset, erant quæ rara modo, et quæ densa relaxat,
> Vertuntur species animorum et pectora motus
> Nunc alios, alios dum nubila ventus agebat,
> Concipiunt."

Marine birds appear to be the most sensible to all these atmospheric variations. Thus the petrel, the storm-bird, the albatross, &c. indicate the approach of the hurricane by their importunate cries and uncertain flutterings near the rocks. We likewise find the majority of birds whose plumage is not as much impregnated with oil as that of the palmipedes and other birds inhabiting shores, suffer very much from rains, and endeavour to avoid them by seeking shelter. In fact, when the water does penetrate their plumage, they remain a long time wet, are retarded in their flight, and often made ill by obstructed transpiration. On the contrary, all birds, the aquatic races excepted, are never better than in dry countries and seasons. They then multiply astonishingly, as we always find they do in the ardent climates of the tropics.

The arrival of the ortolan in our climates marks the presence of severe cold; whence the French term this bird *Ortolan de neige.* The *Ampelis garrulus* of Latham, which comes from Bohemia, announces the first frosts: when the cuckow sings, the leaves begin to germinate. But, in fact, it would be endless to enumerate all the indications which man derives from the feathered race.

The aërial sojourn of birds, and their constant habits of flight, isolates them in some measure from the earth, and in part withdraws them from the influence of climate. The annual migration of many species, rendering them, as it were, cosmopolites, gives them a character totally different from that of terrestrial animals. Less circumscribed in their dwelling, they have more liberty, audacity, and independence. Respiring a purer air; less surcharged with aqueous vapours and terrestrial exhalations, their natural constitution is more fine and subtile, and their sensations more delicate. As men and animals which inhabit low and humid countries have soft fibres, flabby flesh, dull nerves, obtuse sensations, and heavy intellects: and as we see in species inhabiting dry and elevated localities, such dispositions replaced by more active qualities—by tensity of fibre, firm flesh, irritable nerves, a lively sensibility, and acuter intellect,—so the birds inhabiting the wide expanse of air are provided with such qualities in a degree still more eminent. In fact, the muscular fibres of birds are, in general, arid, hard, and very much distended, which contributes in no small degree to the vigour and rapidity of their motions. Do we not observe that slender and even meagre men are much more lively, mobile, and excitable, nay, much more endowed with mental acuteness than the generality of those heavy human masses, which are moved with difficulty, and whose spirit is as heavy and benumbed as their bodily organs? the first partake of the volatile character of the feathered race, and the latter of the complexion of the quadrupeds.

The tension of fibre, the dry temperament, and extreme mobility of the muscles in birds, render their sensibility more energetic. Organs so excitable are put in sudden motion by the slightest impressions. Such animals have need of multiplied sensations. They pass their lives in a perpetual state of agitation and motion. Repose is to them a torment; for, in proportion as their sensations are more lively, so are they more changeable, as is observable amongst mankind. The birds are of an irritable and nervous constitution: everything animates them to excess. They are ardent, choleric, amorous in the extreme, and quick and impetuous in all their actions. They all sleep but little; and as for what has been said concerning the immersion of swallows in the bottom of lakes during the winter, and the retreat of quails into caverns, it appears extremely contrary to the nature of those animals. Emigration is the much more natural and probable mode of accounting for their disappearance.

The extent of sensibility possessed by birds cannot be, as we have already seen, at all traced to the sense of touch, which, from the covering of their bodies, and the hard and osseous character of their beaks and feet, must be extremely obtuse. Laminæ, or very callous scales, invest all the toes; and among a few species only the beak is just barely surrounded at its base with a little naked skin. But from what we have already said concerning the power of vision in birds, it appears evident that their quick sensibility, and extreme vivacity of character, are greatly dependent on the wonderful development of this sense. We may remark, indeed, as a general rule, though perhaps not wholly without exceptions, that animals of very limited power of vision, and still more those which are destitute of sight, are sedentary and inactive. The fishes, which are so lively and agile, have, like the birds, a very extended range of sight; while worms, mollusca, zoophytes, &c. whose gait is groping and slow, are almost all blind.

This extreme vivacity, common to the majority of birds, renders them less capable of education than other more tranquil animals, and produces, in this respect, the same effect as its opposite quality, stupidity. For though they are well organized for the purposes of learning, their boiling impetuosity, the perpetual variety of their motions and sensations, hinder them from fixing their attention, so that ideas shall be permanently imprinted in their sensorium. Still they appear to imagine much in the variety of their operations and migrations ; but this appears to be the result of instinctive feeling, rather than of intelligence. They have, generally speaking, but slight glances of things, which are easily effaced by time. They experience only fugitive impressions, which are speedily replaced by others equally fugitive. They feel, in fact, more than *conceive.* An attention, a steady and reflective character is necessary to penetrate into the knowledge of things. Thus we find the elephant, whose gravity and reflexion are so remarkable, is also one of the most intelligent of animals. Parrots, which are in general less turbulent than other birds, are also more susceptible of instruction ; and if we succeed in teaching Canary birds, goldfinches, linnets, &c., it is only by keeping them imprisoned, and constraining them perpetually to attend and reflect. It has been even observed that birds become blind receive instruction with greater facility than otners, because their attention is less distracted. This observation has given rise to the atrociously-cruel practice of bird-fanciers in burning out the eyes of nightingales and other birds which they keep in cages.

The articulation of the head by a single round condyle is remarkable, as it enables the bird to turn the front of the head full half way round, which no other vertebrated animal can do

The brain of this class is distinguished from that of the mammalia by presenting six visible masses. These are the two hemispheres, the two optic beds, the cerebellum, the medulla oblongata. The two first are without circumvolutions, and there is no corpus callosum, or septum lucidum ; but the most distinguishing character of the brain of birds is, that each of the anterior ventricles is inclosed in a thin partition, which is not found in any other vertebrated animals.

Though the brain of birds is without the corpus callosum, the septum lucidum, pons Varolii, and some other less important parts, still the tubercles called nates, acquire a considerable development, and especially those eminences, which are analogous to the corpora striata, become very considerable. These animals possess, upon the whole, a voluminous brain, even more so than many of the mammifera. This is peculiarly the case with the smaller species, for large birds, as the ostrich, goose, &c., have small heads; but the sparrow, canary, and other small birds, have a cerebellum proportionally larger than that of man himself, sometimes composing even a twenty-second part of their whole body; and accordingly we find such birds, like the parrot, possessing a very considerable portion of intelligence *.

* Our readers will, perhaps, forgive us for once more adverting, in this place, to a subject frequently touched on before in the course of these volumes—we mean the proximate causes of intellectual superiority in man and other animals. We shall not, we trust, be readily suspected of any leaning to the doctrines of materialism; but, setting his spiritual part totally out of the question, we must, in explaining his mental endowments, avoid taking a partial view of the complicated machinery of man. Man's superiority over other animals does not consist, even materially, in the superior development of the brain;—were this the case, the birds above mentioned would be at least his equals. It consists in the admirable harmony and connexion that subsist between all the parts of his entire organization. His hand, as Helvetius has remarked, gives him infinite advantages; but that *hand* was formed for that *head*. The hoof of a horse would have been a very inadequate instrument for performing the actions suggested by the intelligence of a man. In every animal system, the peculiar conformation of one part necessitates the peculiar conformation of every other. There must be a correspondence, a harmony, an unity in the whole system, for the production of a given end. Man is eminently an *intelligent* animal, and accordingly we find that his entire organisation tends to the production of this point. Certain declaimers

The perpetual state of activity in birds has a tendency to develope their muscular system in an extreme degree; and as the labour of the muscles dries, hardens, and fortifies the body, these animals must necessarily be of a complexion arid, but robust. In fact, the flesh of birds is of a substance extremely compact, and almost tendinous. This habit of violent exercise must also engender considerable heat; and as their organs are necessarily much worn by constant labour, so have they need of frequent and copious reparation. We find, accordingly, that the heat in birds is greater, and their appetite more keen, than in the majority of other animals.

Their corporeal heat depends, however, more particularly on another cause, which is the principle of the immense vivacity and force with which they are gifted. For, indeed, what wonderful vigour must a bird possess, to be able to sustain itself in the midst of the air by repeated springs, and to perform such lengthened journeys in so short a space of time? What amazing action of the wings, and what tremendous force in the pectoral muscles are necessary, to enable a heavy bird to proceed at the rate of some hundreds of leagues a day, and to execute such prodigious voyages? The source of this mus-

may flourish as they please respecting the advantages possessed by some animals over man; but no animal is so well organised as man, not only in the brain, but in every other part, for the station which he holds at the head of the animal kingdom. Some may possess greater acuteness of one sense, some of another; some may have greater muscular force, others more agility; but none possess such an union of advantages as man does, to fulfil the peculiar purposes of man's creation. Without such an union, the development of his intellect could only serve to render him miserable; and, we may add, that with many of the seeming advantages that other animals possess, such development could not possibly take place. Paine imagines that man would have been better with the wings of a bird, forgetting that such a faculty would necessitate a covering that must diminish his sensibility, and a volatility of character that would unfit him for reflection. No! man need not envy the pinions of the eagle. Let him content himself with those winged thoughts which can carry him beyond the confines of the earth, and lift him to the heaven of heavens!

E. P.

cular vigour is the powerful and quick respiration which we have already noticed. The immense mass of air continually penetrating into the lungs, and all the aërial sacs and canals of the animal, and being decomposed there perpetually, carries the fire of life throughout the system, warms and reanimates all the organs by continual stimulation. The oxygen gas, flowing into the lungs, and combining with the blood in considerable quantities, communicates its stimulating qualities to this fluid, increases the action of the heart, and propels the tide of circulation with inconceivable rapidity. So prompt are the pulsations in the arteries of a bird, that it is with the utmost difficulty they can be counted. The heat which arises from this great vascular action is more considerable in the bird than in the quadruped. The heat of this last is no more than 32° of Reaumur, and it is the same in man; but the birds have 35°, and even more. Thus they are enabled to sustain with ease the rigour of cold in the elevated regions of the atmosphere; and thus we see even the little wren pass gaily the coldest winters without perishing. If we see sparrows and some other birds die during this season, it is for want of nutriment. It is, therefore, by no means credible that animals of so much heat, and which have a respiration so strong and continual, should lethargise, or even plunge to the bottom of the waters without being drowned, as is reported of swallows.

From this great respiration, and the heat which it developes, two characters are derived which distinguish birds most eminently;—those are their voice and their amorous propensities, between which, as we shall presently see, there is a very close connexion.

If we consider that, of all the animals of the earth, the birds have the greatest extent of chest, and the largest lungs in proportion to their size; that these lungs, attached to the ribs, are not bounded by any diaphragm; that they have pouches or membranous sacs, even in the abdomen; and finally, that the air penetrates into all the parts of their body, we shall cease to wonder at their compass and power of voice. Moreover, they possess a tracheal artery, composed of rings entirely cartilaginous, destitute of epiglottis, and which does not carry its vocal chords towards the pharynx, but which forms a lower larynx towards the bifurcation of this same tracheal artery. The upper part of this canal, which surmounts this lower larynx, serves it in some measure as a speaking-trumpet. Besides, the sound of the voice, coming in collision with the circular fibres and the demi-osseous rings of this tracheal artery, resounds with force, especially in the males, who are often provided with a sort of tendinous drums towards the glottis, while the females are destitute of such appendages. This musical apparatus in birds may be compared to the French horn, for that instrument is formed nearly on the same principles. These organs of song are considerably less perfect in the females, for they are without those demi-osseous and resounding cavities which the males possess, inasmuch as they are not designed for singing. "The bird," says Buffon, "which makes itself heard at the distance of a league in high air, (as do storks, wild geese, &c.) and produces sounds in a medium which considerably diminishes their intensity, and more and more abridges their extension (in consequence of its rarefaction,) must possess a voice four times the strength of those of men or quadrupeds, which can only be heard half a league at furthest on the surface of the earth. This calculation, too, is probably rather under than over the reality; for, independently of what has been now advanced, there is another point which adds weight to our conclusions, and that is, that the sound produced in the midst of the air must, in being propagated, fill a sphere of which the bird is the centre, while the sound produced on the surface of the earth fills only a demi-sphere; and that portion of the sound which is reflected against the earth, aids and furthers the propagation of that which is heard vertically and laterally."

In truth, the song of the blackbird is heard at least at as great a distance as the voice of a man ; and if we consider that the croaking of the raven, the cry of the duck, of the peacock, and of the goose, are perhaps stronger than the bellowing of a bull, or the braying of an ass, we shall find that the bird, in regard of voice, has been more favoured than terrestrial animals. The sea-birds have, for the most part, a voice excessively sonorous ; for, being obliged to call to each other from considerable distances, and in the midst of the roaring winds, they are forced to give an enormous extension to their cries.

But the powerful extent of voice in birds would seem to presuppose a similar excellence and analogous modifications in the auricular organs. This, however, is by no means the case. They are not nearly so well provided in this respect as the mammifera. They are musicians rather by instinct and the perfection of their vocal organs than by the ear. They in some measure resemble in this deaf persons, who call excessively loud, believing that nobody can hear. Besides, the perfection of the voice in birds seems to have been a necessary compensation for the defects of the auricular organ, for they have no external conch to the ear. Instead of interior osselets, there is nothing found but an osseous plate. A species of cone with two cells, and a little arched, represents the cochlea in quadrupeds. The nocturnal birds, which have more need of this sense, have large cavities attached to the cell of the ear. These melancholy birds send forth plaintive accents, as if Nature had established a sort of harmony between their character, the melancholy silence of night, and their funereal cries. In the same manner the complaining tones of the nightingale are still more touching, from their accordance with the decline of day, as the loud concert of the joyous musicians of the fields is in unison with the cheering aspect of the rising sun.

It is easy to distinguish in the tones of birds, a certain language. All animals, in fact, have a language, not indeed articulate, but most undoubtedly comprehensible by cries and signs. The birds perfectly well understand each other, by means of these natural cries. Thus the mothers perfectly comprehend the wants of their little ones, by their piping note of appeal. The swallow chirps in her nest to her young ones, and appears to hold conversation with them. When the hen is alarmed for her chickens, she utters a cry of warning, and they instantly come, and shelter themselves under her wings. This first language is that of nature; it expresses the passions and wants that are felt. It is innate, depends on the organization of the animal; is the result of instinct, just like the accents of grief, joy, surprise, and pleasure, which are equally observed in men and quadrupeds: all animals have this language, which serves, not for the communication of ideas, but of feelings; for their gestures and actions represent nothing but sensations. The principal, and perhaps indeed, the only communication which exists between us and the brutes, is one of feeling, not of thoughts. They do not even understand our articulated language. It is the tone, the action, the physical language, which they comprehend. Menace an animal in the same manner as you caress him, and he will not understand the difference. The case of trained animals affords no exception to this remark. It is by a powerful, too often by a cruel appeal to their sensations, that they become habitually sensible to the meaning of certain sounds. The domestic animals have many more physical relations with man than moral; they study our corporeal movements, the pantomime of our passions, our natural accents. The motions connected with their physical sensations influence them most. They will not trust to the call of pretended kindness, when they see the knife, or the club uplifted to destroy them. They are better acquainted with the heart, than the mind of their masters; because they are, as it were, more material than intellectual, and feel rather than reflect.

Independently of this natural language, which is the mere expression of physical wants, we may observe another sort of language among animals, which may be almost termed acquired. This is the result of the social state in which certain animals live. We find solitary quadrupeds, and birds, uttering sounds but seldom, and almost always of the same character. It is remarkable, that even dogs that have become wild, are said to have lost the habit of barking. We may also observe, that the smaller species, especially among birds, are the most continuously sonorous. The larger species are generally serious. The ostrich has scarcely any cry. The nhandu and cassowary send forth a sound like strong sighing. The pelicans and cranes but rarely utter their clamours, while nothing can stop the eternal prattling of the little songsters of the woods.

This sort of language to which we have last alluded, is closely connected with the necessity of reproduction. The song in birds is nothing but the expression of love. After the time of incubation, the woods are generally silent. The nightingale, which so charms us by the melody of his voice, when endeavouring to attract his mate, utters nothing but a horrible cry, resembling the hissing of a reptile, after the period of his amours. We find, that birds kept in cages never sing so strongly as when deprived of their females; and some have been observed so transported with passion at the sight of a female of their own species that they could not get at, that they sung with a kind of fury, and seemed ready to drop dead. Stimulating and abundant nutriment tends very much to improve the song of birds in cages. Olina pretends that the odour of musk, amber, or civet, has a wonderful effect in stimulating the nightingale to sing. We observe, that the capon does not crow like the cock; and the female birds are totally destitute of this peculiar language of song.

Acquired language, or sounds, is more general among species approximated to each other, than among those which live in an isolated state; on which account, parrots, pies, jays, blackbirds, &c., all the granivorous and insectivorous races which are not mutual enemies, like the carnivorous, have also a greater multiplicity of sounds; and many of them, a melodious song. The polygamous male birds, such as cocks, pheasants, peacocks, ducks, geese, swans, &c., have a sonorous, hard, resounding voice, but destitute of that flexibility of tone, and touching modulation, which distinguish the monogamous races. These latter are forced to adopt the art of pleasing their mates; or, rather it is the order of nature, that they should do so. The others, like the imperious sultans of Asia, command their females with despotic sway. The reason of both proceedings is obviously to be found in the disproportion of numbers between the two sexes.

A peculiar conformation of the beak and tongue in some birds gives a greater or less facility in the imitation of articulate sounds. Thus we remark, that those species with a broad tongue, and a hollow and widened beak, nearly like the palate of man, have the greatest aptness for articulation. The seminivorous birds with thick beaks, as chaffinches, bullfinches, &c., have also a fuller voice than the insectivora, with fine attenuated bill, whose voice is more slender and piping.

As parrots, pies, jays, crows, blackbirds, starlings, and many other species, have a tolerably wide beak, and a thick fleshy tongue, analogous to that of man, they can be taught to articulate some words, to express them mechanically, but without comprehending their meaning, or attaching the slightest idea to them. They understand nothing of human speech, though they articulate it; and if ever they have been known to apply a phrase correctly, it was purely the effect of chance, and by no means the result of intelligence; for their usual application of phrases is quite unmeaning, or in a manner precisely opposite to their sense. It is not at all astonishing, that repeating the same phrases, on a multitude of occasions, they should sometimes make a fortunate hit, and surprise their hearers; thus giving an opportunity for ignorance and credulity to magnify their intellectual powers. They chatter continually, but never speak; for speech is the expression of thought. The simple and almost physical ideas which such animals possess, can have no relation with the abstract thoughts of man, and, no more than with all other animals, can we hold any intellectual intercourse with them; but merely an exchange of affections and of physical sensations.

These animals can never introduce their acquirement of speech among their own species; and this, by the way, is one of the greatest distinctions between man and all other animals. Those animals that are the most successfully trained and educated by man, are quite incapable of communicating their acquisitions to their fellows. All the knowledge rests in the individual, and dies with him. There is no system of mutual instruction among brutes. Under the immediate guidance of man, they are indeed sometimes rendered influential in the training of their fellows; but, of their own accord, they could never become so. The birds of which we speak, even after they are taught our sounds, communicate with their own species only by natural cries and signs. It is only in their relations with us, that they repeat the words which we have taught them. Every thing which comes from without, never enters into the proper composition of the animal. It is only a superficial modification, a fugitive impression, destroyed with the individual, or even effaced by time; the natural bias resumes its ascendant as the tree regains its original position, when the force which bent it is withdrawn.

This imitation of speech, however, presupposes some general aptitude for education, independently of the conformation of the vocal organs. These birds seem to possess a sort of sensibility analogous to our own, a sort of sympathy with man, which is indispensably necessary to all education of the lower animals. The nature of the other species is more harsh and intractable, for we never find them so much tamed as those birds which can learn to talk or whistle. In truth, neither the birds of prey, nor the gallinæ, nor the grallæ, nor the palmipedes, are capable of the same degree of improvement as the small races of birds, the insectivora, the climbers, &c. Still less do they possess any capacity of imitating the human voice. They are more brutal and indocile. They attach themselves to us, not as friends, companions, or guests, but merely as receiving food from us like interested parasites. But these little musicians, the canary, linnet, goldfinch, thrush, and blackbird, exhibit, as do parrots, more attachment and intelligence, more sympathy with man, and more general delicacy of character. They grow more familiar, they approximate more to humanity by their amiable qualities and a sort of fineness of tact; they become friends rather than slaves. Man, therefore, observes a very different conduct to these different species of birds. The first he feeds, and domesticates for his wants, and sacrifices them without compunction. The second, he breeds, and educates almost like children, partaking his dwelling with them, and feeding them with his own hand.

There is little doubt, that the differences of character in the various families of birds, may be clearly traced in the nature of their voice. The piercing cries of the birds of prey; the reechoing clangor of the palmipedes; the harmonious warbling of the small insectivorous and granivorous races; the importunate clamours of the grallæ; the shrill and sonorous call of the gallinæ, all mark the peculiar disposition, constitution, and habits of these different tribes.

The male birds are not only distinguished from the females by their song, their fiercer character, their constitution generally more vigorous, but also by external marks of great importance. The beak and claws, though alike in both sexes according to the species, are nevertheless stronger, and more developed in the majority of the males. These last are also furnished with certain arms, or distinctive parts, by which they can be recognized independently of the beauty of the plumage, or the vivacity of their colours. Thus most part of the gallinaceous male birds (except those of the American continent) have the legs armed with spurs, or horny protuberances, which are never found on castrated individuals, as capons, &c. Among the pheasants, cocks, turkeys, sea-peacocks (*tringa pugnax*, Lin.) poeintades, the males are provided with caruncles, either fleshy papillæ, or crests, more or less large on their heads; others have beards (as some of the *gypætos*), a tuft of hair under the throat, a collar of feathers, like the *tringa pugnax*; a fine tail, like the male peacock; or aigrettes of lively colours, or peculiar forms of plumage, of which all the females are destitute. It is well worthy of observation, that these distinguishing characteristics are never more remarkable than at the periods of sexual intercourse. The peacock loses his fine tail, the *tringa pugnax* his collaret of feathers; in fine, each of these animals is more or less degraded after this period is past.

The young bird has an obscure and dull plumage like the female, when the colours of this last are different from the male. If the plumage of the female be similar to that of the male, then the young bird has at first a covering peculiar to his age. Arrived at the period of puberty, he is invested with more brilliant colours, as if to attract the attention of the female; she is invariably covered with a more sombre plumage, or one of little brilliancy. The females have generally less ardour than the males, except among the partridge kind.

Vivacity, splendour of plumage and colours, and continual loquacity are signs in each species, of ardour, energy and vigour.

The infinite diversity of colours in birds is one of the greatest obstacles to the perfection of ornithology. A female, or a young individual, is often very difficult to recognise, as to species, so uncertain are the shades of plumage according to climate, aliment, migration, age, sex, domesticated, or wild state; insomuch so, that naturalists have often, out of a single species, created many. Besides, birds vary in a manner quite different to quadrupeds, being more numerous in collateral races, in species congeneric, and approximating in mixtures, and finally in the modifications which occur every season, at each moulting of the plumage. Nevertheless, on accidents of such inconstancy, species are determined, multiplied *ad infinitum*, and naturalists imagine that they are enriching science by loading it with dry and useless descriptions of individuals. It may also be questioned whether the publishing of splendid figures at an enormous expense, of rare and beautiful birds, is not more calculated to gratify private vanity than to be generally useful. Well does Lord Bacon remark on this subject: "Industria scriptorum enituit; itâ tamen, ut potius luxuriata sit in superfluis (iconibus animalium aut plantarum et similibus intumescens) quam solidis et diligentibus observationibus ditata, quæ ubique in historiâ naturali subnecti debebant."

As vivacity of colour in the plumage is a characteristic of the male birds, so those which are most particularly distinguished by brilliant colours are of the most ardent character, and *vice versá.* Birds of lively and striking colours abound most in the tropical climates. Those of cold countries have generally a pale and dead kind of plumage, for cold diminishes as much as heat increases this ardour of constitution. Hence it also happens that most males are produced in the warm climates, and most females among the northern species. We find that the aquatic races, the palmipedes, the scolopaces, the grallæ, whose plumage is generally grayish, dull, tarnished, or livid, and which have more females than males, abound principally in the climates approximating to the poles. Whereas, the climbers, the insectivora, the parrots, the woodpeckers, the colibris, the birds of paradise, the toucans, &c., whose plumage is of the most brilliant dye and richest variety of tints, have also in their species more males than females, and inhabit the warm climates almost exclusively. Paleness and whitishness of colour denote effemination and debilitation; and domestication, which degrades the animal, commences almost invariably in the individual by a degeneration of colour, as we find to be the case with canary birds, pigeons, &c.

The birds of cold countries are, in general, polygamous, in consequence of the fewness of males in proportion to females in each species. The birds of warm countries, having many males and few females, are, on the other hand, monogamous. It is singular enough that just the reverse is the case with the human species. It also happens, that among the polygamous families, the males are more vigorous than among the monogamous, a necessary compensation for the defect of number. The polygamous males are also less attached to their females. They abandon to them the care of hatching and the nourishing of the young. It is not uncommon with some of them to break and scatter the eggs; and in such cases a new laying and incubation takes place. Nay, this often occurs more than once during the season.

These polygamous males are moreover jealous tyrants. They use force with the females, and assemble them in a sort of seraglio, of which they must be the sole possessors. Should a rival make his appearance, war is instantly kindled. Cocks, quails, partridges, sea-peacocks, most of the grallæ, and in general all polygamous males, are naturally bold, choleric, and always ready for combat. Nature has therefore provided them, as we before observed, with weapons of offence, independently of their vigorous conformation, and greater development of beak and claws. But the monogamous birds, having each a female which suffices them, combat more rarely. They attach themselves to their companion, assist her to construct the nest, take their turn in the fatigues of incubation, enliven her with their songs, bring her nutriment, feed the young, and, in short, contract an intimate union and form a family where the comforts and troubles are equally shared.

The changes observed in birds at the period of their amours are very remarkable. M. Virey examined two sparrows, one at the period of reproduction, and the other towards the end of summer. The first had a plumage more lively and lustrous than the second: the flesh was more firm, and even coriaceous; the muscles thick and of a blackish red, almost without fat; but more especially the larynx and tracheal artery were fuller and more developed. The abdomen was harder, and the anus more inflated. The tissue, in general, was extremely solid, and the beak black and very much pointed. On the contrary, the plumage of the other sparrow was almost discoloured, and in disorder; the flesh soft, partly withered, and of a pale red;

the glottis was less plump, the abdomen extremely wide, and the testes almost obliterated: the beak was of a leaden colour, and the general tissue of the body relaxed and incompact.

We find, upon the whole, that after the epoch of reproduction, the feathered race are less lively, less robust, and less gay than before. They seldom sing, and their movements are not characterised by the same rapidity and energy which they displayed at the season alluded to;—and, indeed, the same is true of all other animals.

We shall now make a few observations on the nidification and incubation of birds. A remark, which we have had occasion to make before, may with great propriety be repeated here; namely, that in almost all cases, the productions of instinct are more perfect than those which emanate from human ingenuity. The nidification of birds is one of the most striking proofs that can be adduced of this, and is altogether a subject of the most curious speculation. That it is a process depending wholly upon innate impulse in the animal, and not acquired by reason and experience, and transmitted from generation to generation, is evident from the fact that birds, placed under any circumstances, will build their nests as nearly alike as their situation, and the materials afforded them, will admit. Taken when quite young, or even hatched artificially, they will build their nests when they breed in a state of captivity as much as possible upon the model followed by their respective species. This clearly proves that the art is intuitive, not acquired; for in such instances instruction is wholly out of the question.

Among the many pleasures attendant on the return of spring, there are few more delightful to a contemplative mind than to observe the proceedings of the monogamous races of birds which people our groves and fields. They seem replete with happiness, and intent on the performance of what we consider in man some of the highest duties which he owes to society. In the formation of an intimate union of affection and friendship; in providing shelter and food for their offspring, and attending by every means to their comfort and education. A truly philosophic mind sees more in all this than meets the eye; it is raised to the contemplation of that informing soul which breathes throughout all the works of nature—

> "What is this mighty breath, ye sages say,
> Which, in a powerful language, felt not heard,
> Instructs the fowls of heaven, and through their breast
> These arts of love diffuses? What but God?
> Inspiring God! who, boundless spirit all,
> And unremitting energy, pervades,
> Adjusts, sustains, and agitates the whole.
> He ceaseless works alone, and yet alone
> Seems not to work: with such perfection fram'd
> Is this complex stupendous scheme of things."

Every species having an instinct and an industry peculiar to itself, constructs its nest in its own peculiar way. The palmipedes place theirs either on the ground, or among the reeds in the neighbourhood of waters. The grallæ fix theirs near marshy places, or conceal them on the ground among the tufted plants. The gallinaceous birds, in the furrows of the fields, or on the gentle declivities of the lesser hills. But all these fowls being polygamous, and the males abandoning the care of the eggs, which are usually very numerous, entirely to the females, they cannot be, with strict propriety, said to construct any nest, contenting themselves with little heaps of straw, &c. to deposit their eggs in. The ostrich and cassowary expose theirs on the naked sand, leaving them in a great measure to be hatched by the influence of the sun. But the tadorna, a species of duck, some penguins and sphenisci, deposit their eggs in a sort of burrow, which they dig like rabbits. Other water-fowl suspend their nests in rushes at the surface of the water, as the colymbi. Some construct theirs in the clefts of rocks, or in little hillocks, like the cormorant and the sea-mew. The flamingo builds its nest in a sort of clay island in the midst of the water. Some of the ciconiæ place their nests on the summits of buildings, and the herons in the lofty forests.

The large birds, generally speaking, particularly the species which do not usually perch, the gallinæ, the grallæ, and the palmipedes, construct their nests with but little art or industry, placing them most usually on the ground among the herbage. The vulture and eagle tribes generally make choice of the clefts of precipitous and lofty mountains; and sometimes these last prefer the top of the loftiest trees to construct an immense nest in, interlaced with small branches, and carpeted within with grass disposed without much ingenuity. The nocturnal birds of prey, to which nature has refused the means of constructing a nest, lay their eggs in the hollows of a tree or rock, or take possession of some nest abandoned by birds of their own size. The pici, the woodpeckers, the sittæ, the hoopoes, many tom-tits, fly-eaters, &c. lay their eggs in holes of trees or walls, on materials heaped inartificially together. The bee-eaters and martin-fishers do the same in hollows of the earth. Crows, jays, pies, &c. construct their nests on trees, give them considerable solidity with a tissue of roots, fibres of plants and moss, and furnish the interior with wool and hair in abundance. The magpie builds an inaccessible fort, surrounded and covered with thorny branches.

All birds do not build nests. Some make use of such as they find abandoned. Others, as we have seen, deposit their eggs in any place that appears convenient. The genuine cuckow lays her eggs in a strange nest, and leaves to a strange mother the care of hatching and educating the offspring. Wilson has lately made us acquainted with a North American bird, the *passerina pecoris*, (vulg. cow-blackbird,) which does the same. These, however, are the only instances of which we know, as yet, of this deviation from a general law.

The care of constructing the nest more usually devolves on the female than the male, who seldom does more than collect and transport the materials with which she operates. Some males even do not give themselves any trouble about the matter. The female, bending and interlacing the sprigs of dried plants, gives the first form and solidity to the nest; and, in proportion as she furnishes it, pressing on the materials which she has accumulated, separating and arranging them by the movements of her body, she finally puts the entire into a suitable form.

The monogamous species construct by far the most perfect nests, and the most artificially disposed. Our chaffinches, goldfinches, &c., form nests well tissued without, warm and downy within, of an hemispherical form, and fixed with much art between the branches of trees. The bullfinch takes particular care to have an opening only on the side least exposed to the wind. The hoopoe, the pici, the wren, place their nests in the hollows of trees. The loriot suspends its nest on the bifurcations of the branches, and covers it over like a havresack. The swallow is peculiarly admirable in the formation of its nest, which it glues in the angles of windows and chimneys, and cements very solidly with clay, thickened with straws and hairs, and furnished inside with feathers or down. It only leaves a small aperture on the side. The remtz (*parus pendulinus*) has the art of weaving the down of the willow-flower, of the poplar, of the thistle, of the dandelion, and thus fabricating a thick felt, or sort of cloth, the woof of which it strengthens by filaments of plants, and gives it the form of a pear hollowed inside, and wadded within with the same down, not thus manufactured. The aperture is placed on the side, and provided with a ledge, which the bird can close. But, above all, this little being has the address to attach this nest, with the flax of hemp or the nettle, to a moveable branch, suspended over a running stream, so that no animal, such as the rat, lizard, or snake, can destroy its family. Others of the pari, or tomtits, as that of the Cape, the guit-guit, many of the gross-beaks, put in operation all the resources of architecture, to lodge their little ones.

Certain species of orioli attach their nests under the foliage of the banana-tree. Some of them construct in common numerous houses, divided into four chambers, and lodging several families; and to prevent any mutual embarrassment, they trace corridors, winding paths, by which each can repair to its nest. The caciques form theirs after the fashion of a gourd, and suspend them, like numerous girandoles, on the same trees. The anis of the savannahs *(crotophaga)* lay and hatch, in common, in large nests divided into compartments, and covered with foliage. The yapous suspend their nests like alembics, or small lamps, on the trees in South America. The baltimore's nests resemble purses, with two openings. The small fig-eaters, with yellow neck, hang their nests to the flexible branches of willows; and the *motacilla sutoria* sows a leaf detached from a tree, to another leaf placed at the extremity of a branch, in a sort of scuttle shape, to receive its delicate brood. The nest of the baglafecht *(loxia phillipina)* is a sort of sac, twisted spirally like the shell of the nautilus, and suspended to the extremities of the branches. In the same manner are formed those of the toucnam-courvi, nelicourvi, &c.

We find the perfect art of the basketmaker in the nest of the *hirundo acutipennis* of Louisiana. It constructs at first a sort of platform, with little dry branches and briars, cemented with the styrax of liquid amber, on which it places a nest composed of small sticks, glued together with the same gum, and disposed nearly after the manner of the osiers of a basket. It gives to this admirable little piece of workmanship the form of a third of a circle, and fixes it by its extremities to the walls of a chimney.

Among the grallæ, the small water-rail *(rallus porzana,* Lin.) constructs a nest well worthy of observation. This nest is formed like a bark, floats upon the water, and is attached by one of its extremities to the stalk of a reed.

The *motacilla salicaria* constructs its nest round three stalks of reeds, with plants which grow in the marshes. These stalks serve to retain the nest, which ascends or descends along these stalks, according as the surface of the water on which it reposes rises or falls.

The last nest which we shall notice in this place is that very celebrated one of the *hirundo esculenta,* and which constitutes a dainty in great request among the Chinese and Japanese. This swallow constructs its nest in the hollows on the steep shores, or in the caverns of the Molluccas, and many other islands in the Indian ocean. In Java these nests form a considerable article of commerce, and are sold extremely dear, when they are quite fresh, and not dirtied during the process of incubation. These nests are made with the branches of a sort of fucus, discoloured and agglutinated by the swallow. It was for a long time imagined that these nests were formed with the spawn of fishes, or other animal substances, collected by this bird on the surface of the sea. It has, however, been clearly ascertained by M. Valenciennes, that they are made of the branches of a certain fucus, by an accurate comparison of some colourless fucus brought from the Molluccas, with the composition of the nests in question, deposited in the King's Cabinet. This comparison was made by M. Desfontaines, that most expert botanist. This is the less surprising, when we consider that many vegetable productions of the Indian ocean are edible, and that one of them, the *fucus sacchariferus,* contains a large portion of sugar.

Is is also proper to state, that M. Reinwardt, a celebrated professor, who made a long stay at Java, was of opinion that this bird consolidates its nest with a viscous and glutinous humour, secreted in its very large parotid glands. The sums netted by the sale of these nests are very considerable. Near the Goenong-Goetoe, one of the largest volcanos in Java, there is a cavern from which the proprietor derives a revenue of more than fifty thousand Dutch florins per annum.

As a winged animal, like the bird, could not bear about with it its offspring in the womb, like the mammifera, nature has provided for this inability by rendering it oviparous; and that the eggs, which have a shell that does not give way, unlike the eggs of reptiles, which are soft, may be more easily laid, birds have the ossa ischia, and the ossa pubis, remarkably prolonged behind. In this large cavity of the pelvis the eggs acquire their volume, and the white which surrounds the vitellus.

The ovaries of the female are tolerably large, and situated near. the reins. An oviductus receives the vitellus, which is enveloped in an albuminous substance, commonly denominated the white. When the egg descends to the lower part of the oviductus, it begins to be covered with a cretaceous matter, the thickness of which increases in the cloaca, whence the egg is finally expelled by the action of the peculiar muscles of this part. The colour and form of the egg-shell vary in the different species, and form a criterion of distinction, which imperatively claims the attention of naturalists. If the ovule has been fecundated in the act of coïtion, the heat produced by incubation is sufficient for the developement of life. Among our domestic fowls, where the developement has been investigated with the greatest accuracy, it has been observed, that at the end of six hours a small red point appears on the vitellus. This is the *punctum saliens,* which is to be the heart of the chicken. From this *punctum saliens* proceed numerous radiations of vessels, which are, as it were, the outlines of the venous system. A small crescented gray line which surrounds the little red point, becomes the spinal-marrow. It inflates in front to form the brain. The legs, then the arms, and, finally, the viscera, are developed.

The eggs are usually of an elliptical form, more or less elongated, according to the species. There is a large and a small end; the first is rounded, and the other approximates to a point. In the majority of birds, the eggs are of one predominating colour, over which are dispersed spots more or less numerous, and more or less varied. These spots augment in size, and become deeper in colour, according to the progress of incubation; if they then appear more numerous, it is not that they have actually increased in number, but that they have become more sensible to the eye. This is very visible in the green and red eggs, &c. These spots are commonly wider, closer, and more numerous towards the large end of the egg, where they form a sort of zone or crown. Among many birds, however, they have one uniform colour, without any spot.

The eggs of the *diurnal birds of prey* are of a whitish colour, spotted with red, or red spotted with brown. The eggs which border on a red, diminish in tint in proportion as they are laid; so that sometimes the last is merely a light-reddish, or whitish, pricked out with clear red.

The *owls* and *howlers* have white, or whitish eggs, without spots. Among the speckled magpies, the eggs, on a white ground, have, at the broad end, a circle of red, brown, and bluish spots, over which the same colours are sprinkled. Birds which nestle in the hollows of trees, of walls, or rocks, have, in general, eggs of a pure white. Such are those of the *hoopoe*, the *pici with black plumage*, the *torcol*, the *martin-fisher*, the *bee-eater*. The woodpecker's eggs have a few red points.

Birds which nestle to a certain height in the trees, as ravens, crows, pies, &c., have usually green, or greenish eggs, spotted or picked with brown.

It has been remarked, that the white or whitish eggs in the *swimming* birds are short and rounded, while the yellow or greenish and spotted eggs are very much elongated.

The eggs of the *grallæ* have spots on a gray, yellow, yellowish, green, greenish, bluish, red, or reddish ground. They are rarely spheroïd, being mostly elongated, and diminishing very rapidly from the large end.

White is the commonest colour of the eggs of the *gallinacea;* some, however, have a green, greenish, or yellowish ground. It is remarkable that the eggs which certain species deposit on green herbs, partake more or less of this colour.

The *passeres* have eggs, the ground of which is white or whitish, blue or bluish, green, usually spotted with deep colours, such as red, brown, and black.

The *Tomtit* kind, which nestle in the hollows of trees, have eggs altogether white, or white picked with red. The same is the case with the *swallows* and *martens*. The *larks, pipis, &c.*, have the eggs of an earthy hue. The nest is scarcely finished when the bird commences to lay, and if the eggs be removed in proportion as they are deposited, they will lay a greater quantity. But the number, though undetermined, is more considerable among the polygamous species, such as the palmipedes and gallinacea, than among the monogamous.

The *birds of prey*, such as the eagle, the vulture, and the falcon, lay but two, or four eggs at most, each brood. Most of the divers, &c., only one, but which is very bulky.

The rapacious birds are less fruitful than the other species, more particularly so than the small granivorous and insectivorous races. This, indeed, seems a wise provision of Nature in all cases; but, in truth, it must be the infallible result of the peculiar constitution and regimen of animals. Those which derive their subsistence from the vegetable kingdom, must naturally be more numerous, as their food is more plentiful; and the smaller races more especially so, as each individual consumes a smaller quantity.

Hens often lay infecundated eggs, which the Romans called *ova subventanea*, and the Greeks ὠὰ ὑπενέμεῖα, because they imagined them to be produced by the influence of the zephyrs. Many other birds are also liable to lay infecundated eggs.

The attachment which birds exhibit in the process of incubation is very singular in animals of such a volatile constitution. The mother, seated the live-long day upon her eggs, forgets all the necessities of nature; she passes hours, days, and weeks, under the influence of an instinct, whose domination is as imperious as its cause is incomprehensible. Her natural character undergoes a temporary change, and flinging off the timidity which usually characterizes her, she braves every danger, and dares the most unequal conflicts for the safety of her young. Some birds never quit their nests without plucking feathers from their own breasts to cover their eggs; others cover them with dry leaves; and among some species, as the pigeon, the male hatches in his turn, or brings food to the female. But, as we before observed, there are one or two exceptions to this general law of Nature.

The period of incubation varies, not only according to the species, but also the degree of temperature which the eggs undergo. Cold will retard, and heat accelerate, the coming forth of the young. The eggs of the tomtit take but eleven days in hatching; those of the pigeon eight-and-twenty; hens have twenty-one, and many of the scolopaces and palmipedes from twenty-eight to thirty. It is said that the eggs of the *mergus serrator* (Lath.) take even fifty-seven days.

It is well known that eggs may be hatched by artificial heat.

That the chick may be enabled to cut the shell in which it is imprisoned, Nature has provided it with a little osseous eminence on the beak, which falls soon after this operation has been performed: an admirable foresight, which of itself is amply sufficient to indicate the views of an allwise and intelligent Being.

The incubation of birds must be considered as correspondent to the gestation of quadrupeds. Nature has imparted to the females of the accipitres a larger size and greater vigour than to the males, from the necessity of providing living prey for the young. The females of the gallinacea, on each of whom singly devolves the care of a very numerous offspring, could not provide for it, if their chicks were not endowed with the instinct of seeking food for themselves. We find that it is towards the period of the birth of the young that the mothers put in requisition all the resources of their instinct. So much tenderness and trouble lavished without compensation; such a sublime and generous self-devotion in the most urgent dangers, proves that this natural and amiable sentiment is not the result of any mechanical connexion of ideas and sensations, but of a law altogether divine. The swallow, precipitating itself into an edifice in flames to rescue its young; the hen, which hesitates not to brave death in defence of her chickens; the timid lark presenting herself to the fowler, to divert him from her nest; the little colibris, which prefer an eternal slavery with their offspring to liberty without them;—in fine, all these touching evidences of affection for the helpless, in animals so light and volatile, clearly indicate the sacred impulse communicated to all that breathe by the Mighty Being, who has willed the perpetuity and support of every species. Here, indeed, we recognise the workmanship of the Divinity in all its admirable wisdom and surpassing benevolence: *digitus Dei est hìc!*

We also find the birds deserving of the most attentive observation in the education of their young. The assiduity with which they bring them food; the care which they take to adapt it to their tender stomachs; the degrees by which they teach them to fly, calculating with such accuracy the proportion of their growing strength; all these, and many other points of a similar nature, are subjects of the highest interest to the contemplative lover of Nature.

It is a very mistaken idea, to imagine that the rapacious birds, after having reared their offspring for some time, chase them from the nest from the want of parental feeling. Among all carnivora it is a common habit to excite their progeny to seek their prey alone. Already have they fashioned and prepared them for this, by bringing them living victims. It is the useful lesson of necessity, and of the experience of an active and enterprising life, which is thus transmitted from father to son by this expulsion, in all appearance so barbarous and unfeeling. We find that the crow, after driving its offspring from the nest, still leads and directs them for awhile in the search of subsistence.

We discover in the young bird, even in the nest, the germination of the instinct and character which must determine its future life. The eaglet soon exhibits traits of the fierce and sanguinary disposition of its sire; while the humble chick, in issuing from the shell, knows already how to scratch the earth and pick up the grain. The young swallow soon commences to essay its rapid wings, and prepare itself by small excursions for its future long and unwearied migrations. The cygnet aready delights to bathe itself in the crystal wave, and glide along with that instinctive grace which is so amply developed in maturity.

Every species chooses at once its own proper domain, follows the impulse of instinct, puts its little organs into play, and exhibits in its infant efforts all the rudiments of vigour and address. Thus each successive race, among the wild species, is the exact representative of the energy, strength, courage, in fine, of all the qualities of the preceding. Degeneration and change are unknown, except among those favoured species which experience the fostering care of man.

Whether the birds are naturally more precocious, or that the Author of nature, in consequence of the wants and dangers to which their peculiar destiny of existence exposes them, has thought proper to diminish the period of their infancy—certain it is that they acquire their full perfection sooner than quadrupeds. Their short sojourn with their parents does not permit them to receive that developement of intelligence which depends on the association of individuals. The flights of cranes, flocks of partridges, of geese, &c., in fine, all the general assemblages of birds do not constitute societies in which there are sufficient mutual relations for the developement of the internal sense. Birds, accordingly, except in the construction of their nests, do not exhibit the industry and intelligence observable among some quadrupeds, either because they are less happily organized, or have less natural aptitude for instruction. Still, as we before observed, many other birds, as well as the psittacidæ, possess a capacity and a considerable power of imitation.

Goldfinches in cages may be instructed to perform many little tricks. Perroquets were exhibited in Paris in 1803, and some Java sparrows in London a few years ago, which had been taught many amusing exercises. Canaries and other small birds exhibit a considerable degree of familiar attachment.

It may be noticed, that the intelligence of birds is more considerable in proportion as we proceed from the palmipedes, through the grallæ and gallinacea to the perching birds, the accipitres, the passeres, and particularly the picoïdes, the coraces, and the climbers. Accordingly, the cerebellum of these birds is more voluminous. The last-mentioned birds have also a shorter neck, and a head generally more bulky in proportion to the body. Were animals to be classed according to the scale of their intelligence, the psittacidæ should come first among the birds; and then other intelligent and docile species. While the palmipedes, many of the grallæ, and the imbecile ostrich, with its long neck and weak brain, should be removed to the end of the list. If Nature has given to man the first rank among terrestrial animals, not on account of his size, or corporeal strength, which are considerably inferior to those of many others, but by reason of the great superiority of his intellect; doubtless, the species most highly gifted in this respect, deserve the foremost places in their respective classes.

Parrots are capable of being taught a thousand things, which require not only docility and flexibility of organization, but also considerable memory, and some glimpses of reason. The American Indians employ their leisure not unfrequently in instructing these birds, and thus dissipate that *ennui* which is as liable to creep into the hut of the savage, as into the palace of the king.

The jacana, one of the grallæ, is capable of being made a faithful servant to man. It can be taught to watch the flocks, take its regular rounds, call back the sheep when they stray, with a loud voice, and force them to return with strokes of its beak. It is only necessary to hint, in this place, at the capacity and docility of hawks and falcons. In China, cormorants are trained to fish for the advantage of their owners.

There are many birds distinguished by very remarkable habits. Thus the agami, which is a kind of ventriloquist, utters a hoarse and deep sound, which one would suppose proceeded from the anus. The crane, called in French *Demoiselle de Numidie,* gesticulates and makes a motion like dancing. Many of the nocturnal birds make singular and ridiculous gesticulations during the day. The cincle, or sea-lark, buries itself under water and walks there. Many of the magpie tribe spit the little birds and insects which they catch, upon thorns, that they may eat them at their leisure. The vultures are said to have an excellent scent; and ancient writers have informed us that, after the battle of Pharsalia, the vultures of Asia and Africa passed over into Europe, to feast upon the bloody carcasses of the slain. Ravens are also observed to follow armies.

In short, each species has its peculiar mode of life. "Their habits and manners," says Buffon, "are not so free as might be supposed: their conduct is not the result of a freedom of will or choice, but a necessary effect derived from the conformation, the organisation, and the exercise of their physical faculties. Determined and fixed, each in the manner of life which this necessity imposes, none attempt to infringe it, and none can withdraw themselves from its influence. It is by this necessity, as varied as is the structure of animated bodies, that all the districts of Nature are peopled. The eagle quits not the rock, nor the heron the shore; the one drops from his airy height to carry off or tear the lamb, by no right but that of power, and by no means but those of violence: the other, with his feet sunk in mire, awaits, at the command of necessity, the passage of his fugitive prey. The woodpecker never abandons the trunk of the trees, round which he is ordained to creep. The snipe must remain in his marshes; the lark in his furrows; the singing-birds in their groves. Do we not observe all granivorous birds search out inhabited countries, and follow the track of cultivation; while, on the contrary, those which prefer berries and wild fruits, invariably shun the footsteps of man, and in the dense wood, or on the solitary mountain-steep, abide alone with Nature, which has dictated the laws they shall obey, and furnished them with the means of such obedience? She it is who retains the wood-hen beneath the thick foliage of the fir-tree; the *solitary* blackbird (*turdus cyaneus*) in the rock; the loriot in the forest, that re-echoes to his cries; while the bustard haunts the dry fallow land, and the rail the humid meadow. These are the eternal immutable decrees of Nature, as permanent as the forms of her productions. These are her grand and rightful properties, which she never yields nor abandons, even in things which we imagine we have ourselves appropriated altogether; for, let us have acquired them how we may, they are not the less under her dominion. Has she not, for example, quartered upon us such troublesome guests as the rat in our houses, the swallow under our windows, and the sparrow beneath our roofs? And when she calls the stork to the summit of the ruined tower, within whose walls the night-bird has already taken up his abode, does she not seem hastening to resume the possessions which we have usurped for a period, but which she has commissioned the resistless hand of Time to restore to her domain?"

We shall conclude this preliminary essay on birds in general, with a few observations on their molting, migrations, and habitat.

It is a truth generally recognized in physiology, that organized bodies are first developed, and then gradually wear out, both externally and internally, by the action of decomposition, which is antagonist to that of composition. They never remain in a constant state, or in an identical body. The alimentary matter, after being assimilated with the animal substance, ends by being decomposed and excreted. The vital force is perpetually acting on the organs in a propelling direction to the external surface, in proportion as reparation takes place. This *mutation*, or evolution of living beings, is the source of the changes which their external surfaces undergo in the different periods of their existence. These changes are of great importance to study, inasmuch as an ignorance of them has often caused the multiplication of species and confusion of sexes, and distinctions in many instances where there was no real difference.

The first rudiments of the plant are already organized in the grain or seed, and the first rudiments of the animal in the egg. Nutrition, through the interior, augments all the dimensions of the living body, and increases it to a determined point of size. Each individual part of the organised being has its peculiar nutrition, emanating from the general nutrition of the body, because each has its peculiar force originating in the vital principle, common to the whole machine. Thus the body has not only a general evolution, but each of its organs has a particular one, which may take place even independently of the other parts, and augment at their expense.

If each organ has its own peculiar life, it has also, without doubt, its age and duration, independently of what it receives from the whole body. In fact, certain organs grow old and die before the general death; as, for instance, the organs of generation. These are not developed until long after the birth of the living body, and they die before it. Their particular vitality has, therefore, much less duration than the general vitality. It is the same with many other parts, the vital duration of which is very short in comparison to that of the individual. This is particularly the case with several external organs, such as horns, teeth, hair, feathers, shells, &c.

Since each part of the animated body is thus endowed with its peculiar life, it has its period of youth, perfection, decrease, and particular death. This is matter of daily observation in organised productions; for when an organ is completely dead in a being endowed with life, it separates and falls, because a dead substance cannot co-exist with a living. The internal force which should maintain it in its organised state is gone, and destruction follows.

Now moulting is nothing else but this natural death of some part of each animated being, in consequence of the developement of other more interior parts; and this peculiar function is subject to certain laws which are tolerably constant.

In the vegetable kingdom we observe, at the end of each year, the fall of the leaves, flowers, &c., because these organs have gone through all the natural phases of their existence. The defoliation of trees, and the fall of their organs of reproduction, may be considered as their *annual moulting*, which takes place also among other vegetable products, even among evergreens, but in a manner less rapid and perceptible, as one leaf successively replaces another.

Could we doubt that the life of organised bodies corresponded with the revolutions of the terrestrial globe, and that its phases were regulated upon them, we should find a striking proof of this truth in the defloration and defoliation of vegetables, and the moulting of animals. In spring, all living and vegetating nature renews and developes its productions; the earth is clothed with verdure, and the animal tribes become invested in a fresh and more brilliant livery in this season of universal reproduction. The cause of this grand external revolution in all beings is this: during the winter their functions, long compressed by the cold, have gained a superabundance of juices, of sap, of nutriment, which only awaits the return of external heat to assist its propulsion to the surface. Accordingly, at the appointed season, the germs shoot forth with trebled vigour. Everything in our organisation is equally propelled outwards. A proof of this may be observed in the cutaneous eruptions that so frequently appear on the return of spring.

We find, then, the germs of leaves, of flowers, of fruits in vegetables, the hairs, feathers, scales, horns, epidermis, &c., in animals, increasing and developing themselves in spring, to flourish in succession, at least for the duration of the summer solstice in our hemisphere.

But at the approach of the autumnal equinox, plants and animals, being more or less exhausted by the vast expenditure of their vital forces in the great work of reproduction, and also by the increased energy with which those vital forces acted in propulsion to the surface, their external functions begin to be enfeebled, and by so much the more as the heat of the sun diminishes. Then these external parts, these vernal productions, cease to receive aliment through the body; they have, besides, arrived at the full term of their augmentation, and can admit of no further nutriment. They dry up, wither, are detached, and fall. Thus is operated, sooner or later, the fall of flowers, leaves, and fruits, and the change of hairs, feathers, horns, epidermis, scales, &c., when animal and vegetable bodies are brought into this sort of autumnal concentration to prepare them for the winter. In the Austral hemisphere, as our winter is its summer, and reciprocally, the periods of molting every year, must be exactly opposite to ours.

Under the torrid zone, as the sun passes the equinoxial twice a year from one to the other tropic, it produces, in some measure, two summers and two winters, the latter being seasons of perpetual rain. It also determines the moulting and reproduction of animals and vegetables twice a year. Organised beings, in consequence of this, live much more rapidly there than elsewhere; they are continually in a course of production and destruction. New flowers arise by the side of the fruits; the new leaf replaces the old and withered; the bird recommences its amours, and chaunts renewed pleasures by the side of its nestlings of six months before.

The birds, by their brilliant plumage, at the season of coupling, announce most remarkably the changes of the moulting. The females, as we have said before, having pale and dull colours, appear much less to undergo the moulting, the new plumage being not so distinguishable from the old. But the males shine in the richest apparel at the epoch of pairing, a phenomenon, without question, intimately connected with the secretion of the seminal fluid, and more especially observable under those burning skies. The intertropical birds, having usually two broods every year, resume their nuptial dress, when the sky becomes pure and serene. They then seek out the females. But when the rainy season sets in, they lose their beauteous plumage and sonorous voice, at the same time with their sexual desires. Dull, and, as it were, ashamed of their gray or brown dress, they then bury themselves under the thick foliage, as if to escape, during this temporary degradation, the observation of those who admired them in the days of their brilliancy and enjoyment.

In the very cold countries, a different system of moulting is observed in various birds and quadrupeds in winter. The covering, which accompanies the slumber of the sexual organs, is peculiarly proper to secure the animal from cold. Thus the *lepus variabilis*, the ermine, many other mammalia, and a crowd of northern birds, of palmipedes, and grallæ, which in summer have the plumage brown, or shaded to various depths, moult in autumn their hairs and feathers, and change them for white, or pale tints, for winter. This whiteness is caused by the inaction of the rete mucosum, and its colouring matter, from the constriction of cold. An effect altogether similar can be produced on sparrows, by depluming them, and rubbing them over with spirits of wine. The feathers that spring afterwards remain white, because the spirits of wine prevents the developement of the subcutaneous colouring matter. These white animals resume in spring, with their sexual desires, their coloured hairs or feathers. The pen-feathers of the wings and tail do not usually moult at this time, but only the smaller feathers.

The philosophy of moulting in birds (to which we must confine ourselves here, though the principle is the same in all species) may be explained in a few words. In the feather of the bird, at the extremity of the tube, a blood-vessel penetrates, like that under a tooth The dry and slender pellicle of the interior of this tube is at first a gross fleshy canal, receiving vessels filled with lymph, and very multifariously ramified in young birds. These lymphatico-sanguine fluids serve for the nutriment of the feather. Its barbs are at first nothing but a sort of pap, and are rolled cornet-wise, under long membranous tubes. This sort of case for the growing feather, which is analogous to the laminæ of the bud which envelopes the growing leaf of the tree, soon drops off in plates. The feather, like the leaf, is more rapidly developed than the other parts, and the nutriment is carried to it in superabundance, from the necessity of clothing the bird.

When the feather has received its full complement of size and nutriment, it ends, like every other living substance, by drying up. Its saturated canals can admit no further aliment, and it becomes a dead part. It must, therefore, fall; at the same time the nourishment supplied by the body of the animal is carried to the germs of feathers yet in embryo, under the epidermis, and thus a new plumage succeeds to the old.

The habitat of birds is not circumscribed within such narrow limits as that of quadrupeds, because, by means of their wings, they can traverse more space, and even cross the seas. The aquatic birds, by alternate flying and swimming, can proceed to the most remote countries. Nevertheless, each species adopts a country, chooses a climate suitable to its nature, and, when the change of seasons obliges it to seek, under new skies, a country analogous to its former one, it is but for a season. These birds always return to their favourite country at the season of reproduction. The stork, indeed, has two separate broods, one brought forth in Europe, and the other in Egypt.

Birds, generally speaking, appear to belong more to the air than to the earth. They constitute moving republics, which traverse the atmosphere at stated periods, in large bodies. These bodies perform their aërial evolutions like an army, crowd into close column, form into triangle, extend in line of battle, or disperse in light squadrons. The earth and its climates have less influence on them than on quadrupeds, because they almost always live in similar degrees of temperature, passing the winter in hot climates, and the summer in cold. This continual interchange of birds establishes a sort of communication between all countries, and keeps up a sort of equilibrium of life. The bird, passing in summer from the equinoctial climates to the cold regions of the north, and again in winter from the poles towards the equator, knows, by an admirable instinct, the winds and the weather which are favourable to his voyage. He can long foresee the approaches of frost, or the return of spring, and learns the science of meteorology from the element in which he almost continually lives. He needs no compass to direct his course through the empire of the cloud, the thunder, and the tempest; and while man and beast are creeping on the earth, he breathes the pure air of heaven, and soars upwards nearer to the spring of day. He arrives at the term of his voyage, and touches the hospitable land of his destination. He finds there his subsistence prepared by the hand of Providence, and a safe asylum in the grove, the forest, or the mountain, where he revisits the habitation he had tenanted before, the scene of his former delights, the cradle of his infancy. The stork resumes his ancient tower, the nightingale the solitary thicket, the swallow his old window, and the redbreast the mossy trunk of the same oak in which he formerly nestled *.

All the volatile species which disappear in the winter do not, therefore, change their climate. Some retire into remote places, to some desert cave, some savage rock, or ancient forest. Such are many of the starling kind, the loriots, the

* Linnæus tells us, that a starling came regularly to lay during eight years, in the same trunk of an alder, although it emigrated every winter. Spallanzani having attached a red thread to the legs of the swallows which nestled under his windows, beheld them return for many years in succession.

cuckow, &c. &c. They sally from their retreats at the close of winter, and spread themselves through the country

Other families of birds do not, properly speaking, emigrate. They content themselves with approaching the southern climates, in proportion as they are pursued by the cold. The species called erratic, such as the greenfinches of the Ardennes, larks, ortolans, other frugivorous races, and especially parrots, go in troops, begging, as it were, their subsistence on their passage. Others follow the track of cultivation, and spread themselves in proportion with the habitations of men.

Of the birds which emigrate every year, some depart in autumn and return in spring, while others depart in spring and return in autumn. Our insectivorous races, and many granivorous, finding nothing at the beginning of winter but a soil deprived of its productions, presenting every where the image of desolation and death, are necessitated to betake themselves to more favoured climes. At the commencement of this season of gloom, when the fields are denuded of herbage, and all terrestrial animals have retired each to his peculiar shelter, and many species have fallen into a state of torpor, the birds prepare to set out on their voyages. They assemble in troops at the appointed period, and take advantage of the favourable wind which is to aid them in their course. Their proceedings are fancifully and beautifully depicted by a French poet :—

> " Dans un sage conseil par les chefs assemblé
> Du départ général le grand jour est réglé.
> Il arrive. Tout part: le plus jeune peut-être
> Demande, en regardant les lieux qui l'ont vu naître,
> Quand viendra ce printemps par qui tant d'exilés
> Dans les champs paternels se verront rapellés."
>
> L. RACINE, fils.

Those which, through negligence or weakness, remain behind are placed in no very comfortable predicament. They drag out a miserable existence, and constantly perish from famine, in the midst of frost and snow *.

As our summer birds abandon us towards the close of autumn, we receive, at the same time, fresh supplies of feathered hordes from the populous North. When the weather grows dull, we see passing through the misty air large detachments of woodcocks, of lapwings, of plovers: these are followed by triangular bands of cranes, storks, of teal, of wild-geese, and ducks. They alight in inundated fields or reedy marshes, or spread themselves in the glades of humid woods denuded of their foliage. They continually utter clamorous and melancholy cries, in accordance with the bleak and wintry scene around them, like the whistling of the north-east wind through the defoliated forests. It is a most curious circumstance to observe the cranes return and come back every year, on the same days, with the most marvellous exactness.

The palmipedes and grallæ come to us in winter from the northern climates; where they return, in spring, to their cold and humid habitations, whence they had been driven by the ice. The insectivorous and granivorous races come back with the return of the flowers and fine weather. They return from southern regions into their native country, allured by the expectation of renewed enjoyment and abundant food.

It is at the periods of the equinoxes that these great voyages of birds are performed. These are also the periods of great winds, as if nature had intended that the birds should be thus

* The female of the greenfinch emigrates the first into Southern Europe, and comes back in spring to find the male. It is not the rigour of cold which obliges birds to emigrate, for our wrens can support the severest winters, but it is the want of sufficient food. Their longest voyages take place quickly; and when it is necessary to cross arms of the sea, the birds rest themselves in islands. Thus immense numbers of quails are seen every year in the isles of the Archipelago. As to the immersion of swallows under water during winter, it appears totally devoid of all probability.

assisted in their flight. The cold, which drives the birds of the polar regions into more temperate climates, sends those of temperate climates into the hot countries. But on the first indication of summer the hot climates send back to the temperate their aërial inhabitants, and the temperate send back to the cold regions their native tribes. Thus there is a general concentration of birds towards the torrid zone in winter, and a general dispersion towards the poles in summer.

The triangular figure which migrating birds adopt in their flight is the most favourable for cutting through the air. The bird placed at the point is the most fatigued of the entire band: accordingly, each takes this place in turn. The migrations of fishes are conducted in the same manner: the most robust places himself at the head; the other males follow, and the females and young come last. When the ranks of the storks are broken by the wind, they condense into a circle. They do the same when attacked by an eagle.

Whatever the emigrations of birds may be, they yet do all adopt a peculiar country. The palmipedes, such as the penguins, the manchots, the petrel, the albatross, wild-goose, duck, &c., prefer the northern climates and the polar seas. They are entirely aquatic. The grallæ, such as water-hens, colymbi, herons, curlews, woodcocks, teal, storks, cranes, seek out marshy places, covered, humid, and cold countries. These are long-legged birds, and grope in the mud for prey*. They do not bear extreme cold as well as the palmipedes, and consequently they proceed further into the temperate regions. The gallinacea inhabit the fields, dry ground, and even small hills,

* Nature, by a singular foresight, has imparted the faculty of sensation to the extremity of the beak of these birds, by means of a nervous branch from the fifth pair which terminates there. This sensibility was necessary to these races, because their sight could be of no assistance to them in finding their prey in the mire. They are moreover inferior to other birds in the acuteness of this last sense.

warm vallies, and are fond of rolling in the dust. From this circumstance the French call them *pulvérateurs*. The small granivorous and insectivorous kinds, as sparrows, gross-beaks, titmouse, &c., haunt the thickets, bushes, and brakes, and never fly but to a moderate height. The birds of prey, as vultures, owls, eagles, hawks, falcons, kites, and buzzards, delight in rocks, mountains, and elevated and solitary stations in general. Finally, the climbing birds, as peckers, toucans, hoopoes, cuckows, and under the tropics the numerous families of psittacidæ, prefer lofty forests and warm climates.

As the grallæ, or waders, are less tolerant of wet and cold than the palmipedes, so the gallinacea are still less so than the grallæ. But they are peculiarly terrestrial, and natives of the temperate climates. The small granivorous and insectivorous races attach themselves less to earth than the preceding, and bear cold still worse. The birds of prey elevate themselves more in the air, and in general repair towards the warmer climates. Finally, the climbers never attach themselves to the ground, and inhabit principally towards the tropics. There is, then, a marked gradation from the aquatic to the climbing birds, from the penguin, or the manchot, to the parrots. The first remain towards the poles, the second under the tropics. The first remain continually in the waters or on the ground; the second under the most elevated trees. The first have a dusky plumage, and dull and tarnished colours; the second are invested with plumage of the most brilliant dye. The aquatic bird under a hazy sky, in a cold and humid atmosphere, has a heavy and fat body, a dull and stupid character. The climber, under a serene heaven, in a warm and dry atmosphere, has a thin and delicate body, and a lively disposition. The inhabitant of the waters is voracious; its voice hoarse and disagreeable. The inhabitant of the tropical forests is temperate, the voice flexible, and the song delightful. The first is polygamous, and cold in constitution; the second, monogamous, and ardently

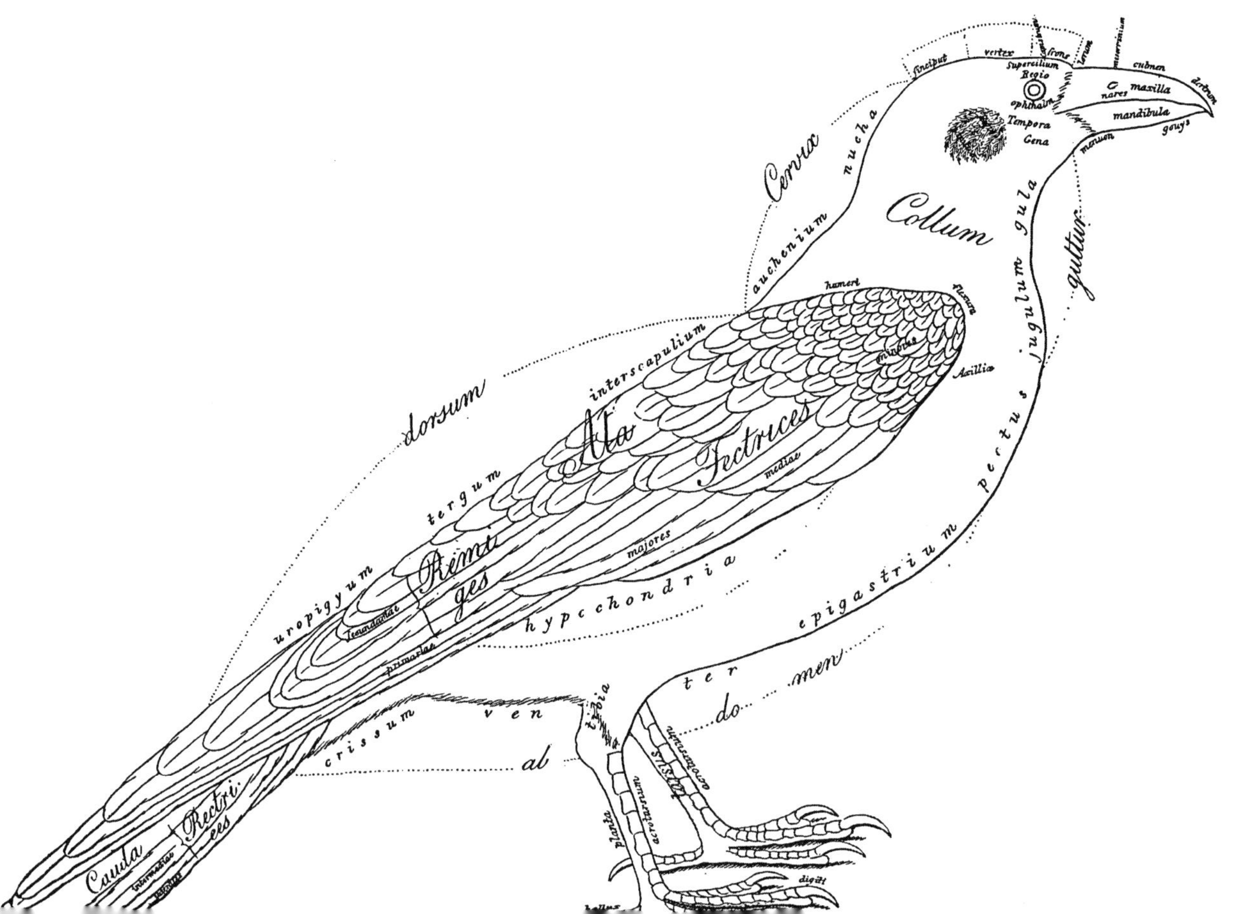

sinciput
vertex
frons
lorum
culmen
Supercilium
Regio
ophthalm
maxilla
nares
dertrum
mandibula
gonys
Tempora
Gena
mentum
Cervix
nucha
auchenium
Collum
gula
jugulum
guttur
humeri
flexura
minores
Axillia
interscapulium
dorsum
Ala
Tectrices
mediae
pectus
tergum
majores
uropigyum
Remiges
primariae
hypochondria
epigastrium
ter
men
do
ven
crissum
ab
tibia
acrotarsium
tarsus
planta
Cauda
Rectrices
intermediae
digiti

attached to the female. The intermediate shades of these two extremes are filled by the families of the gallinacea and grallæ, which approximate more to the aquatic races, and by the birds of prey, and small granivorous and insectivorous races, approximating more or less to the climber. The palmipedes, the gallinacea, and grallæ seldom perch. It is seldom that the others do not do so. In fine, there is an immense difference in favour of the latter on the score of intelligence.

Instead of entering at large into the subject of terminology, we have, as the reader will perceive, given an outline figure of the bird, with the scientific terms for all the various parts of the body. A glance at this engraving will be quite sufficient to point out its utility, as the view of the different parts, with their correspondent denominations, is much better calculated to produce a clear and lasting impression on the mind, than the most minutely detailed description could do without the assistance of such an appeal to the senses.

SUPPLEMENT ON THE ACCIPITRES.

The Accipitres, or birds of prey, also termed rapaces and raptorial birds by some writers, constitute the first order of the class Aves. They are divided, as we have seen in the text, into two families, the diurnal and nocturnal accipitres.

The accipitres, as an order, are very strongly distinguished from all other birds. Their curved and powerful beak, strong limbs, acerated talons, robust head and neck, expansive wings, rapid and lofty flight, compact and solid frame, characterise them as eminently carnivorous. They all subsist by rapine, on living prey, or dead carcasses, and, unlike the granivorous races, they can dispense with water. The females are handsomer, and generally one-third larger than the males. These birds are exceedingly analogous to the carnivorous quadrupeds. The vultures, the griffins, the eagles, the hawks, &c., hold a similar place in the creation with the lion, the tiger, the bear, and all the different feline or canine races. They build their nests on the loftiest rocks and in the wildest solitudes. They seldom lay more than from two to four eggs, and are monogamous. Their temperament, like that of the carnivorous quadrupeds, is sanguinary and ferocious, and their voice is hoarse, shrill, or piercing.

Few birds exhibit so many changes on the type in plumage as the diurnal accipitres, from their birth to advanced age. Accordingly, we find it extremely difficult to determine the species, and even the sexes with precision, during the first two years, except among a few of them, in regard to size, in consequence of the similarity in the liveries both of male and female. In the young, the colours are less pure, and the spots more prominent and numerous before the first moulting, and often before the second. After this last, the tints grow purer, the spots and streaks begin to change; and this takes place more and more in proportion as the bird grows older. In certain species, these spots and streaks undergo so great a change in the course of time, that scarcely any vestige remains of them in the old males, which has often occasioned an erroneous distinction of species. Thus we find the vulture of Malta, passing from brown to white, becomes the vulture of Norway and the little vulture of Buffon: the monachus ends by being the black vulture, quitting its gray and brown plumage to assume a very dark brown. The fulvous vulture, reddish in its youth, becomes successively gray, ashen, and of a uniform gray white in advanced age. We find the osprey become the gray-headed, and, in old age, the white-headed pygargus. This has been contradicted, it is true. It has been advanced that the white-headed pygargus is a distinct species from the gray, which last is found only in North America and the most northern parts of Europe. But M. Vieillot declares that he has seen the osprey, and the gray and white-headed pygargus, in the United States; all three of which he considers, as in Europe, to belong to the same species. Another fact, cited by the same ornithologist, is that a white-headed pygargus in the menagerie of the "Jardin du Roi" was taken in France, and on its arrival there resembled the osprey extremely. The plumage of the pygargus passes more quickly to white on the head in the northern regions of both continents.

The birds of prey are much more numerous, in species, in Paraguay and the neighbouring countries, than in the rest of the world, according to M. d'Azara. There is one species of them to nine of other birds, while in the old world there is but one to fifteen. The birds of prey described by this naturalist are not quite so ferocious or carnivorous as others, for the majority of them live on insects, frogs, toads, serpents, &c., rather than on quadrupeds and other birds.

The first division of our author, on which we shall offer a few remarks, is that of the VULTURES. But it is by no means our intention here, or in any other part of our supplementary observations, to notice all the species which have been enumerated by naturalists. To do so would, in fact, be to dwell for the most part on a series of names, which have been constantly applied to the same species seen under different modifications. M. Vieillot remarks, that, after having observed the living vultures under the various metamorphoses which the difference of age occasions in their plumage, and having most attentively studied the subject, he is fully convinced that few of their genera are composed of as many species as some naturalists have adopted without examination, and others have repeated without reflection. In short, he considers few synonymies in such a state of confusion as theirs*.

Brisson, Gmelin, and Latham have described seven or eight species of vultures in Europe, though it appears more than probable that there are but three or four. As this is frequently the case, though, in our additions to the text, as formerly in our tabular synopsis, we insert all the enumerated species without vouching for their authenticity; we shall be careful, in the supplement, to speak of none that are not pretty well verified, and to give no particulars of any but such as are interesting and important.

Of all the characters drawn from the anterior portion of the body in the vulture tribe, the most distinct is the greater or less degree of nudity of the head and neck. To this may be added, that they differ from the eagles with which they have been vulgarly confounded, by having their eyes on a level with the head, while the eyes of the others are sunk within their orbits.

* A modern author has observed that it would be better not to quote these synonymies, than to attempt the arrangement of such a chaos. This, however, would be as short a way of getting through business, or rather of evading labour, as if a judge, for sake of despatch, should never hear but one side in any cause. A reform in the nomenclature of natural history is loudly called for; and we conceive that a work designed for the use of beginners in zoology should confine itself to two of the most approved names of each species (a popular and a scientific one), and dispense with the eternal business of repetition and reference.

They differ also in their discovered ears, in the form of their claws, (those of the eagle, properly so called, being almost semicircular,) and in the tarsi, which, in the known species, are totally naked. Besides these characters, which are merely methodical, there are others of a more prominent kind which cannot lead into error, nor permit the confusion of the genuine vultures with any of the other birds of prey. Their port is inclined, half horizontal, a position indicating their grovelling nature; whereas the eagle stands proudly upright and almost perpendicular on its feet. On the ground, to which, by the way, they are much attached, their wings are pendant, and their tail trailed along. Accordingly, we find the end of the pen-feathers constantly worn. Their flight is heavy, and they experience considerable difficulty in taking their full soar. Finally, they are the only birds of prey that fly and live gregariously.

Their mode of life, disposition, and habits, exhibit characters still more marked. The vultures are cowardly, disgusting, gormandizing in the extreme, voracious, and cruel. They rarely attack living animals, but when they can no longer satiate themselves on dead bodies. They attack a single enemy with numbers, and tear carcasses even to the very bone. They are attracted by the savour of corruption and infection. The hawks, the falcons, and even the smallest birds of this order, exhibit more courage than the vultures; for they hunt their prey alone, almost all of them disdain dead flesh, and will reject that which is corrupted. Comparing birds with quadrupeds, the vulture appears to unite the strength and cruelty of the tiger with the cowardice and gormandism of the chacal, which likewise joins in troops to devour carrion and root up the dead: while the eagle has the courage, nobleness, magnanimity, and generosity of the lion.

Endowed with a sense of smelling extremely keen, the odour of corrupted flesh attracts the vultures from a considerable distance. They fly towards it in flocks, and all the species are admitted indiscriminately to the disgusting banquet. If pressed by hunger, they will descend near the habitations of men, but they never attempt an attack except on the peaceable and timid tenants of the poultry yard.

The vultures are more numerous in the southern than in the northern parts of the globe. Still, it does not appear that they dread the cold, and seek warmth in preference; for in our part of the world they live in the greatest numbers on the highest mountains, and descend but rarely into the plains. In the hot climates, such as Egypt, where they are very numerous and of great utility, because they clear the surface of the earth of the debris of dead animals, and prevent the ill consequences of putrefaction, they are more frequently seen upon the plain than in the mountains. They approach inhabited places, and spread themselves at daybreak in the towns and villages, and render essential service to the inhabitants by gorging themselves with the filth and carrion accumulated in the streets. In our climates the vultures, during the fine season, inhabit the most lofty and deserted mountains: there, says Belon, they build their nests against shelvy rocks and in inaccessible situations. Authors are not agreed as to the number of their eggs, some stating it at two, others more. They do not carry food for their young in their talons, like the eagles, which even tear their prey in the air to distribute it to their family; but they fill their crop, and then disgorge the contents into the beaks of the little ones. In winter they migrate into a warmer climate.

The *Fulvous Vulture* of the text, which was first properly described by the anatomists of the French Academy of the Sciences, was judged by these gentlemen to be the large species of vulture indicated by Aristotle, the colour of which approaches more to that of the cinereous species, according to the Greek naturalist. Buffon has rendered this somewhat vague conjecture of the Academy more probable; but the want of proper information on some species which it was difficult to procure, led him into a mistake when he imagined the golden and black vulture to be simple varieties of the fulvous, when in reality they are distinct species.

The fulvous vulture, which M. Vieillot calls "*le griffon*," is about three feet and a half in total length, and eight from the tip of one wing to that of the other. Its head is covered with small white and slender feathers ; but those of the occiput and nape form a tuft about an inch long. The neck is almost naked: the short and scanty down with which it is sprinkled does not prevent the brown and bluish tints of the skin from being visible. At the bottom of the neck some long feathers are arranged like a ruff of a dazzling white. There is a large hollow furnished with hairs at the top of the stomach: this is the place of the crop. But notwithstanding this external cavity, there is a bump internally, and a great enlargement in this part of the œsophagus, which raises the skin of the external hollow, and fills it out when the bird has taken plenty of food. The feathers of the body are of a reddish-gray; the quill-feathers of the wings and tail are black; the beak blackish, with some bluish in the middle; the iris of a fine orange; the feet and claws are blackish.

The plumage of this vulture varies with age. In the first youth the body is fawn-colour; in the second and third year, varied with gray and fawn, more or less deep above. In a more advanced age, it is totally of a beautiful ash-colour, nearly blue.

This species, which is seen in numerous flocks on the Alps and Pyrenees, abandons them in winter. It appears also to be considerably spread in Africa, since Le Vaillant mentions having seen it at the Cape, on the Table Mountain, which it never quits except during violent storms from the south-east. Sonnini has also met with it in Egypt and the Levant, where the Turks and Greeks hold its fat in high estimation. They use it as a topical application in rheumatic cases. Its name in modern Greek is σκανια. That of *percnoptère*, derived from the ancient Greek, was adopted by Buffon to distinguish it from all others. The Catalans call it *trencalos*.

This vulture, says Aristotle, has all the vices of the eagle, without any of his good qualities. It allows itself to be chased and beaten by ravens; it is lazy in pursuit, heavy in flight, always clamouring and lamenting, perpetually in search of carrion to allay its sateless hunger. To a vile and ill-proportioned form, this bird adds the disgusting attribute of a perpetual flow of humour from the nostrils, and from two other holes in the beak, from which the saliva runs. The crop is prominent, and when on the ground, this vulture, like the rest of the tribe, has the wings pendant and half developed. When it is digesting or sleeping, the neck is drawn in between the shoulders, and the head buried in the feathers of the nape.

The *Cinereous Vulture* (*Monachus* of Linnæus) is called by some writers the *black vulture*. Brisson and other authors, who have attributed to this bird feet feathered to the toes, were mistaken, for its tarsi are smooth. This error appears to have arisen from the long feathers of the legs sometimes descending sufficiently low to cover the tarsus as far as the toes, as Edwards has well observed in his description of the *black crowned vulture*. If this was not the reason of it, it arose from naturalists referring to Belon, who imagined that all the vultures were thus provided. It is, however, certain that all the vultures of Europe, with the exception of the *vultur aureus*, *barbarus* and *barbatus*, which have been separated from this genus, have the most considerable part of the tarsus naked, as can be verified at the Museum of Natural History in Paris, where specimens of all are to be seen, either in the menagerie, or in the gallery of stuffed birds.

The Cinereous Vulture is nearly the size of the fulvous, (sometimes larger,) and has a collar of long, narrow, and bristling feathers; the naked skin of the head and neck is blue, and garnished with down; the beak blackish; the cera, tarsi, and toes are the same colour as the head; the legs are covered with long and pendant feathers on the sides, which grow down below the articulation with the tarsi. The first remex is shorter than the sixth, and the fourth the longest of all; the tail is rounded at its extremity, and composed of twelve *rectrices*, or tail quills.

In the first year the plumage is varied with brown and dirty gray. The down of the head and neck is, in the second year, gray and brown; the circle round the eye white; the collar ashen; the body is brown, but clearer underneath. In the third year the down becomes totally brown, and the body of a blackish brown. Finally, in the fourth year, the down of the head and plumage are black.

The *Sociable Vulture*, or *Oricou*, received this last name from Le Vaillant, in consequence of a membrane which edges its ears, and is prolonged over the neck, which last is entirely denuded, as well as the head. The crop, which is prominent, is covered with a silky down. There is on the neck a broad and frizzled demi-collar. The under feathers of the body are bristling, and curved like the blade of a sabre. A fine down extends over the legs and a part of the feet, which, as well as the toes, are covered with large scales. The tail is wedged, and always worn at its extremity.

Long black lashes surround the eyes, the iris of which is of a moronne-brown; reddish and violet constitute the tints of the skin of the head and neck; the throat is black; the upper part of the body, wings, and tail, are blackish; the under of a clear brown; the down of the legs white; the beak yellow at the base, and horn colour at the point. The young bird is clothed with a whitish down, and its plumage gradually assumes the sombre tint of the adult.

This large vulture, the height of which exceeds three feet, and which measures from tip of wing to tip of wing ten feet, inhabits the lofty mountains of the south of Africa, principally in the country of the Great Namaquois. The Dutch colonists of the Cape know it under the name of the black carrion bird, and the Namaquois call it *ghaip*. It abides and constructs its nest in the clefts of the rocks. It lays two or three white eggs. The young are born in the month of January.

We give a figure here from Major Smith of a vulture of a distinct, and probably a new species:—that gentleman names it *V. Nubicus*, or *Macrocephalus*, and thus describes it: "It is a bird of the largest size, equal to *Gypaëtos Barbatus*, with the head considerably larger, and thereby also clearly distinguished from *V. Indicus*—bill, cera, and legs, white; head naked, ruff brown; back and wings brownish, ochery, and grayish; white down each side of the neck; breast white, with a few pointed streaks; vent feathers buff; thighs white and brown, feathers downy. Shot in Nubia."

The *King of the Vultures* (*V papa*) is termed *Zopilote papa*, by M. Vieillot; the first name being that given to a genus by that naturalist, embracing the Condor, &c. The various denominations given to this South American bird originate in the idea, that it is so much respected by the *aura* and *urubu*, that they recede from a dead body the moment this vulture descends upon it, and give him place. This, however, says M. D'Azara, is neither the effect of respect nor consideration; it is merely the fear of superior size and strength. It is called in Cayenne, *king of the couroumous;* and the Guaranis of Paraguay call it *iriburubicha*. This species is extended in the New Continent, from the thirtieth degree of north latitude to the thirty-second degree of south latitude; but its numbers increase in proportion as we approach the torrid zone. It is found in Peru, Brazil, Guiana, Paraguay, and Mexico. It must not be confounded with the *coz-quauhtli* of the Mexicans, as some ornithologists, especially Brisson and Buffon, have confounded it. This last bird is the *aura*, which Laertius has described in his *Historia Novæ Orbis*. But the coz-quauhtli of Hernandez and Fernandez (*regina aurarum*), appears, from its Latin denomination, to be the *King of the Vultures*.

NUBIAN VULTURE *of Hamilton Smith.*

V. NUBICUS.

C. Hamilton Smith Esq. del.

Frankfort Museum.

London Published by G.B. Whittaker Nov. 1. 1827.

The King of the Vultures, which the Spaniards of Paraguay call *white crow*, from the colour which predominates in its plumage, flies away quickly when approached on the ground, or on an isolated tree; but is easily killed in the woods when some carrion has been left by way of bait.

We are assured, says M. D'Azara, that it makes its nest in the hollows of trees, and lays but two eggs. We are indebted, for a complete description of this vulture, to this eminent Spanish naturalist. He has described it under the various liveries which it assumes up to the age of four years. The beak is straight for about one-third of its length, then very much curved, and surrounded at its base by a membrane which forms on each side, as far as the eyes, a considerable sinking in, in which are situated the ample apertures of the nostrils; between them arises a sort of crest, which is neither elongated nor retreating, and which falls indifferently on either side: it is of a soft substance, and its extremity is formed by a remarkable group of warts. On the head is a crown of naked skin as red as blood. A bandelette of very short and black hairs, extends from one eye to the other by the occiput; below the naked portion of the neck is a very handsome sort of frill, some of the plumes of which are directed forwards, some backwards. It is so ample, that the bird, in drawing itself in, can conceal in it its neck and a part of the head. Behind the eye are thick wrinkles, which unite over the occiput to a fleshy band, projecting, and of an orange colour, which descends from there as far as the collar. These wrinkles conceal the auditory canal, which is very small, and near which other wrinkles join, which extend as far as the beak. Between these wrinkles some down is perceptible, as well as on the rest of the sides of the head. The remiges, and the large upper coverts of the wings, the tail, a trace on the back, the beak as far as the membrane, and the tarsi, are black. The membrane and the fleshy crest of the beak are orange; the naked skin of the base of the beak is purple; the edges of the eyelids are of a lively red; the naked portion of the neck is coloured by the most agreeable tints: it is carnation on the sides, purple below the head, yellow in front, and a blackish violet near the bands and the wrinkles of the occiput. The iris of the eye, and all the rest of the plumage, are white. Some individuals, supposed to be males, have a weak tint of red over the white of the upper part of the back on one side. The total length is twenty-nine inches and a half; that of the fleshy crest eighteen lines. This description is applicable to a bird of four years of age, complete.

The differences which this bird exhibits at three years old, consist in some upper coverts of the wings, which are black in the middle of the white. At two years of age, the entire head and the naked part of the neck are black, bordering on violet, with a little yellow over the neck; all the upper parts are blackish; so are the lower, but with long and white spots. The black crest falls on neither side, and its extremity only is divided into three very small protuberances. In the first year the bird is altogether of a deep bluish, with the exception of the belly and sides of the crupper, which are white. When the feathers underneath are raised, some white ones are also observable. The tarsus is greenish; the upper mandible of the beak is of a reddish black, the lower orange, mixed with blackish, with long and black spots; the naked part of the head and neck black, and the iris blackish, as well as the crest, which consists, at this age only, in a solid and fleshy excrescence.

This vulture differs from the one mentioned by Bartram in his Travels in the southern parts of North America, though sometimes confounded with it. The tail of the latter is quite white, a colour never found in the *vultur papa* at any age. As we have mentioned this bird, which is called by Bartram, *painted vulture*, *white-tailed vulture*, and *vultur sacra*, we may as well subjoin a short description. The beak is long and straight to the extremity, where it curves very abruptly, and

THE CONDOR.

VULT. GRYPHUS. Lin

Published by Whittaker & C.º Ave Maria Lane. Jan.ʸ 1830.

grows very pointed: the head and neck, almost to the stomach, are naked, where the feathers begin to cover the skin; they are gradually elongated, forming a ruff, in which the bird, by contracting the neck, can even cover the head. The naked skin of the neck is spotted, wrinkled, and of a lively yellow, mixed with a coral red. The lower part is almost covered with thick and short hairs, and the skin of this part is of a deep purple, which clears and grows red in approaching the yellow of the sides and front. The crown of the head is red; some appendages of an orange-red are on the base of the upper mandible. The plumage is usually white, but the quills, and two or three rows of the coverts, of a beautiful deep-brown; the tail is large and white, tipt with dark brown or black; the legs and feet are of a clear white; the eye is surrounded with a gold-coloured iris; the pupil is black. The Creek Indians make their royal standard with the feathers of this bird, to which they give a name signifying eagle's tail. They carry this standard to battle, but then paint a band of red between the brown spots. In negociations, and other pacific affairs, they carry it new, clean, and white. This standard is held sacred by them, and very elegantly ornamented. These birds seldom appear in Florida; but, when the grass of the plains is burnt up, which often happens, either from lightning, or the Indians setting it on fire to rouse up the game, then these vultures come from a considerable distance in great multitudes, and descend upon the plains, still covered with ashes, to pick up the serpents, frogs, toads, &c., which have been scorched to death. They are very easily killed at this time, being so intent on their repast that they will brave every danger.

We now come to one of the most celebrated species of the vulture tribe, and, indeed, of all the accipitres, the far-famed and formidable *Condor*. For the substance of our description we must be indebted to that most eminent naturalist, philosopher, and traveller, the Baron de Humboldt, a name which can only perish with the extinction of science, of letters, and of civilization itself.

It is astonishing, as this most judicious observer well remarks, that one of the largest of terrestrial birds and animals inhabiting countries which Europeans have been accustomed to visit for more than three centuries, should have so long remained so imperfectly known. The descriptions even of the most modern naturalists and travellers concerning this bird, are replete with contradiction, error, and falsehood. By some, the size and ferocity of the condor have been immeasurably exaggerated; others have confounded it with approximating species, or assumed the differences observed in the bird from infancy to age, as the diagnostic characteristics of sex. Baron Cuvier, in speaking of the form of the condor, after a careful investigation of all that has been written on the subject before Humboldt, expresses himself thus: "Some authors attribute to the condor a brown plumage, and a head clothed with down; others, a fleshy crest on the forehead, and a black and white plumage. It has not yet been described with any precision." Of the two drawings given by Dr. Shaw, the second alone bears the least resemblance to the great vulture of the Andes. "But the head," says Baron de Humboldt, "is without character. It more resembles that of a cock, than the head of the Peruvian condor; Buffon has not even risked an engraving of this bird. The one added to the edition of his works, at Deux Ponts, is below all criticism."

The Baron de Humboldt having resided for seventeen months in the native mountains of the condor, having had occasion constantly to see it in his frequent excursions beyond the limits of perpetual snow, has been enabled to render the most essential service to zoology, by publishing a detailed description of this animal, and the drawings which he sketched of it on the spot.

The name of condor is derived from the Qquichua language, the general language of the ancient Incas. It should be written *cuntur*, as other naturalists had previously observed. Europeans, by a corrupt pronunciation, change the Peruvian *u* and *t*, as they change the syllable *hua* into *gua*. They say, for instance, the volcano of Tonguragua, instead of *Tungurahua*, and Andes, instead of *Anti*. Baron de Humboldt thinks, that *cuntur* is derived from *cuntuni*, which signifies to smell well, to spread an odour of fruit, meat, or other aliments. This language is so rich, that it has three neuter verbs, *mucani*, *cuntuni*, and *aznani*, which express, to *smell*, generally, without determining the odour; to *smell well*, and to *smell disagreeably*. The Baron observes, that, as there is nothing more astonishing than the almost inconceivable sagacity with which the condor distinguishes the odour of flesh from an immense distance, the etymologist may be allowed to believe, that both *cuntur* and *cuntuni* come from one and the same unknown root. He has chosen, however, to retain the popular name of condor, lest any doubts might be started respecting the identity of the bird which he describes, with that of which so many fabulous stories have been related.

M. Duméril has separated the condor from the genus vultur, and joined it, and the papa, and the oricou, in a new genus, to which he has given the name of *sarcoramphus*. This appears a very judicious distinction; for the crests, or fleshy caruncles, which crown the beak, present a very distinctive character.

The young condor has no feathers. The body, for many months, is covered only with a very fine down, or a frizzled whitish hair, resembling that of the young ululæ. This down disfigures the young bird so much, that it appears almost as large in this state as when adult. The condors at two years old have not the black plumage, but a fawn-coloured brown. The female, up to this period, has not the white collar formed at the bottom of the neck by feathers longer than the others. This collar the Spaniards name *golilta*. From a want of proper attention to these changes produced by age, many naturalists, and even the inhabitants of Peru themselves, who take little interest in ornithology, have announced two species of condors, black and brown (*Condor negro y Condor pardo*). M. de Humboldt has met persons, even in the city of Quito, who assured him, that the female of the condor is distinguished from the male not only by the absence of the nasal crest, but also by the want of the collar. Gmelin and the Abbé Molina make the same assertion. It is, however, quite certain, that such is not the fact. At Riobamba, in the environs of Chimborazo and Antisana, the hunters are thoroughly acquainted with the influence produced by age on the form and colour of the condor; and for the most exact notions concerning those varieties we are indebted to them.

The vulture of the Andes is much more remarkable for his audacity, the enormous strength of his beak, his wings, and his talons, than for his dimensions from point to point of the wings. A few years previously to his traversing the chain of the Andes, M. de Humboldt lived in the country of Saltzbourg; and has seen at Berchtesgaden, Læmmer-geyers *(V. Barbatus)* fully as large as any condor.

The beak of the condor is straight in the upper part, but extremely crooked at the extremity. The lower jaw is much shorter than the upper. The fore part of this enormous beak is white, the rest of a grayish brown, and not black, as stated by Linnæus. The head and neck are naked, and covered with a hard, dry, and wrinkled skin; this same skin is reddish, but furnished here and there with brown or blackish hairs, short and very stiff. The cranium is singularly flat at the summit; as is the case with all very ferocious animals. Here should be the organ of benevolence according to Dr. Gall; but it is totally wanting in the condor. M. de Humboldt, in alluding to the bold but ingenious system of this philosopher, of which he confesses he was ignorant during his residence in Peru, regrets having lost the cranium of the condor, and having neglected to observe whether it possessed the longitudinal protuberance, which is found in the middle of the sagittal suture in the eagle and the chamois. This, according to the craniological system, is the organ of elevation; and the condor soars above the height of Chimborazo, an elevation six times greater than that of the clouds above our plains. "This," he says, "is a point not unworthy the consideration of future travellers."

The fleshy, or rather cartilaginous crest of the condor occupies the summit of the head, and one-fourth of the length of the beak. This crest is entirely wanting in the female, and M. Daudin has erroneously attributed it to her. It is of an oblong figure, wrinkled, and very slender. It rests on the forehead, and the hinder part of the beak; but at the base of the beak it is free, and almost sloped. In the void thus made, are situated the nostrils; for without this sloping off of the crest, the scent of the animal would be very feeble. The skin of the head in the male forms behind the eye, folds, or rugosities, like barbles, which descend towards the neck, and unite in a flabby membrane, which the animal can render more or less visible by inflating it at pleasure, much in the same way that all turkies do. It is proper, however, to observe, that the crest of the condor does not at all resemble the comb of a cock, or the flabby cone of the turkey. It is very hard, coriaceous, furnished with very few vessels, and cannot be inflated; in an anatomical point of view, it has no analogy with the thick caruncle of the *Vultur papa*. The ear of the condor exhibits a very considerable aperture; but it is concealed under the folds of the temporal membrane. The eye is singularly elongated, more remote from the beak than in the eagles; very lively, and of a purple colour. The entire neck is garnished with parallel wrinkles; but the skin is less flaccid than that which covers the throat. These wrinkles are placed longitudinally; and arise from the habit of this vulture of contracting its neck, and concealing it in the collar, which answers the purpose of a hood.

This collar, which is neither less broad, nor less white in the adult female than in the male, is formed of a fine silken down. It is a white band, which separates from the naked part of the neck the body of the bird furnished with genuine feathers. Linnæus, and after him Daudin, have both asserted, but without foundation, that this collar is wanting in the female. In both sexes, the hood is not entire; it does not close exactly in front, and the neck is naked as far as the place where the black feathers commence. It is necessary, however, to look very close to perceive that the down of the collar is interrupted towards the breast; for the naked band is very slender. Molina tells us, that the female has a small tuft of white feathers on the nape of the neck; but M. de Humboldt met with nothing of the sort in all the numerous specimens which he saw in the Andes.

The rest of the bird, back, wings and tail, are of a black slightly grayish. It is false, that the back of the male is white, as many naturalists have pretended; and among the rest, the Abbé Molina. It appears so, when the bird is seen hovering below you; but this is occasioned by the reflection thrown by the wing-coverts, which form a white spot. The plumes of the condor are sometimes of a brilliant black; most frequently, however, this black borders on a gray. They are of a triangular figure, and cover each other mutually, like tiles.

The primores of the wings are black; the secondaries, in the male and female, have the exterior edges white; the difference of sex is much more visible in the *tectrices*. In the female, these quills, which cover the *remiges*, are of a grayish black; but in the male condor (and this character is strongly marked) the points, and even one half of the quills are white, so that the wing of the male appears adorned with a beautiful white spot. The tail is cuneïform, rather short and blackish, in both sexes.

The feet are very robust, and of an ashen blue, ornamented with white wrinkles; the talons are of a blackish colour; they are not much crooked, but remarkably long. The four toes are united by a very flaccid, but very perceptible membrane. The fourth toe is very small, and its talon is most curved.

The dimensions of a female condor, killed at the volcano of Pichincha, were as follow: (the measures are French):—

Length of the female from the point of the beak to the end of the tail, 3 feet 2 inches.

Length of the beak, 1 inch 2 lines.
Diameter of the eye, 6 lines.
Thickness of the head, 3 inches 1 line.
Breadth of the hood, or white collar, 2 inches 1 line.
Extent of the wings at full stretch, eight feet 1 inch; for each wing was 3 feet 8 inches, and the diameter of the body, 9 inches.
The longest feather of the wings was 2 feet 2 inches; the *pennæ secundariæ*, 14 inches.
Length of the tail, 1 foot 1 inch.
Naked part of the feet, 10 inches.
Diameter of the tibia, 8 lines.
Length of the intermediate toe, 5 inches 2 lines.
The two lateral toes, 2 inches 6 lines.
The fourth toe (the smallest), 1 inch 6 lines.
Length of the claws of the three great toes, 11 to 12 lines.

The dimensions of a male condor, taken on the eastern declivity of Chimborazzo, were as follow:—

Length of the head from the occiput to point of beak, 6 inches 11 lines.
Length of the beak, 2 inches 9 lines.
Breadth of the beak closed, 1 inch 2 lines.
Length of the crest, 4 inches 9 lines; breadth, 1 inch 5 lines; thickness, half a line.
Length of the bird from point of beak to tail, 3 feet 3 inches, 2 lines.
Height of the animal perched, having the neck but moderately elongated, 2 feet 8 inches.
Breadth of the collar, or white hood, 2 inches 2 lines.
Envergure of the wings, 8 feet 9 inches.
Breadth of the tibia, 11 lines.
Length of the intermediate toe, without reckoning the claw, 3 inches 11 lines.
Length of the claw of the same, 2 inches.
Length of the two lateral toes with claw, 3 inches 7 lines; without claw, 2 inches 3 lines.
Length of the smallest toe with claw, 1 inch 8 lines.

Naturalists, says M. de Humboldt, who shall attentively observe the dimensions here given, will no doubt be surprised to recognize a bird merely of the European size. He has seen no condor, the envergure of which, or measurement of wing from tip to tip, exceeded nine feet French measure. Many persons in Quito and the Andes, worthy of the highest credit, assured M. de Humboldt, that they never killed any that exceeded eleven feet in the envergure. Even on a careful examination of the narratives of travellers, who visited these regions previously to M. de Humboldt, it will appear that, among the naturalists who have measured the vulture of the Andes, there are but few who assign to it a very extraordinary size. Father Feuillée, whose exactness in matters of natural history is quite exemplary, killed in Peru, in the valley of Ylo, to the south of Arequipa, a condor whose envergure was only eleven feet four inches. The measurement which he gives of the different parts of the bird perfectly accords with the dimensions given by M. de Humboldt, with the exception of the length of the beak. The condor of Feuillée appears to have been a female, for he says nothing concerning the crest. The male condor measured by Fresier had an envergure of only nine feet. From his own observations in Peru and Quito, M de Humboldt thinks differently from Buffon, that the condors measured by Feuillée and Fresier were not young ones. He also doubts very much whether any condor ever surpassed fourteen feet in the envergure. Dr. Strong, quoted in the *synopsis* of Ray, killed in Chili, near the island of Mocha, a condor, whose extended wings measured twelve feet two inches. The individual described by Dr. Shaw, from the Leverian Museum, had an envergure of fourteen feet English. The Abbé Molina himself seems to regard this as the maximum of the size of the condor. On the other hand, old travellers, less interested in the progress of natural history, have given the most exaggerated dimensions. Père Abbeville, for instance, assures us that the condor is twice the size of the most colossal eagle. Demarchais tells us, that its extended wings measure eighteen feet; that the enormous size of its wings prevent it from entering the forests; that it attacks a man, and can carry off a deer. Such exaggerations are not to be wondered at in naturalists who, instead of observing for themselves, did nothing but copy and compile the traditions of the natives. Marco Polo informs that the Roc, a bird of Madagascar, carried up elephants into the air *. Herodotus was acquainted with ants, which were smaller than some dogs, but larger than some foxes. We must always be on our guard, even in the present age, against the exaggerated accounts of form and size. Were we to trust to the rash assertions of the inhabitants, we might easily believe that in Egypt and South America there existed crocodiles from thirty to forty feet in length. Nevertheless, those who have actually measured these animals have not found any that exceeded twenty-eight. From every authentic account of the dimensions of the condor, it appears that this bird is not larger than the *vultur barbatus*, or Læmmer-geyer, which inhabits the central chain of the mountains of Europe, and with which both Buffon and Molina have confounded it. It has been with the condor as with the Patagonians and so many other objects of descriptive natural history,—the more they have been examined, the more have their enormous dimensions been found to diminish. The average length of the condors, from the point of the beak to the end of the tail, is but three feet three inches. Their usual envergure eight or nine feet. Some individuals, from a superabundant supply of aliment or other causes, may have attained an extent of wings of fourteen feet. The læmmer-

* This eagle-roc, of which Marco Polo speaks, exists, according to him, in the islands to the south of Madagascar. A domestic of Cublai Khan, who was taken prisoner by the inhabitants of these islands, related that the roc had feathers more than twelve paces in length. "*Avis vero ipsa tantæ fortitudinis, ut sola, sine aliquo adminiculo, elephantem capiat, et in sublime sustollat, atque rursus in terram cadere sinat, quo carnibus ejus vesci possit.*" Marco Polo adds, that he believed for a long time that the roc was a griffin, which, *as everybody knows*, is a sort of winged lion, with the head of an eagle. The word *roc*, under which name the old naturalists have placed all vultures, comes from the Persian *rhoc*, and signifies hero. These birds were obviously the creatures of mythological fiction.

geyer of the Alps of Switzerland and the Tyrol, from beak to tail, is usually four feet long. Its common envergure, seven or eight feet according to M. Bechstein, nine or ten according to Gmelin. Some individuals have been seen with an extent of fourteen feet. M. Salerne relates, that in France, at the chateau of Mylourdin, a vulture of this species was shot, measuring eighteen feet in the envergure. If this be true, our European vultures exhibit specimens of colossal size fully equalling the most exaggerated accounts of the most credulous or lying travellers concerning the condor.

The nature of the regions inhabited by the condor has, without question, contributed to produce these marvellous notions concerning the conformation of its body. These animals are considerably larger than the *vultur aura*, the *vultur papa*, and other rapacious birds inhabiting the chain of the Andes. They are generally beheld perched in the most solitary stations, often on the crest of the naked rocks which border on the lower bounds of the everlasting snows. Isolated and remote from every living being to which it is possible to compare him, the condor presents himself in contrast only with the blue depths of the horizon. This remarkable station, and the large crest of the male condor, makes the bird appear much larger than in reality he is. M. de Humboldt himself was long deceived in this way during his visits to the desert summits of these volcanos. He believed the condors to be of a very gigantic size; and it was only by a direct measurement of the bird when dead, that he could be convinced of the effect of this optical illusion.

If the læmmer-geyer of Switzerland, and the condors, be the largest animals which nature has endowed with the faculty of elevating themselves in the air; and if, in their habits, audacity, and strength, these two species bear a striking mutual resemblance; they are yet very different from each other in physiognomical characters. The *vultur barbatus* has neither the naked head, the nasal crest, nor the collar of white down.

It was from doubting the existence of this extraordinary crest, that Buffon was led to join the condor with the læmmer-geyer of Europe. The engraving given of the first in the small edition of Buffon published at Deux Ponts resembles any vulture of Europe more than the object it was intended to represent. It is still more singular that the Abbé Molina, a native of Chili, should have known so little about the condor. After having pointed out the false characters used to distinguish the two sexes, he concludes by assuring the reader that the condor differs from the *vultur barbatus* only in colour. This naturalist does not even mention the crest of the male condor.

The condor, like the lama, the vicunna, the alpaca, and several alpine plants, is peculiar to the chain of the Andes. The region of the globe which he appears to prefer to every other is of an elevation of from 1600 to 2500 toises. Whenever the Baron and his friend M. Bonpland were led, in the course of their herborizing excursions, to the limits of perpetual snows, they were always surrounded by condors. There they used to find them, three or four in number, on the points of the rocks. They exhibited no distrust, and suffered themselves to be approached within a couple of toises. They did not appear to have the slightest inclination to attack. Baron de Humboldt declares that, after the utmost research, he never heard a single example quoted of a condor having carried off a child, as has been so frequently reported. Many naturalists have asserted that condors have killed young persons of both sexes of from ten to twelve years of age. These assertions are not less fabulous than the report concerning the tremendous noise made by the vulture of the Andes in his flight, of which Linnæus observes "*Attonitos et surdos fere reddit homines.*" M. de Humboldt does not, however, doubt that two condors would be capable of depriving a child of ten years of age of life, or even a grown man. It is very common to see them attack a young bull, and tear out his tongue and eyes. The beak and talons of the condor are of the most enormous force.

Nevertheless, all the Indians who inhabit the Andes of Quito are unanimous that this bird is not dangerous to man. M. de Humboldt even hesitates to believe that any well-authenticated instance of a child having been attacked or carried off by the læmmer-geyer of Switzerland can be produced. People not unfrequently dread misfortunes, merely because they believe them possible; simple probabilities are elevated in their credence to the rank of historical facts. M. de la Condamine, a writer of the utmost credibility, relates, that the Indians present to the condor, by way of bait, the figure of an infant, composed of very viscous clay, on which it immediately darts with a rapid flight, and in which its talons become engaged so that it is unable to extricate them. But that gentleman prudently adds the qualifying terms, "on prétend." The figure of some small quadruped would appear to be much more likely to attract the presence of this formidable vulture. Nothing is more common than to see the little children of the Indians sleeping in the open air, while their fathers are employed in gathering the snow to sell it in the neighbouring towns. Yet who, asks M. de Humboldt, has ever heard of any of those children, thus surrounded by condors, having been ever attacked or killed?

Though the condor exclusively belongs to the chain of the Andes; though it prefers situations more elevated than the peak of Teneriffe or the summits of Mont-Blanc; though, of all animals, it is the one which removes to the greatest distance from the surface of our planet; it is yet not less true, that hunger will sometimes induce it to descend into the plains, and more especially into those which border on this mighty mountain chain. Condors are to be seen even on the shores of the southern ocean, especially in the cold and temperate latitudes of Chili, where the chain of the Andes may be almost said to border on the margin of the Pacific. Still it is observed that this bird sojourns but a few hours in these lower regions. It prefers the mountain solitudes, where it respires a rarefied atmosphere, in which the barometer does not rise above 16. On this account, in the Andes of Peru and Quito, many small groups of rocks, and platforms elevated 2450 toises above the level of the sea, bear the names of *Cuntur-Kahua*, *Cuntur-Palti*, *Cuntur-Huachana*, names signifying, in the Inca language, watch-tower, brooding place, or nest of the condors.

M. de Humboldt, during his travels in America, never saw the condor except in the kingdom of New Granada, in the province of Quito, and in Peru. He was informed, however, that it follows the chain of the Andes, from the equator even into the province of Antioquia, to the seventh degree of north latitude. The western Cordillera, or that branch of the Andes which by Choco extends towards the isthmus of Panama, appears to be elevated too little to be the habitation of the condor. Connecting under the same point of view the geography of plants with that of animals, it may be said that the condor proceeds no farther towards the isthmus than the *quinquina*, the *befaria*, the *escallonia*, and other alpine plants of the higher Andes. M. de Humboldt is ignorant whether this bird is found to the north of Panama. M. Sonnini has ventured to assert that the condor has been seen in Mexico; but this is extremely doubtful, for the *cozcaquauhtli*, a bird which plays no inconsiderable part in the mythology of the Aztecs, is the *vultur papa*, and inhabits, by preference, warm, or at least very temperate climates. Travellers, for a long time, were in the habit of giving the name of condor to every bird of prey of extraordinary size. It has even appeared in print, that condors have been killed in Africa, in Asia, nay, in the very heart of France—at Chateauneuf on the Loire.

As the eastern branch of the Andes extends through the mountains of Pampelona to those of Merida, which are covered with eternal snow, it would be interesting to know if the condor extended its migration to the neighbourhood of the sea of the Antilles. It is certain that it is found on the eastern declivity of the central chain of Quindiu in the environs of Ibagué, but it is not ascertained whether it exists in the chain of Summa-Paz, and Chingasa, to the east of Santa-Fé de Bogota. Neither is M. de Humboldt aware whether it is found in the colossal group of the mountains of Santa-Marta. Birds, like plants, are often circumscribed within certain limits, beyond which they are not found, even though the nature of the country and the climate are the same. The condor and the guanacos mutually accompany each other through the entire chain of the Andes, from the straits of Magellan to the northern frontiers of Peru, over an extent of above nine hundred leagues. But the guanacos and the vicunna, which inhabit the austral hemisphere exclusively, are no longer found to the north of the ninth degree of latitude, while the condor follows the Cordillera beyond the equator at least three hundred leagues farther than the vicunna.

Alpine plants present a curious example of identity of species, notwithstanding the immense distance which intervenes between many of the great mountains of the world. On the Silla of Caraccas, the same *befaria* is found which adorns with its purple flowers the mountain declivities of the kingdom of New Granada. How the seed of this beautiful plant came to be dropped on this projecting peak, the only part of the chain on this coast which, from its elevation, is sufficiently cold to permit the existence of the *befaria*, would be a useless and unphilosophic question, for the first origin of things can neither be a problem of history, nor an object of research to the naturalist. It is, however, remarkable that in animals this identity of forms in situations remote from each other, but analogous in climate, is much less observed than in plants.

The Indians of the Orinoco often mentioned to M. de Humboldt, during his navigation up that river, certain large birds of prey, which unfortunately he had no opportunity of seeing. He is of opinion that these may be the two large eagles discovered by M. de Sonnini in French Guiana. This naturalist confesses that at first sight he took these birds for condors, but in the sequel he rectified this error. The condor is not known in the elevated mountains on the coast of Venezuela, nor in the chain of Dorado, nor in Brazil. The *Ouira-Ouassa* of the Brazilians, which Buffon conceived to be synonymous with the condor, is a very different bird, although, as the story goes, it is large enough to devour apes, and even attack men. It is, perhaps, doubtful whether the condor is extended over the entire chain of the Andes to the most southern extremity of the New Continent. In the account of Cordoba's voyage to the Straits of Magellan, the only case in which men of education made any stay in that quarter, among the animals observed in Terra del Fuego, and on the coasts of Cape Victoria, are mentioned colibris, American ostriches, guanacos, and wild dogs; but there is not a word about the condor. It is, however, certain that it exists there; for the condor described by Dr. Shaw was killed at the Straits of Magellan. It was brought into Europe by Captain Middleton on his return from the South Seas. Although the figure of this bird from the Leverian Museum is not much like Humboldt's, yet this writer is of opinion that it was the true male condor, and not a different species or variety. Dr. Shaw, whose description is very exact, thus characterises it: "*Saccum in gula, seu pellis quædam dilatata a basi mandibulæ inferioris longe per collum ducta. Prodeunt etiam a latere colli appendiculæ septem quasi carneæ, seu carunculæ, semi-circulares et cærulescentes. Collum et pectus nuda et rubentia, pilis raris nigricantibus aspersa. Crista capitis sinuata,* ALTERA AD NUCHAM, AMBÆ NIGRICANTES, CÆRULEÆ, *et nonnullis in locis rubentes. A collo inferno dependet tuberculum pyriforme. Dorsum atrum, remiges albæ secundariæ, cauda atra, pedes albi.*" The two crests, the white feet, and the white secondaries, might certainly lead us to believe that the bird of Dr. Shaw differed from the true condor. But these differences may result from the animal not having been described in a living state, or well preserved. The other vulture, from the Leverian Museum, would appear to have been a young female condor. This also came from the Straits of Magellan; but it is remarkable that all the other very large specimens have been from Chili, or the most southern part of Peru. M. de Humboldt queries if there be a larger race of condors in the cold or temperate climates than in the torrid zone? The temperature of the lower regions of the air must, however, be a matter of indifference to a bird which can chuse its climate at will, in the various elevations of the Andes. But it is probable, that the proportion of food, and other local circumstances, may be influential on the development of its organisation. It is impossible to indicate, with any certainty, the causes which determine what naturalists have thought proper to designate by the vague denomination of the distribution of *races*.

The condor advances to the east in the mountains of Santa-Cruz, of the Sierra, and of Cochabamba. As these peaks seem to join those of Mattagrosso, it is possible that the bird may exist in Brazil. But the group of mountains called *Cerro do Frio* and *Cerro das Emeraldas* appear not to be sufficiently elevated or sufficiently cold for the habitation of the condor.

It appears very doubtful that the condor has ever been transported alive into Europe. A bird was exhibited in London some years ago under this name, but it was uniformly brown, and without the white on the wings which distinguishes the true condor. It was said not to be young, and therefore the less likely to differ from the common condor in the mere evanescent peculiarities of age and sex. In fact, it would not be a very easy matter to bring a bird like the condor to Europe. It might, however, be brought by four different routes,—by Cape Horn, the isthmus of Panama, down the river of the Amazons, or the Madeleina. The first would be the best way, according to M. de Humboldt; because, though the animal suffers captivity very well, it is probable that its stay in very hot countries, and subjection to great barometrical pressure, might prove injurious to the health of the animal. The condor prefers a temperature of two or three degrees above congelation. It often remains, to be sure, for many hours in the hot vallies, where the centigrade thermometer rises to 30°. Still, there would be every reason to apprehend that the constant heat it would experience in the isthmus of Panama, in the province of Jaen de Bracamorros, or in the river Madeleina, from Honda to Carthagena, would cause its destruction.

Among the birds of prey, as with the insects, the female is generally larger than the male. This difference, however, is not very sensible in the condors, though there is variety enough in the magnitude of different individuals of both sexes. Inhabiting solitary situations, and having no other enemy but man, who does not greatly occupy himself in destroying it, it appears not unlikely that the condor attains an advanced age. They do not, however, multiply greatly. While the vultur aura is observed in numerous bands of forty or fifty at a time, more than five or six condors are never seen together. Of all the rapacious birds, however, of America, the vultur papa seems the least numerous.

M. de Humboldt was assured that the condor builds no nest; that it deposits its eggs on the naked rock, without surrounding them with straw or leaves. The eggs are said to be altogether white, and from three to four inches in length. It is also reported that the female remains with the little ones for the space of an entire year. When the condor descends into the plains, it prefers alighting on the ground to perching in the trees, like the vultur aura. The talons of the condor are very straight; and it is a remark of Aristotle, that birds of prey with very crooked talons are not fond of settling upon stones or rocks.

The habits of the condor are similar to those of the læmmergeyer. If it is not larger than the latter, it appears to be superior in strength and audacity. Two condors will dart upon the deer of the Andes, upon the puma, the vicunna, and the guanaco. They will even attack a heifer. They pursue it for a long time, wounding it with their beak and talons, until the animal, breathless and overwhelmed with fatigue, thrusts out its tongue, bellowing. The condor then seizes the tongue, a morsel to which it is much attached. It also tears out the eyes of its victim, which sinks to the earth and slowly expires. In the province of Quito, the mischief done to cattle, but more especially to sheep and cows by this formidable bird, is immense. In the savannahs of Antisana, 2101 toises above the level of the sea, bulls are constantly found which have been wounded in the back by condors.

The condor, when satiated with food, remains perched, phlegmatically, on the summit of the rocks. In this situation the bird has an air of sombre and sinister gravity, and will not give himself the trouble to escape the chase. But when stimulated by hunger, the condor will fly to a prodigious height. He hovers in the air for the purpose of taking in at a glance the vast extent of country which is to furnish him with his prey. On those days in which the sky is peculiarly serene, the condor is usually to be seen at its most extraordinary degree of elevation. It appears attracted on such occasions by the transparency of the atmosphere to review a large extent of territory, which in duskier weather would be concealed even from its piercing view.

In Peru, Quito, and the province of Popayan, they are in the habit of taking the condor alive with nooses. The inhabitants are fond of this sport, and particularly delight in entertaining their European guests with it. The mode is thus:— A cow or horse is killed; in a short time the odour of the dead animal attracts the condors, whose scent is remarkably fine. They soon appear in great numbers in places where nobody had previously any suspicion of their existence. They devour with incredible voracity. They always begin with the eyes and tongue, their favourite morsels. Then they attack the carcass through the anus, that they may arrive more quickly to the intestines. When the condors have thus satiated themselves, they are too heavy to fly, and the Indians pursue and take them with facility. The bird is said, under these circumstances, to make tremendous efforts to elevate itself in the air. Scarcely has it succeeded in so doing, than it begins to vomit most abundantly. In such efforts the condor contracts and elongates the neck, and approaches its talons to its beak. This motion, purely accidental, has given rise to a report that the condor makes use of its claws to assist the operation of vomiting; but it seems by no means likely that it could even touch the throat slightly with the claw. The condors thus taken alive by the Indians are subjected to the most cruel torments.

At Riobamba, it is said that the natives put poisonous plants into the belly of the animal used as a bait, which produces an effect on the condor similar to that of intoxication.

The condor, when taken alive, is melancholy and timid for the first hour, but soon after grows extremely mischievous. M. de Humboldt had a living female in the yard of his house at Quito for eight days. Fear had rendered her so savage, that it was quite dangerous to approach her.

The condor appears to have more tenacity of life than any other bird of prey. M. de Humboldt was present at certain experiments on the life of a condor at Riobamba. They first attempted to strangle it with a noose. They hung it to a tree, and dragged the legs with great force for many minutes: but scarcely was the noose removed, than the condor began to walk about as if nothing had been the matter. Three pistol-balls were then discharged at him, within less than four paces distance. They all entered the body. He was wounded in the neck, chest, and belly, but still remained on his feet. A fifth ball struck against the femur, and rebounding, fell back on the ground. This ball was for a long time preserved by M. Bonpland. The condor did not die for half an hour after of the numerous wounds which it had received. Ulloa informs us, that in the cold region of Peru the condor is so closely furnished with feathers, that eight or ten balls may strike against his body without one piercing it.

It is worthy of observation, that the condor prefers carcasses to living animals. It subsists, however, on both, and seems to pursue small birds less than quadrupeds.

We forgot to notice in the proper place the *Angola Vulture*, of which we have given a figure, from a specimen which is in the British Museum.

We shall now notice two of the principal species of the subdivision PERCNOPTERI.

The first is the *Percnopterus* of Egypt, (*Vultur Percnopterus* of Linnæus, and *leucocephalus* of Gmelin.) In the illuminated plates of Buffon it is entitled *Vautour de Norwege*. The plumage of this bird is perfect in this drawing, but the figure is not exact as to the tail, which is represented equal at its extremity, whereas it is wedged. The variety noticed by Latham is of the same species, but of a different age. Buffon, who gave to the vulture described by Mauduit the epithet *petit*, was deceived in saying that the *brown* and Egyptian vultures of Brisson must be separated, the second not being a vulture, but a bird of another genus, to which Belon has thought proper to give the name of *Sacre Egyptien*. Of the identity of these two birds there is now no doubt, and of their belonging to the species of percnopterus, which we are about to describe. He was again deceived in saying, that his Norwegian or White-headed Vulture was of a different species from the brown and Egyptian vultures of Brisson; that it differed in having the feet naked, while those of the two others are covered with feathers. This even appears to be a typographical error, though repeated in many editions of his works; for Brisson, whom he seems to have consulted, gives to the two birds the naked feet which they really have.

It is proper to notice here an unaccountable inconsistency of M. Sonnini. In his article on the *petit vautour*, just

ANGOLA VULTURE,

F. ANGOLENSIS, Gm.

Sowerby del & sc
Mus. Brit.

London. Published by G.B. Whittaker Nov. 1. 1827.

mentioned, he says, in a note, " I do not think this bird is the same with the *vulture of Egypt*, or the percnopterus of Linnæus and Hasselquist:" while in another article, on the vulture of Egypt, he assures us that this last has many relations with the *little vulture*, or the *vulture of Norway;* and sets down among its synonymes the *percnopteri* of Linnæus and Hasselquist. It is the more necessary to notice such errors, as they have crept into a number of publications. The reader, indeed, cannot be too often cautioned in this department of our work, against the mistakes of authors, of individuals for species, founded on the differences which characterize age and sex only.

This vulture, known by the Europeans who frequent Egypt by the name of *Pharaoh's Chicken*, is called by the Turks *Akbobas*, which means *white father*. The Egyptians and Moors call it *rachama*. These names have been erroneously applied to many birds of a totally different genus, such as the pelican, the stork, and the swan.

The individual described by Bruce under the name of *rachamah* has a very strong and pointed beak, the end of which is black for about three-quarters of an inch; the remainder is covered with a yellow and fleshy membrane, which envelopes it above and below; and the front of the head and under part of the neck is covered in like manner by this membrane, which terminates in a very sharp point at the bottom of the neck. This membrane is very wrinkled, and the under part of it is thinly set with a few hairs. The apertures of the nostrils are very large, and so are the orifices of the ear, which are not covered by any sort of feathers. From the middle of the head, where the yellow membrane ends, to the tail, the body is perfectly white; but the large feathers of the wings are black, and six in number. After these come three small ones, of an iron gray: these again are covered by three others, smaller still, and similar in form, but of a rusty gray. The coverts of the great feathers of the wings are iron-gray in the tips for about four-fifths of an inch, and the rest is perfectly white. The fourteen pen-feathers of which the tail is composed are wedge-formed, which makes it seem to terminate in a point, as Bruce affirms, and it does exceed the end of the wings by more than half an inch. The thigh of the rachama is covered with a very soft down, as far as the articulation of the leg. The leg is of a dirty and almost flesh-coloured white, and is covered with black and fleshy tubercles. The claws are black, very strong, and crooked. The female is brown.

This animal incessantly hunts after the most stinking carrion; it exhales from its own body a most infectious odour, and putrefies the moment it is dead. In Cairo it is considered as a crime to destroy these birds.

Sonnini tells us that these vultures are not ferocious in Egypt: they are to be seen on the terraces of houses, in the midst of the most populous and noisy cities, perfectly quiet, and living in complete security amongst men, who feed and cherish them with the utmost care. They also frequent the deserts, and prey upon the carcases of men and animals which have perished in those immense wastes, consecrated, as it were, for ages, to nakedness, desolation, and sterility. Those which inhabit Egypt are not known to quit it, but some of the same species are to be found in Syria and Turkey; less numerous, however, because they do not enjoy the same prerogatives, nor is their existence protected in these countries by ancient superstition, as in Egypt: for they were considered sacred among the old Egyptians, whose opinions on this point, as on many others, have been transmitted to their successors, even to this day. In truth, they perform very considerable services to this country, in sharing with other birds, equally sacred in ancient times, the task of destroying the rats and reptiles which abound in this fertile and slimy region. They also clear away the carcases and filth which under a burning sky, and on a soil kept in a frequent state of humectation by the inundations of the Nile, would otherwise fill the atmosphere with pestiferous exhalations. The fields of Palestine would remain uncultivated and abandoned, if these vultures did not clear them of a prodigious quantity of rats and mice, which breed there superabundantly.

The *Ourigourap*, described by Le Vaillant among the African birds, whose name signifies *white crow* in the language of the Great Namaquois, is a bird of this species. The plumage of the one figured by Le Vaillant was not yet perfect: the Hottentots call it *hoa-goop*, and the Dutch colonists *white krai*, which words have the signification aforesaid.

The forehead, circle round the eye, and cheeks as far as the ears, are naked, and of a saffron colour, more lively towards the base of the beak. The throat is furnished with a scanty and fine down, which allows the skin to be seen, which is yellowish, wrinkled, and capable of great extension. The top of the head and all the neck are covered with long and slender feathers: the plumage is in general white, tinted with fawn colour: the primaries of the wings are black, the secondaries fawn colour on their external side, and blackish on the interior. The tail is wedged, and of a reddish white. The end of the beak and claws are blackish: the feet of a yellowish brown.

The young ourigourap has all that portion of the head and neck which is naked in the adult, covered with a grayish down. During the season of reproduction, the beak of the male is redder than during the rest of the year. The number of eggs generally laid, according to the report of the Hottentots, is from three to four.

The ourigouraps do not appear in flocks, except when attracted and assembled by the immediate expectation of prey; at other times, they are only found in pairs. The male and female never quit each other. They construct their nests in the rocks.

These vultures are rare at the Cape, but very common in the country of the Little Namaquois. In still greater numbers are they found on the banks of the Orange River, and among the Great Namaquois. They are not very wild, and will suffer themselves easily to be approached: the natives never hurt them, because they clear their encampments, &c., from the abundance of filth with which they are generally encumbered.

The vulture of Angola, the percnopterus with black wings, and the vulture of Malta, (*V. Fuscus*), are referred to by M. Vieillot, and most probably belong to this species.

In Cuvier's division of the GRIFFINS comes the *Bearded Vulture, Læmmergeyer* of the Alps, (*Vultur Barbatus et Barbarus*, Linn. and Lath.), *Phene Ossifraga* of Savigny. The German name signifies, *Vulture of Lambs;* and this bird is, in fact, a very formidable scourge to the flocks which pasture in the Alpine valleys. It wages cruel war on sheep, lambs, she-goats, and even calves: the chamois, the hare, the marmot, and other wild quadrupeds, also become its victims. Its force corresponds with its corpulence, which according to some writers is immense, and is equal even to that of the condor. Fourteen and even eighteen feet in the extent of wings have been attributed to the Læmmer-geyer. Gesner reports that the eyrie of one of these birds was discovered in Germany, placed upon three oaks, and constructed of branches, &c., so far extended, that a waggon might have been completely sheltered under it. In this nest were three young birds, already so large as to measure seven ells in the envergure; their legs were already thicker than those of a lion, and their claws as large as the fingers of a man. In this nest were found several skins of calves and sheep. The eggs are white, and spotted with brown.

It would appear, however, that exaggeration has had a good deal to do with recitals of this kind. A very distinguished naturalist, who has observed this species in the Pyrenees, Picot la Pérouse, has described it very carefully, and considerably reduced the magnitude attributed to it by others. He gives to it the following dimensions:—envergure, eight feet and a half; total length of the animal, three feet ten; weight, about ten

THE BEARDED VULTURE.

V. BARBATUS.

London Published by G.B. Whittaker March. 1829.

pounds. The beak is four inches long; it is covered above on the base, as far as its centre, with numerous long and black hairs, directed forward; underneath hangs a tuft of similar hairs, forming a true beard, an inch and a half in length. There are more of these hairs scattered at the corners of the beak and over the throat, near the eyelids and brows. The tail, three inches wide and six long, is rounded, and composed of twelve quills: the wings have two-and-thirty.

The upper part of the head is white among adults, and more especially in old subjects: in the young it is black. The occiput, the neck, and the under part of the body are white, washed with red or orange, a difference occasioned by age in the males: these colours are deeper on the throat and breast, weaker on the belly, legs, and feet. The under part of the wings is gray: the feathers of the tail, upper coverts of the wings, and croup, are of a clear gray, and bordered with black: the wing coverts at the end are spotted with orange: all the rest of the plumage is of a very deep brown. The beard is black.

The Alpine and Pyrenean chains, in their loftiest and most inaccessible regions, constitute the principal asylum of the Læmmer-geyer. From these towering heights, where

> ———— Horror wide extends
> His desolate domain,

this formidable bird descends, on rapid wing, into the fertile valleys of Switzerland and the smiling plains of the South, to pounce upon his prey. Equal, perhaps, at his fullest growth, to the condor, equal in ferocity, and scarcely inferior in strength, he spreads devastation far and wide among the peaceful tenants of the fold, and the wild, but timid inhabitants of the hills, the meadows, and the lawns. The swiftness and activity of the hare, the chamois, or "the nimble marmazet," afford them no security against their winged foe; nor can the smallest quadrupeds escape his piercing ken. It is even reported that this rapacious animal does not confine his attacks to the brute creation, but sometimes succeeds in carrying off children. This relation, perhaps, is no better verified by facts than similar stories of the condor: we certainly, however, have no reason to doubt the capacity of the bird to perform such a feat, nor do we suppose that so much "divinity hedges" the young princes of the creation as to deter him from the attempt. Fortis has beheld the læmmer-geyer on the precipitous rocks which border on the Cittina in Dalmatia, and Pallas on the granite ridges of Odon-tschelon in Siberia, where it constructs its nest. It arrives there in the month of April, and passes the summer there. It is also found in Mongolia, where it receives the appellation of *icello*.

It is probable that the fabulous stories of the *roc*, so celebrated in the tales of Oriental enchantment, originated in some eastern variety of this gypaëtos; that they cannot be referred to the condor has been sufficiently proved.

The Gypaëtos of Africa, described by Bruce, is considered by some ornithologists as a distinct species, and by others as but a variety of the Læmmer-geyer. It was seen by that celebrated traveller on the highest part of the mountain of Lamalmon, near Gondar. The natives call it *Abou-Duch'n*, or Father Long-beard, from the tuft of divided hair which hangs beneath its beak. Mr. Bruce imagined it to be one of the largest birds in existence: it measured eight feet four inches from wing to wing; from the tip of the tail to the point of the beak four feet seven. Its weight was two-and-twenty pounds. The legs were short, and the thighs extremely muscular: the aperture of the eye was scarcely half an inch across: the crown of the head, and the forehead where the juncture exists between the beak and the skull, were bald. We extract Mr. Bruce's account:—

"This noble bird was not an object of any chase or pursuit, nor stood in need of any stratagem to bring him within our reach. Upon the highest top of the mountain Lamalmon, while my servants were refreshing themselves from that toilsome rugged ascent, and enjoying the pleasure of a most delightful climate, eating their dinner in the open air, with several large dishes of boiled goat's flesh before them, this enemy, as he turned out to be to them, suddenly appeared: he did not stoop rapidly from a height, but came flying slowly along the ground, and sat down close to the meat, within the ring the men had made round it. A great shout, or rather cry of distress, called me to the place. I saw the eagle stand for a minute, as if to recollect himself, while the servants ran for their lances and shields. I walked up as nearly to him as I had time to do. His attention was fixed on the flesh. I saw him put his foot into the pan, where there was a large piece in water prepared for boiling; but finding the smart which he had not expected, he withdrew it, and forsook the piece that he held.

" There were two large pieces, a leg and shoulder, lying on a wooden platter; into these he thrust both his claws, and carried them off; but I thought he still looked wistfully at the large piece which remained in the warm water. Away he went slowly along the ground, as he had come. The face of the cliff over which criminals are thrown took him from our sight. The Mahometans that drove the asses were much alarmed, and assured me of his return. My servants, on the other hand, very unwillingly expected him, and thought he had already taken more than his share.

" As I had myself a desire of more intimate acquaintance with this bird, I loaded a rifle-gun with ball, and sat down close to the platter, by the meat. It was not many minutes before he came, and a prodigious shout was raised by my attendants, 'He is coming, he is coming!' enough to have dismayed a less courageous animal. Whether he was not quite so hungry as at his first visit, or suspected something from my appearance, I know not; but he made a short turn, and sat down about ten yards from me, the pan with the meat being between me and him. As the field was clear before me, and I did not know but his next move might bring him opposite to some of my people, so that he might actually get the rest of the meat, and make off, I shot him with the ball through the middle of the body, about two inches below the wing, so that he lay down upon the grass without a single flutter.

"Upon laying hold of his monstrous carcass, I was not a little surprised at seeing my hands covered and tinged with yellow powder or dust. On turning him upon his belly, and examining the feathers of his back, they also produced a dust the colour of the feathers there. This dust was not in small quantities; for upon striking the breast, the yellow powder flew in full greater quantity than from a hair-dresser's powder-puff. The feathers of the belly and breast, which were of a gold-colour, did not appear to have any thing extraordinary in their formation; but the large feathers in the shoulder and wings seemed apparently to be fine tubes, which, upon pressure, scattered this dust upon the finer part of the feather: but this was brown, the colour of the feathers of the back. Upon the side of the wing, the nibs or hard part of the feathers seemed to be bare, as if worn; or, I rather think, were renewing themselves, having before failed in their functions.

"What is the reason of this extraordinary provision of nature it is not in my power to determine. As it is an unusual one, it is probably meant for a defence against the climate, in favour of birds which live in those almost inaccessible heights of a country doomed, even in its lowest parts, to several months excessive rain."

M. Sonnini thinks that this African Gypaëtos ought to be considered as a species distinct from the Alpine or Læmmergeyer; but, certainly, the description from Bruce affords no sufficient characters on which to ground such a description. The differences of five or six inches in length, the differences resulting from age or sex, as the upper part of the head being white (an attribute of the adult), and the throat and lower parts of a golden tint (the distinctive character of the male), cannot be considered as sufficient. As to the powder of which Bruce speaks, it is by no means, even according to Sonnini himself, a remarkable singularity, or one of the multifarious modifications of nature, but a simple effect of the moulting, more perceptible in consequence of the bulk of the animal. In fact, this powder comes from the pellicle which envelopes the feathers at their first production, which follows at first their progression, being elongated with them, and finally dries up as the barbs shoot forth, and becomes divided into very fine light parcels, the quantity of which depends on the number of feathers which are developed at the same sime. This pellicle is usually of the same colour as the feathers, as Bruce has well remarked.

THE FALCONS.—Linnæus has comprehended under the denomination *falco*, the eagles, balbuzzards, kites, and many other rapacious birds, as well as the falcons properly so called, and which subsequent naturalists have found the necessity of separating from that division. Notwithstanding, however, these separations, the species of which the genus Falcon remains composed, undergo in the course of years so many variations in their plumage, that they are scarcely yet distinguished with any great degree of exactitude. But the generic characters have gained a greater degree of precision: they consist in a beak curved from the base, the upper mandible of which, crooked at its extremity, is armed on each side and towards the end with one or sometimes two teeth, more or less projecting; the lower one of which, being convex underneath, is sloped at the point. From the centre of their circular nostrils arises a pliant and conical tubercle: the tongue is fleshy, sloped, and canaliculated: the tarsi are short: the feet are provided with strong toes, of which the external have a membrane at the base, and curved claws, acerated, and nearly equal. The three external pen-feathers of the wings are narrowed and pointed at the end. The second is the longest, and the others, from the fourth to the tenth, are regularly wedged.

Between the falcons proper and the gerfalcons, there are differences which have determined the formation of two sections. The first are distinguished by a tooth more strongly defined on each side of the upper mandible, which, among the others, is a mere festoon: the lower mandible is also much more sloped at its point in the true falcons.

In the species of both sections we find the general characters of the great genus Falco of Linnæus. The head and neck clothed with feathers, the brows forming a projection which makes the eye appear sunk; and the female one-third larger than the male, which occasions the latter to be called in French *tiercelet*. But the falcons, more courageous in proportion to their size, and, therefore, termed *noble birds of prey*, have peculiar habits resulting from the length of the wings, which, in a calm air, renders their flight very oblique, and forces them, when they want to rise directly, to fly against the wind., They are also more docile, and fitter for the purposes of falconry, being more easily taught to pursue the game, and to return when called. Daudin remarks, that the larger species of falcons have, like the eagles, pentagonous and hexagonous scales on the tarsi, and that the smaller species, such as the merlins, have, on the front of the tarsus, half-rings, divided in the centre. M. Savigny has also observed, that the tarsi of the falcons have larger scales on the internal side in front.

The falcons subsist exclusively on living prey, which they seize adroitly, or tire down in pursuit; and they nestle generally in rocks, or very elevated trees.

The *Common Falcon* is about the size of a hen. Buffon has given two figures, a male and female; the former was from a bird one foot six or seven inches in length, and the latter about four inches more. A young one, represented by the same author, has the upper part of the body covered with brown feathers, edged with reddish, and those of the lower part are whitish, with longitudinal brown spots, of an oval form, occupying their centre. These spots are successively transformed into transversal blackish lines, and the plumage of the back becomes more uniform, and of a brown colour, radiated crosswise with dark ash colour: the throat and bottom of the neck become whiter. The caudal quills, brown above, with pairs of reddish spots, exhibit below pale bands, which diminish in breadth with age. The cere and feet are sometimes yellow, sometimes a greenish-blue: but a triangular spot on the cheeks is the sign by which this species is known at all ages. M. Savigny adds to this the white extremity of the tail.

The common falcon, which is usual enough in France, is also found in Switzerland, Germany, and Poland, in Italy, Spain, Rhodes, Cyprus, Malta, and the other islands of the Mediterranean. Wherever it exists, it prefers mountainous and rocky countries. It is, perhaps, of all birds the most courageous in proportion to its size: it does not approach its prey sideways, like the hawk, and some other accipitres. It drops perpendicularly upon it; devours it on the spot if it be large, or carries it off, rising perpendicularly, if it be not too heavy. It frequently attacks the kite, either to exercise its own courage, or deprive the latter of its prey. Such are the habits which have always been regarded as peculiar to the falcon. It appears not to descend from the mountains in summer, except in search of food, when it is not to be found on these elevations, and it never removes from them in winter to hunt in the plains, but when constrained by famine and the rigour of the season. M. Vieillot, indeed, quotes the authority of one of his correspondents, whose observations, made in the plains of Champagne, where the falcons arrive in the month of August, are somewhat different. He reports that he has seen these birds hunting singly, or in couples, and darting with extreme rapidity from a hillock of earth, or the low branch of a tree, the instant they perceived a flock of partridges: the falcon follows this flock, crosses it, and in passing, endeavours to seize a partridge in its claws, or gives it so violent a shock with its breast, as to stun, and even kill the individual. It returns sometimes after this shock, with so much agility, as to catch and carry off the partridge before it has fallen: if it does not reach it until it comes to the ground, it generally eats it on the spot, or takes it behind an adjacent bush. This gentleman adds, that the falcon does not follow the partridges on foot, like the goss-hawk; and also says, that it does not descend perpendicularly on them, but endeavours to make them rise by shaving the earth, and making a noise like the whistling of a bullet. Though it passes and repasses many times, it does not always succeed in its attempts, the partridges squatting down, or concealing themselves in the bushes. The falcon also gives chase to other birds, as pheasants, thrushes, larks, pigeons, and even ducks, which dive the moment they see him. The observer just quoted also remarks that the falcon almost always passes the night in the same place on the thick branch of a tree near the trunk. But as most of these facts do not agree with what the generality of authors inform us are the peculiar habits of these birds, we must entertain some doubts respecting the identity of species.

It is in the cliffs of the most rugged rocks exposed to the south, and in high mountains, that the falcon most frequently establishes its eyrie, where the female lays three or four eggs of a reddish-yellow, with brown spots. In France the little ones are born towards the middle of the month of May, and as soon as they are able to procure their own nutriment, the parents not only drive them from the nest, but force them to quit that particular district, which they reserve exclusively for themselves.

The falcon is very long lived. A falcon belonging to James the First, in 1610, with a gold collar bearing that date, was found in 1793, at the Cape of Good Hope. This bird, though more than one hundred and eighty years old, was still considerably vigorous.

As we do not at all intend to pursue the enumeration of species, or follow any very severe method in this part of our work, which would be totally unnecessary after what has been done, we must confine ourselves to what is most interesting to general readers.

The two species which approach nearest to the common falcon are the *falco frontalis* and *falco tibialis.* The former bird was discovered by M. Le Vaillant at the Cape, and has a very apparent tuft extending from the front to the back of the head, which erects itself when the falcon experiences any agitation, and especially during the season of reproduction. This tuft is bluish, and the whole upper part of its body of a slate-coloured gray: the throat, neck, and breast, are of a dirty white, and the lower parts on this ground have transversal bands, which are also observable on the tail. The beak is bluish at the base, and black at the point: the lower mandible is dentelated, and squared at its extremity: the toes and tarsi are yellow: the eyes orange-yellow, and the cheeks furnished with brown mustachios.

The tufted falcon frequents lakes, rivers, and the sea-shore: it does not hunt, but fishes, subsisting on small fish, crabs, echini, and other shelled mollusca, the envelope of which it breaks easily by the force of the beak. Its nest is on trees in the neighbourhood of rivers, or on the rocks on the sea-coast. The female lays four eggs of a reddish-white: the male brings her the produce of his fishing, and partakes the cares of incubation. As these birds are not exposed to the want of subsistence, they keep the little ones near them a long time: they do not separate until the latter are capable of procreation.

The young have no tuft until they are able to fly They are also distinguished from the old by the fawn-coloured tint of the plumage, and by spots of red and grayish-brown spread over the throat, neck, and chest.

The *falco tibialis* is stronger than the preceding, and has also shorter wings: the beak is yellow at the base, and horn-colour in the remainder: the tibial feathers of the male are of a blackish-brown, like those of the head: the alar and caudal quills are of the same colour, but bordered with white: the back and wing-coverts are gray-brown: the lower parts of the body are reddish-white, with long brown spots: the tarsi and toes are yellow, and the claws black. This bird, a specimen of which was killed by M. Le Vaillant, in the country of the Great Namaquois, appears rare.

The *Hobby* (*falco Subbuteo*). This bird is common in France, Germany, and other countries of Europe, and is found even in the deserts of Tartary and Siberia. It is reported to leave England and some other countries in winter: woods, in the neighbourhood of fields, are its usual places of abode. These birds usually prey on larks; but they also pursue greenfinches, bulfinches, sometimes quails, and according to M. Temminck, some small river-birds. They nestle on very elevated trees, and the female lays three or four whitish eggs, unequally spotted with olive-coloured points, and black spots somewhat larger. For descriptions we must henceforth refer to the text, except where there is any thing peculiarly remarkable, which may have been omitted there.

The *Kober Falcon*, is the gray hobby of Cuvier. This bird hunts in the evening, and even at night: it is very common in Russia, Poland, Austria, and Switzerland; but seldom seen in France. It subsists on larks, and other small birds, and even on insects, especially the coleoptera.

The *Common Merlin* and the *Rock Merlin* seem now to be considered as one species, but some confusion exists regarding their respective habits. According to some writers, these birds inhabit forests, and nestle on rocks or in trees. Others, particularly Lewin, say that they are found in the hedges, along which they fly low, in search of small birds, and nestle on the ground, particularly in the furze. The courage of the merlin is very great, and it attacks birds larger than itself, as partridges, and often kills them. It remains with us only during the winter, though some have averred that it has been known to breed here. It is met with on the continent of Europe, but no where very common, and seems to be perpetually changing place.

The *Kestrel* is a bird very common in almost all parts of Europe: it frequents the open country, woods, old towers, and destroys a great number of small birds; it frequently darts on partridges and field-mice; also common mice, frogs, and even insects form a portion of its nutriment. The female is bolder, and less wild than the male, and will come into gardens, and close to habitations. These birds hover at very great elevations, describing a circle, and sustain themselves for a long time in the same place by beating the air with their wings in an almost insensible motion. They repeat, frequently, and with a sharp sound, a cry resembling the syllables *pri, pri, pri.* When they perceive their prey, they dart upon it with the directness and rapidity of an arrow. If they do not succeed in destroying it at the first attack, they continue to pursue it with extreme velocity and inveterate perseverance. They deplume the birds before they feed upon them; but they swallow the small mammifera with their skin, which they disgorge afterwards through the beak.

Though they are often seen in the neighbourhood of old towers and ruined buildings, they most usually nestle in the woods on the loftiest trees, or in the cavities of such as have been perforated: their nests consist of twigs and roots intermingled; sometimes they even content themselves with the old nests of crows. The female lays from five to six eggs, of a ferruginous colour, pale, and marked with deeper spots, irregularly distributed, and of different forms and sizes. The young are at first fed with insects, and afterwards with flesh brought by the parents.

Considerable variations take place in the plumage of this

species; sometimes the upper parts are reddish, spotted with black; sometimes the top of the head is shaded, more or less with a clear blue, and sometimes it becomes entirely white.

The *American Sparrow Hawk*, or *Falcon Malfini* (*Sparverius*, Lath.), is a Transatlantic bird: it is found in Carolina, Cayenne, St. Domingo, and the Antilles. Lizards, grasshoppers, &c., form the principal aliment of this bird: it also attacks young chickens; it is more sociable in the Antilles than in North America. It nestles in forests on the tops of the largest trees. In Paraguay its nest has been found in the hollows of trees, and even in the galleries of churches. It is remarkable enough that, in the first places mentioned, it lays four eggs, and in the second but two; and M. d'Azzara adds, that the number of eggs is less in South America than in North.

The *Rufus-backed Kestrel*, or *Mountain Falcon* (*rupicolis*), is a native of the Cape. This bird, which often utters the syllables *cri, cri, cri*, passes the entire year in the most rocky mountains, where it lives on small mammifera, lizards, and insects. It constructs a nest on a level on the rocks, composed of twigs and grass. The female lays six or eight red eggs.

The Bohemian falcon inhabits the loftiest mountains of that country; subsists on mice and field-mice, and only hunts in the evening.

The *Maritime Falcon* would seem, from its habits, to be a vulture. It is found on the coasts of the island of Java, and subsists on fish and rotten flesh. We pass over a number of species which have been named and described, but of whose habits nothing is known, and the correctness of whose allocations, in many instances, may be deemed more than doubtful.

On the Gerfalcon, we shall say a few words more at large.

Besides the tooth, very marked, and sometimes double at the upper part of the beak in the true falcons, being almost wanting in the gerfalcons, the slope in the lower mandible of the latter is less defined. They have also one-third of the tarsi furnished with feathers, and the tail exceeds the wings in length, although the latter are very long.

Etymological affinities, which are so often found to throw light on many subjects, seem to have contributed to obscure the natural history of the gerfalcon. Belon traces the origin of this name to the word *gyps*, a vulture, and *falco*; and the word *gyrfalco* seems immediately formed from *geyer*, the German for a vulture, and *falco*.

This association of terms so incompatible, designating birds of different genera, might seem extraordinary, did we not consider the state of natural science at the time when it was formed, and if we had not plenty of examples of names indicative of the uncertainty of naturalists respecting the proper allocation of certain animals in the scale of being. The vagueness, however, of such terms can be easily rectified, by a more intimate acquaintance with the true characters of species. But an inconvenience of another nature has resulted from Belon's exclusive application of the Greek term *hierax*, equivalent to the Arabian word *saqr*, to a species which, perhaps, has no existence, or is, at all events, doubtful: neither of these words was restrained in its acceptation to a single bird. They were used in a general way, to designate a class of birds venerated by the ancient Egyptians, who moreover distinguished the *hieraces* (falcons, hawks, and gosshawks) from the vultures, which were held in equal veneration, but from different motives. An attentive examination of the Egyptian monuments has proved that it was the common gosshawk which was represented on the temples, obelisks, and particularly on the Isiac table, where even the distribution of its colours is observable. If, then, the *hierofalco*, the *falco sacer*, the *sacre* of Belon and others, can be considered as forming the peculiar type of any one species of falcon, there is no reason why these denominations should be applied to the gerfalcon, rather than to the common falcon. Indeed, it seems much less natural to admit them into the synonymy of the first, as probably this bird, a native of Northern Europe, was not known in Egypt.

It remains, perhaps, yet to be verified whether the white gerfalcon and the gerfalcons of Iceland and Norway be particular races, simple varieties, or mere individual differences of age and sex. It is, however, safer to stick to the specific characters of the gerfalcon, as given in the text, and applied to all of these, than run the risk of adding to errors and confusion already far too great.

The gerfalcon is one of the most esteemed of rapacious birds for the purposes of falconry. When at liberty, it preys on nothing but birds, and it will attack very large ones, as, for instance, the heron and stork. It kills hares by dropping perpendicularly on them, and is so ardent in pursuit of its prey, that, after having torn one in pieces, it often abandons it to give chase to another. Pallas relates, that in the north of Russia they take the gerfalcons with nets, above which they suspend waving feathers to packthreads extended from one tree to another, at the same time fastening pigeons on the ground to serve as a bait.

Though, perhaps, strictly speaking, it is not a subject of natural history, we cannot help subjoining a few observations on the ancient and celebrated art of FALCONRY.

This term is given to the methods of instructing and training birds of prey to the chase, and is extended to the amusement itself. The great trouble and expense attendant on this exercise has caused it to be relinquished since the invention of gunpowder, which has rendered it superfluous; and few occupy themselves with it at present, except as an historical monument of the extent of human industry. It does not appear that the earlier hunting nations knew any thing of this art. The most ancient authors who have mentioned it are Aristotle and Pliny; Elian, who reduced it to principles; and Firmius, who developed more at large its practical details. After these came a crowd of authors on the subject, with an account of whose names and works we shall not trouble the reader. We must confine ourselves merely to what is necessary to the understanding of the practice of falconry, and avoid, as far as is possible, the usage of terms as useless, for the most part, as they are barbarous. Technical terms are often unavoidable in the exposition of many arts, but their intemperate usage is a silly and pernicious affectation.

The ancient authors have only treated of the mechanical parts of falconry; but M. Huber, in a work published in 1784, entitled *Observations sur le Vol des Oiseaux de Proie,* has entered into the theory of the art. In this, as in most other matters, practice has preceded theory, details have been carried into operation before principles were examined; and though we might well imagine that the means employed by rapacious birds in seizing their living victims, must form the natural foundation of the art of falconry, yet we apprehend that M. Huber was the first writer who paid any attention to this part of the subject.

This author divides the wings into *rowing* and *sailing* wings (*rameuses et voilières*). The birds provided with the former sort he calls *rowers,* birds of high flight, or, as in the old French, *de leurre;* the latter he calls *sailers,* birds of low flight, in the hawking jargon *de poing.* The wing of the first is slender, attenuated, not much convex, and, when unfolded, subject to very considerable tension. The first ten quills are entire, and their barbs touch each other without discontinuity, in their entire length. The motions of this wing are easy, rapid, and strong: accordingly, we find the rowers fly against the wind, with the head straight, and raise themselves without difficulty into the highest regions of the air, where they sport in all directions. The wing of the sailers is thicker, more massive, and arched, and less stretched in the act of flying. The first five quills, of an unequal length, are sloped from the middle to the extremity. Thus that portion of the wing which is most important for the purposes of flight, presents an interrupted surface to the air, and the wing itself, actuated by forces of less energy, fails of producing so perfect an effect. We find, therefore, that these birds can only fly with advantage when the wind is in their rear. They keep their heads low, and seldom rise but for the purpose of discovering their prey. The French term *planer* (to hover) very appropriately depicts this mode of flying, in which the wings are extended and motionless, and the body is carried along by the course of the wind. In fact, it is, strictly speaking, a sort of sailing. The quills of the rowing wing are also, in general, more firm than those of the sailing. This is indicated, according to M. Huber, by the lively and marked variegation which predominates in the first from one end to the other; while, in the last, a deep, uniform black wash prevails from the sloping of the feather to the point, and a white equally uniform from the origin of the quill to the commencement of the sloping.

There is, likewise, a different conformation in the talons of the *rowers* and *sailers*. These talons the falconers call *hands (mains)*. The toes in the former birds, or in the *noble* division of birds of prey, are longer, finer, and more supple. They embrace a more extended surface, and being moved by a longer lever, they are capable of a more powerful retention than those of the sailing or ignoble birds, which are thicker and shorter. The claws of the rowers, also, being more curved and acerated, penetrate more easily, and inflict a more dangerous wound.

The rapacious birds employ the weapons with which nature has provided them with the most admirable dexterity. The rowing birds seize at once their intended victims, when the latter are more light of body than rapid in their movements. When the prey is of greater weight, and more activity, they strike it to weaken and diminish its strength and speed. With an instinctive precision the most extraordinary, they instantly attack the vital part, which in the birds is at the hollow of the occiput, and between the shoulder and the ribs in the mammalia. It is also remarkable that the smaller species are the most instantaneously destructive; the merlins scarcely touch the place just mentioned before immediate death ensues.

The sailers do not strike with so much precision; their grand resource is to seize their victim and compress it to death. When they cast themselves upon a hare, they seize it by the neck with one of their talons, and strangle it. Their beak, not being indented, tears the skin and flesh, but seldom breaks the bones, except when they are so situated that its point can manage them in its curvature. In the thickest woods these birds exhibit extraordinary address in seizing their prey; and probably the length of the tarsi may prove of considerable utility to them on such occasions.

The rapacious birds of elevated flight perceive, the moment their hood is removed, not only the various birds which are, as it were, immersed in the luminous expanse of air, but also their peculiar kinds, and their natural disposition and means of defence. Accordingly, they instantly select the object of their pursuit, against which they steadily proceed, without being in the slightest degree distracted by the motions of any other birds which may happen to be about it. The low-flying birds also, when they quit their master's hand, fix their quarry with unerring eye, in the darkest obscurity of the forest, either among the birds which circle with such rapidity through the thick coverts, or the smaller mammalia whose almost imperceptible motions would elude a duller ken.

Among the particular resources which the birds of prey derive from the varieties of their conformation, M. Huber does not take the tail into consideration. This part, in fact, does not, as the ancients imagined, serve for a rudder to the bird, to enable it to turn itself to one side or the other, but simply as an assistance in ascending or descending. Even Borelli has long since remarked, that individuals accidentally deprived of their tails, performed all the movements to which this part had been supposed indispensable The first-mentioned writer has, moreover, added to the characters of the rowing birds a dentelated beak and a black eye, while the beak of the sailers is without indention, and the eye is clear. Among the rowers he classes the gerfalcon, the common falcon, the hobby, the merlin, but not the kestrel; and among the sailers, the gosshawk and the hawk.

The birds which are not rapacious may be considered, according to the nature of their flight, either as rowers or sailers. But it would be impossible to establish a marked division in this way. The birds of prey, however, whether from instinct or experience, are at no loss to distinguish these characters where they exist, and to direct their plan of attack and pursuit accordingly. The raptorial sailer will suffer a bird eminently endowed with the rowing capacity to pass without attempting to put himself in motion, well knowing that he would be unable to overtake him. Not so the raptorial rower, who shoots upon his victim without such discrimination, equally capable of assailing him on high, or pouncing upon him below.

If we united the considerations of anatomical structure to the inductions of M. Huber from external characters, we might institute a comparison between the motive forces of these two different raptorial groups, to which, in imitation of him, we have given the denominations of rowers and sailers: those, for instance, which actuate their talons; the texture and insertion of the muscles which put the levers in action; the disposition of the tendons, and the augmentation of force, produced by the re-acting pullies round which they circle. This comparison might be even extended to the organs of respiration, to the degrees of natural heat in those beings, some of which sustain the rigorous cold of the more elevated atmospheric air without detriment to health or respiration, while the others, though to all appearance similarly constituted, rise but seldom, and for a short period, above the lower regions.

We shall now proceed to a slight sketch of the practical part of the art of falconry, commencing with the mode of procuring the birds employed therein.

When it is possible to take the young ones, as yet covered only with down, from the nest, the education of these birds, which are, in the language of falconry, then called *niais (simple)*, is comparatively easy. They have little bells attached to their feet, and are placed on what is termed an *eyrie*, which, for a bird of high flight, is a cask staved at one end, rested on the side, lined with straw, and placed on a low wall, or a hillock of earth, within reach of the master, with the opening turned towards the east. For a bird of low flight they use a kind of hut of twisted straw, set upon a tree of no great height, within reach of the hand. Certain planks are placed near the openings of these, on which the birds perform their first exercises and receive their food. The food consists of beef or mutton, from which the fat and membranous parts have been withdrawn, and which is cut into slender and oblong pieces. This aliment is given daily at seven in the morning and five in the evening, and the bird is excited to partake of it by an uniform cry, which he soon learns to recognize. On those planks, which serve as a table, they always place the food for the high-flying birds, but for the others the food is set on the ground as soon as they are strong enough to descend and re-ascend. Both kinds exercise their strength gradually. They first reach the places which are near them by jumps, and then by a heavy sort of flying, which the French call *monter à l'essor*. At six weeks old they can catch bats, swallows, and other feeble animals, which, when they come near them, are sure to fall their victims. At this period they are deprived of their liberty, being taken in snares or nets, and covered with a thick cloth, that they may be chained down in darkness. The *jesses*, which are attached to the tarsi, are manacles of supple leather, to which is fastened a ring and cord, by which the birds are fixed on a log of wood on a level with the ground, surrounded with straw. They also cover their heads with a hood, which hinders them from seeing, while it allows them to eat. The training is then commenced.

The birds which are taken after they have left the nest, and can only hop from branch to branch, from which they are called *branchiers*, receive the same education as the *niais*. They are more difficult to train than these, though less so than adults, with which, however, the falconers are obliged to content themselves when they can get no others, and which are taken in the following ways:—

The hawk, the merlin, and the hobby, are taken in projecting nets, laid as if for larks. They immediately descend upon the *calling* birds, which are placed in the centre. Falcons and gosshawks are also sometimes taken in the same manner; but as this never happens except when these birds are very hungry, and in the immediate neighbourhood, the fowler desirous of taking them provides himself with a tame shrike attached by a buckle. This bird, which recognizes from a great distance the various raptores hovering on high, and is but slightly agitated when he sees a buzzard, rushes into the hunter's lodge when he perceives a falcon. The hunter then slips a pigeon under his net, also held by a long cord, to leave him the power of fluttering and exciting the falcon, which, when he attacks his prey bitterly, suffers himself to be drawn after it within the fall of the net. Should this plan not succeed, the fowler (if he has one) takes a tame falcon, which age and infirmity have rendered useless, and attaches it to the end of a long and pliant twig, by the feet, and fixes the other end of the twig in the ground. A cord, beginning from the point where the bird is retained, passes through the pulley which occupies the centre of the nets. The hunter, who holds the extremity of it in his box, on a signal given by the shrike, draws it, and the twig bending, obliges the falcon to extend its wings as if about to pounce on a prey. The wild bird then directly precipitates himself on the other, and falls into the snare.

The great horned owl is also employed in taking birds

intended for falconry. The falconers teach this bird to fly from one end to the other of a long cord attached to two logs of wood, on which the owl rests after his flight. To accustom the bird to this exercise, they shut him up in a chamber, in which is placed, at a little distance from each other, two logs of wood, separated by a tight cord, through which a ring is passed; to this ring another slacker cord is attached, which also joins the cord of the bird's jesses: food is presented to the owl on the side opposite to where he is, so that to come at it, he is obliged to cross the interval by flying, without touching the ground. This operation is repeated again and again, until gradually the owl acquires a habit of crossing from one side to the other, merely to change place. When the owl is thus disciplined, they form, in a copse, a sort of saloon, in the midst of which they place a log of wood, and another opposite at about a hundred paces distant, having cleared away the intermediate space. The top and sides of this place must be covered by branches, which, while they suffer the inside to be seen, will not permit a bird of prey to enter with unfolded wings. Nets, called *spider-nets*, are suspended to the top and sides, only leaving that part free which is opposite to where the owl has been placed on the log: the fowler then retires into a lodge or box prepared for the purpose, and judges that the owl sees some rapacious bird in the air, by his lowering his head, and turning the globe of the eye upwards. When the enemy approaches, the owl passes from the log he is on to the other in the centre of the saloon, and draws the rapacious bird after him, who, on whichever side he comes, is embarrassed in the nets, and seized by the fowler before he has time to disengage himself.

As soon as the bird of prey is taken, his legs are passed into very strong manacles, the ring of which is crossed by a cord which serves as an attachment, and little bells are hung to his feet. The person charged with training him fortifies his hand with a glove, and taking the bird on his fist, fatigues him as long as possible in an obscure place, without allowing him to take food, so that his strength being exhausted, he may be the better prepared for submission. When the bird agitates himself very much, and attempts to use his beak, they throw cold water on his head, and even plunge it into a vessel of that liquid. When by these means they conquer his spirit, which is usually done in three days and three nights, they cover his head with a hood, which is taken off and put on, according as he accustoms himself to take food uncovered, which they present to him from time to time. To weaken the bird more speedily, they make him swallow little pellets of hemp, which produce a purgative effect: these are called *cures*. Having thus succeeded in making him take food easily, they carry him into a garden, where he is uncovered, and showing him the prepared meat, which we have already mentioned, and which is held a little elevated, they accustom him to leap upon the hand. When he does this with facility, they place the meat on a representation of a bird, formed by an assemblage of wings and legs, which is called *lure* (*leurre*), and to which they attract him successively from a greater distance, holding him always by the cord. When he has had so much training that he will pounce upon the *lure*, from the whole length of his tether, they accustom him to know and examine the game which he is destined to hunt. This is done by attaching the game to the lure, and allowing it to run or fly near the bird; first attached by a packthread, then at liberty, until they think they can trust to him free of all restraint.

When it is possible to choose birds for training, the falconers prefer those whose shape is the most easy and elegant, glance the proudest and most assured, toes the most elongated, grasp the most ample, and whose plumage is the deepest, and least charged with spots. Neither is the education exactly the same for the *rowers* and *sailers*, and it also varies according to the species: but it may be observed, generally, that the larger the species, the older the individual, and the more northerly its habitat, the greater is the difficulty of training.

This is the case with the gerfalcon of Norway. The first care with respect to him, is to weaken his strength without exposing him to fall into a decline: this is done by reducing his allowance of food one half, and steeping the meat which is given him in water. This regimen is continued for about six weeks, after which they tie down one of his wings with a thread, and throw water over his body with a sponge: they touch the fore and hind part of his head without removing the hood, they rub him with a pigeon's wing, and if they find the movements of his head supple and obedient to the hand, they loosen the hood, and uncover by degrees his eyes, always leaving the beak engaged, and removing and restoring the light by turns. These operations are commenced in the morning, in a solitary and gloomy place, and continued all day long, and in the evening the bird is sufficiently mild, to be carried, though uncovered, into another place, where several persons appear before him, taking care not to go behind, lest they might frighten him. They repeat the exercise of removing and putting on the hood from time to time, and making him feel the pigeon's wing until the middle of the night, for the rest of which they allow him to take his repose; still, however, two months are requisite to complete his education.

The above-mentioned lessons are repeated for fifteen days, leaving the bird, by little and little, a longer time uncovered; and accustoming him to noise, to motion, and to the sight of dogs, which are held at a little distance in a leash. They give him small portions of food, first holding the hood half closed, then removing it altogether: finally, they give him his full allowance. They then carry him into another chamber, having placed upon the table an ox's tail, towards which they draw him by presenting him with the hand a pigeon's wing all bloody, on which he falls furiously, and which they let fall when he is near the tail, which he then seizes, but without being able to eat it. They present him the wing again, raise the hand, giving the cry of *lure* (*leurre*), at first in a low voice, and cover him again gently with the hood. This exercise is repeated the following day, and, in the evening, they add the presence of a light, to which he becomes accustomed in an hour or two. The preceding lessons are renewed during fifteen days in the open air, on the turf, taking care gradually to slacken the cord or thong: they gradually remove the *lure* farther, and, at last, to the distance of 150 or 200 toises, and accustom the bird to the full cry, as it is made in the chase. The ration is all this time diminished, and they administer two or three times a laxative, composed of garlic and absinthium, in an envelope of tow. For two days running, they then set the gerfalcon against a hen, pointing it out at first within five or six paces, and warning him by the cry of *lure;* and on the second day they allure him to feast upon it, talking and shouting about him the whole time he is eating, to habituate him to motion and noise. The following day, they give him but little food; and the day after, they *lure* him at two hundred toises distance, without the string.

From fifteen to twenty days are employed in instructing the gerfalcon in the pursuit of a prey which attempts to escape, and in the choice of that to the chase of which he is designed. If a hare be the object, they enclose a chicken in the skin of this animal, and its head is passed through a hole made for this purpose: this skin is fixed on a plank, as if the hare were lying on its belly. At the distance of three or four paces, they show this hare to the bird, who goes to it: the pullet draws back its head, but its cries and movements animate the bird, who attacks the skin furiously, which is covered with some bloody food to excite him still more. They then draw him off, cover him, and the exercise is recommenced at five or six paces distance. The skin is removed farther and farther on the following days, and to give more motion, they cause it to be drawn along by a huntsman, who gradually augments his pace, and ends by mounting on horseback, and dragging off the skin in full gallop. The bird at first reaches it with the beak open, and out of breath; but, on successive exercise, he gains wind, and comes in with the beak closed. They always take care to give him his repast on this skin.

When they wish to teach a gerfalcon to pursue the heron, buzzard, &c., they lure him with the skin of one of those birds, flinging it daily farther and farther, and habituating him to seize it in the air while falling. They end by employing in these exercises a hen of obscure plumage, or even a real buzzard, attached to a stake, or a kite whose beak and claws have been blunted. When the gerfalcon has seized them at thirty or fifty feet of elevation, they then make him do so at a more considerable distance, which terminates his education.

The instruction of the proper falcons does not require so much care, and may be terminated in a month, or even in fifteen days when they are taken from the nest. The operations for weakening the falcons which have left the nest, or as they are called *haggards*, are of the same nature as those used with the gerfalcon: they give them two or three hempen pellets, and as many baths, which they will take of their own accord when they are fastened near the edge of the water; otherwise they throw them in, and keep them there a sufficient time. In about three days, they manage what is called *making the falcon's head*, that is, accustoming him to the hood: they then teach him to jump from the hand on the table, and from the table on the hand. The lessons of the lure are soon practised in the open air, and there the bird is habituated to leap from the turf on the hand, which the falconer first lowers, and afterwards presents standing at distances more or less considerable. Then comes the exercise of a pigeon attached to a stake; then the pigeon is held by a thread, and the falcon left free; and finally, a black hen is attached to the stake, to teach the hunting of crows, a red hen for the kite, and a grey turkey-hen to represent the heron. On the five-and-twentieth day, the crow, the kite, and the heron themselves are attached to the stake, having the claws blunted, and the beak surrounded with a sort of case, to prevent such resistance as might revolt the falcon. On the twenty-eighth and twenty-ninth days, they teach him to know his game at greater and greater elevations, which is called *demi-escap*, and on the thirtieth they do this at the highest point, leaving the bird at full liberty, which is called *grand-escap*.

The merlins being by far the most familiar and docile of the birds of prey, their training is much less tedious and difficult. It is not necessary to use the hood with them. When the falconer has carried them on his hand for a few days, and enticed them with little pickings of meat, they fly to him the moment they see him. Then shut up in a room, the window of which is only closed by a drawn curtain, they soon accustom themselves to leap upon his hand. When the bird can do this at twenty paces in the open air, they attach a lark to a packthread at that distance: the merlin soon seizes it, takes it in his beak, then in his talons, and carries it off. It is necessary to prevent his doing this, which is the only difficulty in his education. For this purpose they begin by drawing the packthread with a jerk. Frequently the lark does not escape from the merlin, and his head remains in the beak of the latter. In all cases, the body of the lark is quickly passed into a little crook dug in the earth for that purpose; and the merlin returning with fury to devour his prey, at his master's feet, but without being able to take it away, he gradually comes by reiterated exercises, assisted by the voice and gesture, to lose this habit, and never resumes it with small birds of any species. The merlin is employed to hunt, not only larks, but blackbirds, quails, and partridges.

The hobby is much less docile than the merlin, and his training a matter of much greater difficulty; but it is needless to mention it, as it does not differ in kind from what we have already related.

The goshawks, and hawks, are *sailers*, or birds of low flight, and the education of the first is very easy and very short. They use no hood with the goshawk, which, nevertheless, torments himself very much at first, refusing all sustenance. But from the fifth or sixth day, these birds lose all terror at what is going on about them: they seize the food greedily, which is given to them in very small quantities. They are soon habituated to jump on the hand of the falconer, who can carry them in this manner with a thong, in the most frequented places, and amidst all kind of bustle and noise, without inconvenience.

At the end of eight days, having bathed the goshawk in the morning, they lure him in the evening with a cord, several times, at eight, ten, and twelve paces distance, and the following day at twenty and thirty, after which, they leave him at liberty to attack a pigeon fastened to a stake: when he has taken this bird by the head, they pull away the body, and hold it in the hand, so that when the goshawk has eaten the head, he jumps upon the hand, to devour the rest. In the afternoon of the same day, they call him back, from greater and greater distances in the woods; and if he returns readily, they can employ him the following day in the chase, having first carried him for some time on the hand. But if he is designed for any other chase than that of partridges and rabbits, to which he is instinctively prone, it is necessary to habituate him to the particular game, like the falcon and gerfalcon, by means of *lures*.

The hawks are trained like the goshawks; but, although weaker in appearance, they are more fierce, and their education takes more time, especially after they have left the nest. Before they are fitted for the chase, many lessons must be repeated in an orchard, and they must be *reclaimed*, as it is called, until they seek the falconer of their own accord, who conceals himself purposely. Even those which are already educated must be exercised daily, or they would soon become indocile for want of action.

We may see by the system of education pursued with the birds of prey destined to falconry, that the objects of this art are, to teach them to obey man, to bear the hood, to return on the hand from the end of their tether, to accustom them to the *lure*, to rise when desired, even against the wind, to be ready to drop the prey for which they are trained, and not to carry it off without returning.

Falconers train the rapacious birds for seven different sorts of sport; for the kite, the heron, the crow, the pie, the hare, for open fields, and for rivers. Birds of prey, in health, should be fed with beefsteaks, and legs of mutton cut in slices, and the fat and tendinous parts removed. In general they are fed but once a day, but the food is divided into two moderate portions during the moulting time: the evening before a hunt, the portion should be smaller than on other days, and sometimes on such occasions a laxative is administered. During the season of reproduction in the month of March, a custom prevailed of making those birds swallow flints about the size of a nut, with the intention of rendering the females unfruitful, and deadening the desires of the males. Such a plan, however, could not be otherwise than dangerous, and detrimental to digestion in birds whose stomachs are more delicate than those of the granivora. The same result might probably be obtained with less danger, by giving them less nutritious or less abundant food.

In summer the birds of prey are kept in cool places, where pieces of turf are laid, on which they like to repose. A bucket is also placed there, in which they bathe, and if they are observed not to do so of themselves, they are taken and plunged in every eight days. The baths soften the skin, and render the moulting more easy. In the evening these birds are fixed on their perches, in such a manner, as to prevent them from hurting each other. Care must be taken to clean their hood very scrupulously, to prevent an accumulation of dirt, which would injure their eyes. A light is left in the place where they are kept about an hour, to allow them to clean and polish their plumage. In winter they are kept abroad during the day, and at night falconers are in the habit of shutting them in warm rooms. This practice is objectionable; for, as these birds are natives of cold, or, at all events, of temperate climates, it would be sufficient to keep them in sheltered places, without contributing, by too much warmth, to augment the debility, which domestication of itself is calculated to produce.

Authors who have written on falconry have entered into long details concerning the maladies of birds of prey, and the modes of their cure. But their treatment of internal cases was, as may well be supposed from the infant state of the medical art in their days, for the most part exceedingly arbitrary. Their prescriptions merit no attention, except in the case of accidental wounds; and, even in this point of view, it would be equally irrelevant and uninteresting to take any notice of them here *

* We shall avail ourselves of the present opportunity, to offer, in the shape of a note, a few remarks on the education of animals. This is a very curious and interesting subject, and, perhaps, not less important than curious and interesting. The education of animals has not always met from philosophers the degree of attention it deserves, nor has it, in our opinion, been carried as far in practice as it might have been. We may add, that the mode of conducting it has, in most cases, been extremely erroneous. This is the less to be wondered at, when we recollect who the persons have been who have generally undertaken this important task; men, for the most part, ignorant and vulgar, obstinately wedded to old methods, unwilling, therefore, to question their merits, and incapable, were they ever so willing, to appreciate their defects, and substitute better systems.

After what we have said in a former part of this work, on the instinct and intelligence of animals, it is unnecessary to premise that we concede a portion of the latter faculty to the brute creation. Animals, like man, are governed by two grand springs of action, *pleasure* and *pain*: it is by a judicious management of these, in reference to the intelligent faculty of animals, that their education must be conducted. It is thus that *attention* is excited and sustained, and attention is the *sine quâ non* of all

Next come the grand division of the Eagles. Pursuant to our plan, we shall here avoid a repetition of, or enlargement on, the generic and specific details of the text, and keep clear of the thorny path of nomenclature. Linnæus comprehended the eagles, with many other groups, under his genus falco;

instruction. Every method of securing and concentrating this attention must be adopted. This is the object of hooding the falcons before and after they receive their lesson, to prevent distraction; but the coercive, and often cruel measures resorted to with animals, are calculated to produce a direct contrary effect. Chastisement, moderately used, may be sometimes necessary, to fix the desired association in the sensorium, but if carried too far, it produces too strong an image of itself, to admit of any other. The animal is occupied with nothing but the violence of his immediate sensations, and cannot attend to the idea with which you mean to impress him. But, in fact, experience proves, that mild methods are the best in general. The docility of the Arab horse, which is the companion and friend of his master, and never ill used, is an eminent proof of this. The same observation is applicable to dogs. One of the principal reasons of the distrust, and want of docility evinced by cats is, the general ill treatment they receive. I am aware that some animals require a more severe discipline than others. (Indeed, nothing is so requisite in the education of animals, as a profound study of specific and individual peculiarities, and few points are less profoundly studied.) But I am certain that the worst discipline is the discipline of blows: judicious privation will answer all purposes much better. The account which we have given of the training of the gerfalcon is a good illustration of this point.

There is no doubt that education might be much more extended in the animal kingdom than it is. We have seen, in the case of rapacious birds, what the industry, perseverance, ingenuity, and judgment of man, is capable of effecting in this way. Had he a sufficient motive to exert these qualities in the instruction of other wild animals, many more might be reclaimed, and rendered subservient to his purposes. In short, I believe, that all vertebrated animals that can at all be brought under the control of man, are susceptible of instruction: instances of this are not wanting even among fishes. The only thing is, to hit on right methods, which can alone be done by long and partial observation. But, to pronounce an animal untameable, because we cannot tame him, by the hacknied, and, in many respects, injudicious systems pursued with domesticated races, is unphilosophical and absurd in the highest degree.—E. P.

this was certainly embracing too many species, strongly inter-distinguished, under one head. But if Linnæus has erred in crowding too many species into one genus, it is equally certain that some subsequent naturalists have not offended less, by the conversion of species into genera*.

The eagle holds, among the feathered race, the foremost rank, and his station is analogous to that of the lion among the mammalia. There is a general resemblance between the character of the two animals: in both the qualities of ferocity and strength are adorned with a daring courage, and redeemed by a generous magnanimity. The vulgar notions of cruelty, rapine, &c., usually attached to the carnivorous tribes, are, to say no worse of them, exceedingly silly. They may serve to embellish declamation or poetry, when sounding words are found a convenient substitute for just ideas; but they are calculated only to mislead the understanding, and have no place in philosophical investigation. If the eagle, like other carnivora, subsists on flesh, it is because he cannot help it; the structure of his stomach and intestines precludes the use of other food. Unprovided with internal organs to reduce other aliment to a nutritive consistence, he does not violate, but fulfils the laws of nature, by the employment of those destructive weapons with which she has armed him. Neither do these carnivorous propensities constitute a bye-law, or an exception to the grand code of the universe. It is the fiat of

* It is too much the fashion now, especially among flippant sciolists, to depreciate the merit of Linnæus, who was one of the most eminent men of his times, and the greatest of systematic writers. The "Systema Naturæ," with all its defects, is a magnificent specimen of ingenuity, industry, and judgment. Its utility, too, is far from being superseded, and the young Zoologist cannot do better than begin by making himself perfect master of it, before he proceeds to the study of any other work on the subject; otherwise, his notions respecting *natural methods* will, for a long time, remain confused; he will be unable thoroughly to appreciate the great improvements of Cuvier, and to discern the full extent of mischief produced by the mania of everlasting innovation.—E. P.

nature that life must subsist on life: the modes, indeed, are different, but the principle, the result, and the object are the same. The peaceful herds and flocks which graze on the plain, or browse upon the mountain slope, are no less destroyers of life, than the sanguinary rangers of the forest and the air. Even vegetation itself is sustained by what once was animal existence, to which its own origin is in all probability posterior: for lifeless matter could never have produced life, nor the green herb have sprung from the naked bosom of the primæval granite.

We shall not have recourse here, like some writers, to the vague hypothesis of final causes, to explain all that appears contrary to our conventional ideas of right and wrong in the great system of nature. The fact is, that of final causes we know very little: all we know is, that things are so, and we may conclude that they must be so. There are certain conditions of existence without which existence could not be. Wherever we turn, we find indubitable marks of that imperious necessity, to which the highest intelligence must bow, as well as the meanest worm. It is no compliment to the Divinity to laud his wisdom in the provisions he has made for the preservation of any being, when we know that, without such provisions, the being could not exist at all; and it is the height of presumption to pretend to justify his operations, by arguing from an imaginary and an impossible hypothesis.

But without pretending to unravel the mystery of final causes, or to assign a reason why certain animals are endowed with a sanguinary instinct, we may simply observe, that the mischief operated by carnivorous animals in the creation is comparatively very small. The wolf may occasionally abstract a lamb from the numerous flock, the lion kill one buffalo out of the immense herd, the eagle strike a solitary kid, or the gerfalcon a single hare; but the number of victims bears no sort of proportion to the numbers which escape. The benevolent lord of the creation executes more destruction among his peers in one glorious campaign, than all the carnivora from one end of the earth to the other among all the living tribes.

Among the lower animals, as in savage and uncivilized nations, where the intellectual faculties are but slightly developed, strength and courage are the surest titles to supremacy. If, then, the pre-eminent possession of the characteristic faculties of its class, and the resistless exercise of them in the element which constitutes its domain, give any animal a claim to exclusive superiority, the empire of the eagle cannot be disputed by any of the denizens of the air. Shooting impetuously on untiring wing to an incommensurable distance, or sailing majestically above the mountain and the cloud, he assumes his native place among the feathered tribes; and none can escape his pursuit, or rival his elevation. No other bird can cross his path on high; all remain humbly in the lower regions, forming a graduated scale down to the penguin, which is provided only with the rudiments of the organs essential to the capacity of flight. The eagle is distinguished by a lofty mien, an eye of piercing vivacity, a bold assured gait, and a general expression of commanding nobleness. That this magnificent bird should be classed among the *ignoble*, by the professors of falconry, because he disdains a subservience to the caprices of man, is one proof among many of the proneness of human selfishness to the perversion of words.

The eagles are monogamous: they ordinarily subsist on living prey, and never touch the dead, except when ready to perish with hunger. Their admirable power of vision enables them to distinguish their prey at an immense distance; they rush upon it with the velocity of an arrow, tear it instantly, and carry it off in their talons, except when its weight is unusually considerable.

The broad and flat nest constructed by the eagles, between rocks and large trees, is called an *eyrie*. The female usually lays two, and but seldom, three eggs, which she hatches for thirty days. This nest remains, and continues to answer the purposes of the eagle during life, except some accident should destroy it.

In the eagle tribe, as among all the other birds of prey, the female is larger than the male, and in a state of freedom appears to possess more assurance, courage, and subtlety: she appears, in some species, to have a mutual understanding with the male for the purposes of the chase, and, except when she cannot quit her eggs or little ones, she and the male are generally observed at no great distance from each other.

The eagle, especially in a state of captivity, can go a long time without food. Buffon knew one of these birds, of the common species, which had been taken in a snare, to live forty days without any nourishment, and it showed no symptoms of exhaustion but for the last eight days, at the end of which it was killed. This bird, which can quench his thirst with the blood of his victims, can also remain a long time without drink; but it is a vulgar error to suppose that he never drinks at all. When water is presented to him, he will bathe his plumage in it, and drink like other birds.

Spallanzani has made a singular remark on the conformation of the internal canal of the eagle. The capacity of the crop to that of the ventricle is as thirty-eight to three, which explains why a single repast is sufficient for these birds for many days; for, if a large animal becomes their prey, they fill their crop, and digestion proceeds successively, according as some portion of this nutriment passes from the crop into the ventricle or stomach.

The eagles love to haunt the mountains and the deserts. They are not very frequent in islands, and more especially in those of small extent, because they are less peopled with animals than the terra firma. Such as are more frequently found there, and which build their nests on the shore, are the sea-eagles, which subsist more on fish than game. It was observed that the first eagle seen in the island of Rhodes, perched upon the house of Tiberius as a presage of his future empire.

Professor Reisner, of Germany, has published a pamphlet, the object of which is to prove that eagles may be employed to direct a balloon. He states the number of these birds, which he deems necessary, according to the dimensions of the machine, and gives the mode of training, harnessing, and guiding them.

The *Great Eagle (Aquila Chrysaëtos)* also called the *royal* and *golden* eagle, is not confined, as Buffon imagined, to warm and temperate climates, but is also found in colder regions. He lives solitarily in the mountainous regions of Europe, as in the Pyrenees, the mountains of Silesia, Ireland, &c.; also in Tartary and the various parts of Asia, in Western Russia, Kamtschatka, and Siberia. It is also met with in Barbary, but apparently only in the chain of Mount Atlas, for it is by no means certain that the eagles seen in Africa generally, by many travellers, belong to this species. It does not exist in North America, where the common eagle is found.

This bird appeared so redoubtable to the ancient poets, from his bold glance, proud air, the elevation of his flight, and the strength of his limbs, that they consecrated him to Jupiter, and deposited the thunderbolt in his talons. He was termed the celestial bird, and the augurs esteemed him as the messenger of the gods. The Persians and Romans adopted the eagle as their standard of war. Modern potentates have followed their example, and we have ourselves beheld the greater part of Europe tremble at the elevation of this imperial standard. This bird has also been considered the emblem of genius. It is this species which may particularly be compared to the lion as to physical and moral analogies. Full of the consciousness of his strength, the eagle disdains the smaller animals, and despises their insults. He desires nothing but by the right of conquest, and will have no prey but what he takes himself. His temperance is extreme, and he scarcely ever finishes the entire of his game. He leaves the fragments to other animals, and though ever so hungry, will never touch a dead carcass.

Retired, like the lion, in some wilderness, he banishes every other bird which might partake in his prey, and when two pairs of the same species settle in a forest, they keep sufficiently apart to find ample sustenance in the place they have chosen, without interfering with each other. Even the colour, the form of the talons, the terrific cry, the ferocity of character, the erect and imposing attitude, in this bird, all serve to approximate him to the first of quadrupeds. Buffon has added to these qualities, the powerful odour of his breath; but Spallanzani, who kept one of these eagles tame for a long time, has ascertained, by numerous trials, that the breath of this bird emits no disagreeable effluvia whatever.

Notwithstanding the want of docility in the great eagle, it appears that he was formerly employed in the East for the purposes of hunting. But he was found unfitted for falconry, both by reason of his great weight and capricious and irritable temper. Some people of the north, however, still train this bird for the chase. The Kirguis, whose country is situated eastward of the Caspian Sea, judge by certain marks of the disposition of these eagles, and purchase from the Russians of Samara, at a very great price, eaglets taken from the nest, to train them to hunt the wolf, the fox, and the gazelle.

The scent of this bird being feeble, he hunts only by sight. Though he elevates himself in the air above all other birds, yet he rises from the ground with difficulty, especially when overloaded, from the want of suppleness in his legs; yet he can carry off geese, cranes, hares, young lambs, and birds: it is even pretended, that in Scotland children have been found in his nest. When he attacks calves and fawns, he only satiates himself on the spot with their flesh and blood, and carries off the pieces to his eyrie. This nest, which is usually placed in the clefts of rocks, lasts the eagle, it is said, during his life. It is made with sticks of from five to six feet in length, crossed by supple branches, and then covered with rushes and weeds, and has no shelter but some projection of the rock. The female lays there annually two or three eggs. It is pretended that this barbarous mother occasionally kills the most voracious of her young: but, if scarcely ever more than two eaglets are found, and frequently but one, it is no doubt owing to the infecundity of the eggs. The philosophers of final causes find in this a wise provision of nature against the multiplication of destructive beings, as if the occasional infecundity of eggs was not a common phenomenon among all the volatile tribes. Why produce these destructive beings at all, or if a certain number only are necessary, why not limit the production of germs? Why produce any thing superfluous? These are questions the philosophers of final causes cannot answer. But we can:—such is the order of nature.

If it is true that the young eagles are chased from the nest as soon as they are able to fly, this habit would appear derived from the difficulty with which birds of prey procure subsistence. Yet it is well known, that when a mountaineer has discovered an eagle's nest, he can supply himself for some time with an ample store of provision by substracting the game he finds there during the absence of the old ones. It is even pretended, that by tying down the young, he can prolong the period of his robberies. These facts but ill agree with the precipitate expulsion, or rather with the above solution of it. Smith, too, in his history of Kerry, relates a story as little in accordance with it. A poor inhabitant of that county provided for his family abundantly for an entire year, by taking from an eagle's nest the food brought there by the parents: and that he might prolong their attentions beyond the ordinary period, he contented himself with clipping the wings of the eaglets, to retard their voluntary departure.

Perhaps the circumstance of which we are speaking is as philosophically explained by our own poet Thomson, of whose eloquent lines on this subject we shall avail ourselves:—

High from the summit of a craggy cliff,
Hung o'er the deep, such as amazing frowns

On utmost Kilda's shore, whose lonely race
Resign the setting sun to Indian worlds,
The royal eagle draws his vigorous young,
Strong-pounced, and ardent with paternal fire.
Now fit to raise a kingdom of their own,
He drives them from his fort, the tow'ring seat,
For ages, of his empire; which, in peace,
Unstained he holds, while many a league to sea
He wings his course, and preys in distant isles.

The great eagle, though a very lascivious bird, lives for above a century. Klein mentions one which lived at Vienna one hundred and four years in a state of captivity. Some writers have pretended that the death of this bird is accelerated by the great increasing curvature of the beak, which prevents him from taking his food any longer. But this assertion seems founded on no great degree of probability.

The great eagle is tamed with much difficulty; but he can be fed on all kinds of flesh, even on that of other eagles. He will also, in default of other food, eat serpents, lizards, and even bread, according to Buffon. Spallanzani, however, declares that the eagle has a great antipathy to bread, which he will not touch even after a long fast, though he can digest it well enough if he is forced to swallow it.

In proportion as this eagle grows older, the colour of his plumage becomes lighter: whitish tints become visible, and even some places turn entirely white. These changes are likewise produced by diseases, hunger, and long captivity.

The *Common Eagle*, whose species is more numerous than the foregoing, is found all over Europe and North America. It is very common in the high mountains of France, Switzerland, Germany, Poland, and Scotland, and descends into the plains in winter. It has been seen in Barbary, and it would appear that it also exists in Arabia and Persia. It has been found in Louisiana, the Floridas, Carolina, and at Hudson's Bay. During summer it never quits the mountains, but when it descends in winter the forests become its asylum during the rigour of that season. The flight of this eagle is so high, that it is often completely lost sight of. From this great distance, however, its cry is still audible, and then resembles the barking of a small dog. This eagle builds, on the most rugged rocks, a flat nest about five feet square, where it rears the young, whose operations it also directs during their adolescence. Its eggs are of a brown red, with blackish stripes. It is particularly fond of hares, which form its principal food. It also preys on various birds, and even on lambs. The male eagle never hunts alone, except when the female cannot quit the eggs or young. At other seasons they always hunt together; and some mountaineers pretend that one beats the bushes, while the other remains in some elevated place to stop the prey on its passage. According to Marco Polo, the eagle is employed in Tartary to hunt hares, and even wolves and foxes, but this probably applies to the great eagle: the common eagle was of no use in falconry. Spallanzani has observed, in relation to this bird, that when it swallows pieces of meat, two streams of fluid spring from the apertures of its nostrils, run down the upper part of the beak, and uniting at its point, enter it and mix with the food.

The *Martial Eagle*, sometimes called the *griffard*, is a large species discovered in Africa by Le Vaillant. It inhabits the country of the great Namaquois, between the twenty-eighth degree of south latitude and the tropic, and probably exists in other parts of Africa. When perched, it emits sharp and piercing cries, mixed with hoarse and lugubrious tones, which are heard at a great distance. It flies, with the legs pendant, and, like the common eagle, rises so high that it is lost sight of, though its cry is still audible. Highly courageous, it never suffers any great bird of rapine to approach within its domain. It hunts gazelles and hares.

The griffards, like the other eagles, are usually observed in couples, but during the hatching time the male alone provides for the subsistence of the family. The nest is formed between precipitous rocks, or on the summits of lofty trees. Its basis is constituted like that of the other eagles' nests, but it is covered with a large quantity of small wood, moss, and roots, which give it a thickness of about two feet. This bed is again covered with small bits of dry wood, on which the female lays two eggs almost round, entirely white, and more than three inches in diameter.

We have engraved a figure of an eagle exhibited for some time in Mr. Cross's valuable and extensive collection at Exeter Change, said to be from Africa. It seems intermediate between the eagles properly so called, and the Morphni, or eagle hawks of Cuvier. We cannot satisfactorily refer it to either of the known species, and have adopted the name given to it by Mr. Cross.

The Wedge-tailed Eagle, *A. Fuscosa*, is so named on the foot of the stand in the Museum at Paris. Its size is about that of the Golden Eagle, and its principal character is in the shape of the tail.

We now come to the section of the Fisher Eagles.

The *Osprey*, or *Ossifrage*, is so named, because fragments of bones of considerable magnitude have been found in its stomach. It is found in the different countries of Europe and North America. Though it appears generally to prefer cold and even frozen regions, such as Russia, Siberia, and Kamtschatka, Poiret has seen it in Barbary. From its usual habitat on the sea-shore, on the banks of great rivers and lakes, over which it is continually hovering, it has received the denomination of the great sea-eagle. Fish is the principal article of its subsistence, which it seizes by darting on it when it is on a level with the water, and sometimes even by plunging after it. It also preys on sea-birds, young seals, hares, and even lambs. It hunts and fishes both by night and day, having the double advantage of seeing better in daylight than the nocturnal birds, and by night than the diurnal. The morning and evening, however, are the principal times which it devotes to this exercise. Its flight is neither as elevated nor as rapid as that of the great eagle, and not being so long-sighted, it does not pursue its prey so far.

C. Hamilton Smith Esq. del.

Exeter Change.

IMPERIAL EAGLE OF AFRICA.

MORPHNUS?

London Published by G. B. Whittaker March 1828.

C. Hamilton Smith Esq.^e del.^t

Mus. Paris.

WEDGETAILED EAGLE.

F. FUSCOSA.

London. Published by G. B. Whittaker Nov. 1, 1827.

The osprey builds its nest in the rocks which border the sea-coast, or in very lofty oaks. It lays two round and very heavy eggs of a dirty white. It nurses its young with the greatest affection; but as one of the eggs is often unfruitful, the species, though considerably extended, is not very numerous any where.

The Pygargus, which is now ascertained to be the same species as the osprey, though formerly separated, is found in the northern parts of both continents. Pallas beheld a prodigious quantity of them in the mountains of the Volga. This bird frequents the sea-coasts, and lives on fish, young seals, ducks, &c., and the carcasses of animals cast on shore by the waves. To make itself master of the diving-birds, it perches on the point of the rocks, and, judging from the agitation of the water of the place where the bird will reappear, it seizes it at the very instant of its rising to the surface. When it has possessed itself of a prey too heavy to be raised out of the water, it drags it to the shore, flying backwards; but when its talons have entered the body of some large seal, and it cannot disengage them, it is drawn into the water by the animal, and is heard to utter the most piercing cries. Aristotle says, that this bird also preys on fawns, deer, and roe-bucks. It has been observed that the pygargi which frequent inhabited places, hunt only for some hours in the middle of the day, and rest in the morning, evening, and night.

This bird builds its nest in rocks, and composes it of small branches arranged in a circular form: the interior is furnished with weeds, grass, moss, and feathers. Buffon informs us, after Willoughby, that this nest is also found on large trees, whose foliage constitutes its only shelter above. The female lays two whitish eggs of the form and size of goose eggs. Incubation takes place in April, and frequently but one young one is hatched. These birds feed their young by throwing pieces of flesh into the nest, which the latter quit as soon as they are able to fly, and accompany the parents to the chase.

The *Balbuzzard* is one of the most numerous of the accipitrine tribe, and is pretty generally spread through France, Germany, and most of the countries of Europe from north to south. It is also found in Barbary, Egypt, Louisiana, and even in the island of Pins in the South Sea. The balbuzzards of the reeds in Carolina and Cayenne, appear to be only varieties of the same species, which equally inhabits Pennsylvania, and is sometimes called *piravera*.

The places which the balbuzzard prefers to frequent, are not the shores of the sea, but low lands bordering on ponds and rivers, from which habit it might be termed the fresh-water eagle. Perched on a lofty tree, or hovering at a considerable elevation in the air, it watches the fish from afar, descends upon it with the rapidity of lightning, seizes it at the moment it appears on the surface of the water, or even plunges in completely after it, and carries it off in its talons. But this prey, the weight of which renders the flight of the bird slow and laborious, does not always remain the portion of the balbuzzard. On the banks of the Ohio, where it goes to fish, when the *perca ocellata* quits the ocean to enter the river, dwells also the formidable pygargus. When he sees the balbuzzard arrived to the height of his eyrie, he quits his own, pursues him closely, until the fisher, convinced of his inferiority, abandons the prey; then this fierce antagonist with folded wings shoots down like an arrow, and with the most inconceivable address, seizes the fish again before it reaches the river. The right of the strongest is the sovereign arbiter of small and great events, and governs throughout the universe with resistless sway, in the air, on the earth, and under the waters.

But as a corsair, whose booty has been taken by an enemy in sight of port, undertakes a new expedition in the hope of being more fortunate, so the balbuzzard recommences his operations, and possessed of a fresh prey, he usually succeeds, if it be not too heavy, in escaping with it from his redoubtable foe. These scenes continually occur as long as the fish above-mentioned remains in the river. When it returns to the ocean, the pygargus retires to his mountains, to pursue game, and the balbuzzard betakes himself to the sea-shore, where he is no longer obliged to pay tribute for his plunder*

The balbuzzard builds its nest on the lofty trees of thick forests, or in the crevices of rocks. According to Lewin, it is also constructed on the ground in the midst of reeds. Two or three white eggs are generally laid, sometimes four, and spotted with red.

These birds are almost always in pairs; but when the waters are frozen, they separate in search of milder climates and a more facile subsistence; they are usually very fat, and the flesh savours strongly of fish. It is said, that they might easily be trained for fishing as other birds are for hunting, and it appears not improbable.

In Siberia, where they are very common, an opinion prevails that they carry a mortal poison in their talons, and the super stitious inhabitants are dreadfully afraid of a single scratch.

The *Great Harpy* is a bird which has been described under various synonymes, in consequence of the variations which result from age and sex, in its magnitude and plumage. It is found in Brazil, New Granada, and Guyana, where it particularly inhabits the forests of the interior. It is also found in other countries of America, and is peculiar to that continent. It is said to be the most robust and powerful of the feathered

* A still more extraordinary circumstance is related of the pygargus, by M. de Buch, in his travels in Norway and Lapland; and notwithstanding the respectable authority on which it rests, we can scarcely credit it. The pygargi of the isles of the interior sea, known under the name of *Loffoden*, not being able to attack the oxen with open force, have recourse to this stratagem. The bird plunges into the waves, and coming out all wet, rolls himself upon the shore until his plumage is all covered with sand, he then hovers over his victim, shaking the sand into his eyes, and striking him at the same time with his beak and wings. The ox blinded, and rendered desperate, runs here and there, to avoid an enemy who attacks him on all sides, and he falls at last, exhausted with fatigue, or precipitates himself from the summit of a rock. The eagle then drops upon him, and devours his prey in tranquillity.

race. If the stories told of it be true, the benefits of nature seem, in this way, to be pretty equally distributed to both worlds. While the old can boast of the most terrible of quadrupeds, the fiercest and strongest of birds has fallen to the inheritance of the new. Travellers have assured Mauduyt, that the harpy makes its usual prey on the aï and the unau, and that it often carries off fawns and other young quadrupeds. It also attacks the aras, and the larger parrots.

It does not appear very clearly, why this eagle should come under the section of the fisher-eagles, a denomination to which, in many cases, we must not attach much importance, and which is generally applied to those eagles whose thick and short tarsi are altogether or in part naked. The places inhabited by the harpy, and all we know concerning its mode of life, is confirmatory of this observation. Sonnini is persuaded that this bird does not fish, and describes, under the appellation of the great eagle of Guiana, an individual whose size exceeds the usual magnitude of the harpy or destructive eagle. There is every probability of the identity of species in this case, and the individual in question may be the female of the harpy, on the sexual differences of which no well-authenticated observations seem hitherto to have been made. Sonnini has measured and described the individual which he killed, and the only material difference between it and the destructor consists in relative size. It also frequents the hot and humid countries of America. But we cannot expect for a very long time to gain any precise notions respecting a bird whose solitary abode, in the depth of almost impenetrable forests, is so far removed from the habitations of man.

It is not our object to spin out our observations by extending them to all the species, or even by dwelling much on several of the subdivisions of this order. Where nothing interesting in structure or habits is known concerning them, we shall pass them over in silence here. The text, with its additions, it is hoped, will amply answer the purposes of those who delight to

BRAZILIAN KITE, *Var? Male?*

PANDION CARACARA?

London. Published by G. B. Whittaker, Nov. 1. 1827.

GREAT HARPY OF AMERICA.

F. HARPYA Lath F. DESTRUCTOR Daud.

London. Published by G. B. Whittaker. Nov 1 1827

unravel the tangled web of synonymy, and to dwell on the description of external characters. In this part of our work, it behoves us to generalize our views as much as possible, and to reject everything which has no bearing on the philosophy of the subject. In our former supplementary parts we have certainly entered more into the kind of details to which we now allude; but they were better authenticated, and more important in themselves than most of the same sort that can be offered in the department of Ornithology.

In fact, the conflicting accounts of naturalists in this department of Zoology are almost beyond belief. What with errors of many and the corrections of more, they have made " confusion worse confounded." An immensity of labour and research is still requisite to rectify the very defective nomenclature of the eagles and of the birds of prey in general. How, indeed, considering the different appearances according to age and sex, can we presume to pronounce affirmatively on foreign species, when it is recollected how long a period elapsed before the identity of the osprey and pygargus was ascertained, birds constantly found in Europe? A complete and judicious monograph of these birds would be of the highest utility to the science, but it would require a continued series of observations for many years, a thing impossible with regard to beings which live at such a distance from our dwellings, and whose spoils exhibit only variable signs, more calculated for the multiplication than the detection of errors. To form an idea of the extreme difficulty of such a task, it is sufficient to consult the *Observationes Zoologicæ* of the profound Hermann, who, notwithstanding his very careful and painful description of numerous individuals, has left us little but his own personal uncertainties and doubts upon the subject.

The figure is from a specimen in the Museum at Edinburgh. It seems likely to be the male of Daudin's Falco destructor.

The figure of the Brazilian Kite, *Pandion Caracara?* appears to be the Caracara of Jacquin. The specimen was shot at Curaçoa, and was drawn by Major Hamilton Smith before its death; it appeared to be a male bird. The female is larger, and less elegantly marked.

Prince Maximilian's Crested Hawk, *Falco?* is from a drawing also by the Major of a beautiful specimen in the valuable collection of Prince Maximilian, belonging to the tribe of crested short-winged birds of prey. It is about the size of a Goshawk.

The Urubitinga is from the same collection. The specimen differs from the Baron's short description of this species in the intensity of the colour, which is a dark brown.

We shall now take a rapid survey of the HAWKS, KITES, and BUZZARDS. There are two sections of the HAWKS. The HAWKS proper and the GOSHAWKS. The denomination of *accipiter* which has been applied to the whole order of raptorial birds, is the original Latin term for a hawk. But in consequence of this application of it, naturalists have reserved the term *nisus* for the hawks, and *astur* for the goshawks, whose habits are similar, and whose external differences are but trifling. M. Savigny has formed a new genus comprehending the hawks and goshawks, to which he has given the name of *Dædalion*. And M. Vieillot has called these birds *Sparvius*.

The generic characters of this subdivision of Accipitres we shall briefly recapitulate, because from their structural importance they should be impressed on the mind of the student. The characters common to both sections are, a beak greatly inclined from the base, and compressed laterally; the upper mandible greatly crooked, with a very marked tooth; the lower shorter, and obtuse; the cere smooth: the nostrils a little oval; the commissure, or division of the mouth, extending as far as below the eyes; the tongue oblong, thick, and sloped; the tarsi reticulated, principally on the sides, with a rank of lozenges in front; the four toes long, but considerably exceeded by the intermediate one; the talons crooked and acerated; that of the

THE URUBITINGA _ *male*.

F. URUBITINGA.

C. Hamilton Smith Esq. del.

Mus. Pr. Max. of Neuwied.

London Published by G.B. Whittaker Nov. 1.1827.

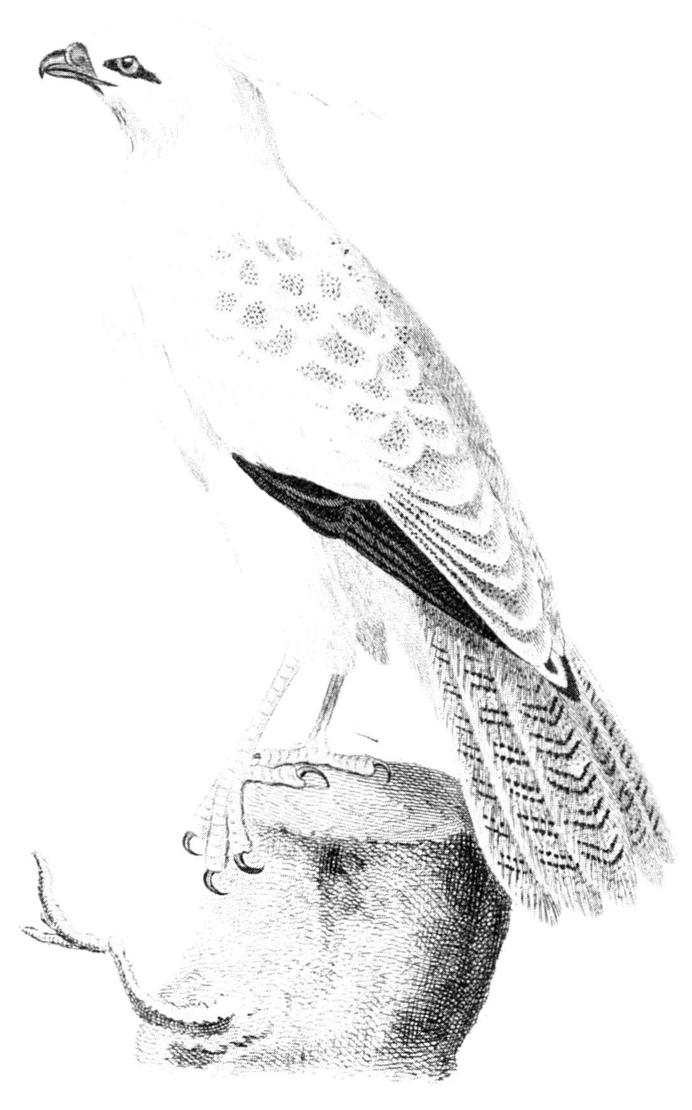

PRINCE MAXMILIAN'S CRESTED HAWK,

FALCO ?

C. Hamilton Smith Esq. del.^t

Mus. Pr. Max. Neuwied

London Published by G. B. Whittaker, Nov. 1, 1827.

lower toe the longest of all; the first remex the shortest, the fourth the longest; the wings scarcely reaching half the length of the tail, which is rounded.

The differences between the two sections consist in the respective proportions of the beak, tarsi, and toes. The hawks have the beak shorter, the tarsi more elongated and slender than the goshawks; they also have the last phalanx of the intermediate toe passing the talons of the lateral toes; the tongue is also more sloped than that of the goshawks, and the latter in general are stronger built, and of a less elegant shape.

The hawks and goshawks have also, in their plumage, a character which distinguishes them from other birds of prey; when adult and past the second moulting, they have transverse stripes on the lower parts of the body, where, previously to this age, there were longitudinal bands.

With respect to the natural habits of these birds, the conformation of their wings does not permit them to fly so high, nor so long, as some of the other Accipitres, which have longer wings; and it obliges them to employ stratagem in the procuring of their prey, while the other raptores fall upon it almost perpendicularly. Their flight is low and horizontal, and they dart sideways on the birds which pass within their reach. When obliged to repose, they fix in the midst of tufted trees, from which they watch partridges, fringillæ, field-mice, and other small mammifera. This mode of hunting naturally removes them from the open fields which are bare of trees. The goshawks, being stronger, attack hens and pigeons. All these birds deplume their feathered prey, and tear it in pieces before they eat it. But they swallow the small mammifera entire, the skin of which, rolled up, is rejected by the mouth. It is only during summer, and the back season, that hawks are seen dispersed in the fields, where they are frequently observed alone, though the two sexes are usually at no great distance from each other; but the male and female, to avoid interfering with each other, are generally perched on separate trees. Sometimes, however, entire families are met with, hunting together. But such assemblages only take place during the early age of the brood, while the parents are instructing them in the exercise necessary to procure subsistence: a proof that there is a natural education among animals, and that all is not instinct.

During a considerable portion of the year, the hawks and goshawks remain in the forests, where they build in the largest trees a nest, in which the female lays usually four or five eggs. Lewin says, that these nests are sometimes constructed in ancient ruins, or on rugged rocks.

The *Common Hawk* is found in almost all parts of the world. Kæmpfer has seen it in Japan, and M. Poiret in Barbary. In Egypt it comes into the towns, and is a sacred bird. Mauduit has found it at Cayenne, and D'Azara in Paraguay. Its usual food consists of moles, mice, thrushes, larks, quails, and other small birds. It also eats lizards and snails. Though many hawks remain constantly in Europe, others traverse the seas to pass the winter in milder climates. The mariners of the Mediterranean call them corsairs, as, during their voyages, they prey on all the weaker species they can find. Notwithstanding their boldness and intrepidity, they are easily, as we have already seen, rendered docile for the purposes of falconry, and were employed in hunting thrushes, quails, and partridges. They are termed *royal* when they have undergone their training. The voracity of these birds renders them easy to be caught, and they are taken in such snares as are usually set for sparrows. Belon was witness to the catching of these birds near the Strait of the Propontis. A fowler concealed behind a bush took a dozen of them per hour, without any other artifice than causing small birds to flutter about, attached to a cord under suspended nets, into which the imprudent hawk precipitated itself impetuously.

We insert here another figure from the magnificent collection of Prince Maximilian. It is nearly allied to the *Acoli* of

BRAZILIAN LONG-LEGGED HAWK *of Prince Maxmilian.*

F. WIEDII BRAZILIENSIS.

C. Hamilton Smith Esq. del.

Mus. Pr. Maxmilian of Nieuwied.

London Pub. by G.B. Whittaker Nov. 1, 1827.

DELAFON'S HAWK,

ACCIPITER DELAFONSII.

C. Hamilton Smith Esq. del.

Mus. Plymouth

London. Published by G.B. Whittaker. Nov. 1 1827.

Le Vaillant, the long-legged Falcon of Shaw. It has the bulk of a pigeon, but is larger. The female is barred only from the abdomen downward.

Our Accipiter Delafonsii was caught by the crew of Captain Delafons' ship in the strait between Bileton and Borneo. Major Hamilton Smith has dedicated the species, as new, to his friend Captain Delafons. It seems allied to *Accipiter torquatus*. It is thirteen inches long, bill black, dentated cere small, of a dirty white colour, head one inch long; a streak of mottled white passes from the forehead to the nape; cheeks ashy ochre, throat white, neck long, and slender behind, darkish sepia in front, and down the throat white mottled with regular oblique rows of rufous drop-like spots gradually assuming the form of bars toward the abdomen; vent white; thigh feathers long, loose, white, crossed with rufous streaks; the mantle white, with broad sepia bars nearly concealing the white; wings short, first primary very short, fourth the longest, all entirely sepia, paler at the edge, inside of the wings whitish-gray transversely barred with ashy sepia. In the tail twelve feathers equal ashy, with three dark broad bars above, pale ash beneath, with six narrow bars, the last being the broadest; legs yellow, faintly aculeated; claws black.

The *Pigeon-Hawk* of America, is, as it name imports, remarkable for the continual war which he wages with pigeons and doves, and for nothing else. This bird is found near Hudson's Bay, and through all North America.

The *Goshawk* is larger and stronger, as we have said, than the common hawk. It remains all the year in France, and is also common in Germany, Russia, Switzerland, and Ireland, but more rare in England and Holland. It is also found in Asia from Kamschatka to Persia, in Africa, in Barbary, and in North America. It inhabits, by preference, forests of fir-trees, and those which are situated on the mountains. Young pigeons, and other small fowl, leverets, squirrels, mice, and moles constitute its principal food. This bird, whose cry is hoarse and frequent, builds its nest in the largest trees, and the female lays four or five eggs, a bluish white, with brown stripes and spots.

The goshawk is often taken with cloths which are used for taking larks, or sometimes, by placing in a space surrounded by four nets, a white pigeon, on which the goshawk precipitates himself. Very frequently he does not attempt to disengage himself until he has devoured his prey. Falconers, according to Belon, prefer, for the purposes of training, the goshawks which are brought from Greece, which are not so indocile as the individuals procured in the Alps and Apennines. The goshawks being birds of low flight are employed in the chase of partridges, pheasants, ducks, wild geese, hares, and rabbits. The principal care of the trainers, independently of what we have stated under the head of falconry, is to feed the young goshawks by the hand with the flesh of fowls, to accustom them to the noise of horses, to expose them every morning to the sun, to make them hunt only when the heat is not too strong, sheltered from the wind, and giving them time to watch the partridges and intermit their own pursuit on the wing. They must not be kept too long without making them fly, and those which hunt lowest are the best. When the trainers wish to teach them to hunt wild ducks, they commence with tame ones. Then they take them to some pond or river where the wild ducks are found, and the moment the latter take wing, the goshawk darts upon them, and seizes the most lazy. For rabbits, after the bird has been accustomed to see them, they take him morning and evening through some warren, and he shoots equally on all he sees.

Particular care must be observed, in the education of the young goshawks, not to make them too well acquainted with hens and pigeons; for this being an easy chase, they would speedily destroy all the poultry-yards and dove-cotes in the neighbourhood.

But few birds appear really to belong to the genus of the

goshawk. Those which have been attached to it are noticed in the text and additions, and we have nothing interesting to add upon them here.

The generic characters of the KITES are, a beak inclined from its base but feebly, and forming a hook only in the middle. The back of it is contracted and angular; the cere smooth and convex; the edges of the upper mandible are dilated, and the lower is straight, obtuse, and shorter than the upper; the nostrils are elliptical, situated obliquely, and marked with a fold at the anterior edge. The tongue is oblong, fleshy and rounded below, and its point is entire and thick. The tarsi are short and slender, and have the upper part covered with feathers. The toes are short, the exterior of the three front ones united by a membrane, the intermediate little exceeding the lateral; the claws moderate, and weakly acerated. The wings, very long, reach the extremity of the tail, which in one species belonging to Australasia is forked or wedged.

The *Common Kite* is extended through Europe, Asia, and Barbary. It is found in France in mountainous districts, and is equally common here, where it frequents marshes and fresh waters, and pursues ducks and other aquatic birds. It is also reported to attack hares and rabbits; field-mice, moles, rats, reptiles, and large insects constitute its ordinary food, and it will devour the dead fish which float upon the surface of the waters. It also approaches habitations to attack the young chickens, but if the hen perceives it in sufficient time, her cries and resistance are sufficient to drive it away. This bird shoots with rapidity from an elevated station in the air, and it hovers so lightly that the motion of its wings is not perceptible. By means of its piercing sight, it soon discovers its quarry, and stoops upon it as if it were only sliding on an inclined plane.

The kite is considered as the emblem of cowardice. It is as voracious as the crow, and yet will suffer itself to be pursued by the latter, and will fly before birds of prey of a much smaller size than itself. This species was formerly called the Royal Kite, because it contributed to the amusement of princes who were wont to send the hawk to attack and vanquish it.

Buffon, though usually so judicious an observer, has drawn, with exaggerated severity, a picture of the cowardice of this bird. Mauduit regards its qualities and defects more with the eye of a philosopher. Though the beak of this bird may not be much inferior in form or dimensions to that of some of the more courageous raptores, yet the weakness of its talons will account for its excessive pusillanimity. These form in fact the principal weapons of the hunting birds ; with these they strike, arrest, seize, carry off, and retain their prey. It is by the form of the talons that we must judge of the extent of capability in birds of this class, and it is because he is badly armed that the kite is cowardly. He flies before the hawk, because his talons are short and of little flexibility, while the latter can reach him from a distance with a supple weapon which imparts facility to all his movements.

The nest of the Kite is usually situated in the hollows of rocks, or on large and ancient trees of the forest tumbling into decay. It is very ample, but is artificially constructed with small branches interlaced with dry grass and herbs. Two eggs are generally laid ; sometimes three, and even four, according to M. Temminck. They are white, with some spots of yellowish red.

The *Black Kite*, of the text, is the *falco parasiticus* of Shaw, and is described by Le Vaillant. It is common in South Africa, and is named, at the Cape, Kuyken-dief, which literally means chicken-thief. There is scarcely a habitation where it does not pay a visit at certain hours of the day, and, bolder than our kite, the sight of man will not prevent it from darting on the young domestic fowl. Even shots did not prevent these kites from returning to the waggons where M. Le Vaillant was preparing his repast, to carry off some pieces of meat.

C. Hamilton Smith Esq.[e] del.[t]

Mus. Philadelphia

MISSISSIPPI KITE.

F. MISSISSIPENSIS.

London Published by G.B. Whittaker, Nov. 1. 1827.

These birds will plunge down into rivers to carry off the fish, and also hunt all kinds of small game. They fight with crows for pieces of carrion, and force them to let them go. They frequent marshy grounds in preference, and build the nest on some bush in the midst of reeds. They also build in rocks and trees like the common kite. The eggs, four in number, have red spots.

It appears not improbable that this kite is but a variety of the common, and also that the Etolian kite of Savigny, *falco Egyptius*, Gm., is the same.

The *Blac* of M. Le Vaillant is the *couhieh* of the Arabs, and is found in Barbary, Egypt, and Africa generally. It is usually on the top of trees or the most elevated bushes: it continually sends forth piercing cries, both when perched and flying. It does not attack small birds, and pursues the shrikes and crows only for the purpose of driving them away from its habitat. Though daring and intrepid, its usual food is grass-hoppers, and some other insects, from which it is thought to derive a certain odour of musk with which its body and excrements are impregnated. As it is exceedingly savage, one cannot easily approach it. It builds a tolerably spacious nest in the forks of trees, which is furnished within with feathers and moss, where the female lays four or five white eggs.

Our figure of the Mississipi Kite was drawn by Major Hamilton Smith at Philadelphia. It is the same specimen as is figured by Wilson, the Carolina Kite of the text.

A species called *Yetapa* is placed among the kites, and described by d'Azara. He calls it *faucon à queue en ciseaux*, for in hovering it opens and closes its tail like a pair of scissars. It is about twenty-one inches long. The upper part all white, with the exception of the anterior portion of the back, which is black. Wings partly black and partly white; cere and tarsi blue.

This bird arrives in Paraguay in spring in flocks of from ten to twenty individuals. Its flight is usually circular; and when descending near the earth, it sees any one approach, it rises promptly, tracing spiral figures in its flight, and is soon out of reach. Grasshoppers appear to be its only nourishment.

The birds known under the denomination of Buzzards have all the general characters of the accipitres. Like the falcons their wings are almost as long as, and even in some instances exceed the tail; but they differ from these, in having the first quill-feather very short, and the third or fourth the longest.

There is also a secondary character in the buzzards, consisting in the relative length of the tarsi. The true buzzards have them thick and short,—in those called *busards* by Cuvier, and a division which may be called sub-buzzards, the tarsi are long and slender. The first also have the head broader, the neck shorter, and the body more clumsy than the second, whose shape is generally finer and more elegant. In all, the female is larger than the male.

The buzzards, as well as the kites, are in general regarded as cowardly birds, and are also considered as the emblem of folly. But this notion, though apparently justified by facts, seems a little exaggerated. Nature, to preserve the species, has given to each being the consciousness of its strength and resources, and we are always exposed to the danger of false judgments, when we decide on results without carefully investigating causes. We have above made some observations on this subject relatively to the kites, and shown that the weakness of the talons is the principal reason of cowardice in those birds. The buzzard, though better organized in this respect, still appears to be equally devoid of courage; but its sight is so extremely delicate, that open day-light dazzles it, and this circumstance naturally explains its habits, which could not be different without ceasing to be in accordance with its organization. If, then, the buzzard prefers ambush to open war, and has the patience to wait for entire hours for his prey among the branches, on which he pounces in its passage, it is because his defective sight will not permit him to pursue it in the upper regions. That sort of tranquil indifference with which this bird will suffer itself to be approached, M. Dumont declares *not* to proceed from the want of perception of approaching danger. But that gentleman has not thought proper to inform us what it does proceed from. If it does not arise from the organic deficiency just noticed, or the obtuseness of some other organ, or the absence of general sensibility, we have no idea to what cause it can be assigned. Certain we are, from the ordinary indications of character in the buzzard, that it does not proceed from an intrepidity of disposition, which M. Dumont has antithetically denominated *tranquil audacity*. However, as to the question of cowardice we perfectly agree with M. Dumont, who well observes, that there can be no true cowardice except in individuals, who, provided by nature with offensive or defensive weapons, have not the courage to employ them. We may also add, that the employment of them is not always a proof of true courage. This quality can only be exhibited against an adversary, equal or superior in strength, a sort of courage comparatively rare among brutes. Their courage is for the most part, if not proportioned to their actual quantum of strength, at least determined by their resources for attack and resistance. With the exception of the demonstrations of maternal instinct, and the cases of the horse and dog, especially the latter, we find but few examples of true courage among the lower animals. It is only in cultivated man that this virtue is to be found in perfection, for the courage of savages and barbarians approximates very closely to that of brutes. True courage consists neither in insensibility to danger from nature or custom, nor in the confidence inspired by strength, activity, or skill; but in a habit of the mind, induced by intellectual discipline, which bears its possessor calmly through scenes of peril and death, conscious of his risk, and conscious of his weakness.

We cannot be surprised to learn that falconers have attempted, without success, to teach these birds an art for which nature has so totally unfitted them. As the weakness of their eyes approximate them to the nocturnal accipitres, we also find in them that air of stupidity, and other similar effects, always produced by short sight.

The buzzards proper, usually establish their abode in cultivated grounds, and in the neighbourhood of habitations, where they feed on fowl, small game of all kinds, moles, mice, and other small mammalia, and even on insects. The sub-buzzards have similar habits. The *busards* (as they are called by Cuvier) are wilder, and prefer the neighbourhood of marshes and watery grounds, where they feed on aquatic birds, fish, reptiles, &c.

The *Common Buzzard (Buteo Vulgaris*, Lacep., and *Falco Buteo*, Linn.*)* was called by the Greeks *triorches*, from an erroneous opinion that it had three testicles. It is a little more bulky than the royal kite. The plumage of this bird is so subject to variations, to so great an extent both in intensity of shades and proportions of white in the different parts, that it would be impossible to give a description that would agree with all or the majority of individuals.

This species is very much extended in Europe: it has been seen in Barbary, and probably exists in other countries of Africa. Quails, partridges, leverets, rabbits, are in summer its most usual prey, and in the same season it plunders the nests of other birds. When food of this description is wanting, moles, field-mice, frogs, grasshoppers, and other insects, supply its place. In this way it renders some service to agriculture, and young buzzards, when tamed, may be employed in the destruction of worms and hurtful insects in gardens; but they will also destroy the small birds, many of which do no mischief, and serve by their presence and song to embellish such places.

The buzzard often hovers heavily over small coppices to discover the minor game. In the fields it fixes by preference on a tree or bush, or a clump of earth, to watch its prey, and dart instantly upon it when within reach. It constructs its eyrie on some elevated tree, and composes it of small branches, and lines it within with wool, or other soft materials. It often takes possession of the nest of a crow, which it enlarges. The female lays two or three whitish eggs, with yellow spots. It nurses the young for a longer time than the other accipitres. According to Ray, if the mother be killed, the male will continue his attentions until the young ones can dispense with them. When the latter have first taken their flight, they are heard perpetually to send forth sharp and plaintive cries.

We shall now speak of the *Honey Buzzard*. Though said to have been very common in France in the time of Belon, it is now rare enough in the different countries of Europe. It is usually found in plains on the trees and bushes: its flight is low and of short duration. It is said, without the assistance of its wings, to be able to run as fast as a cock. Its principal food consists of lizards and field mice, frogs, and insects. Its nest, composed of interlaced twigs, is closely covered within, with wood or other analogous materials. It usually lays but two eggs, which, according to Buffon, are ash-coloured, and marked with small brown spots. In Lewin's figure they are of a rust colour, with deeper spots of the same hue. It feeds its young with the chrysalides of insects, and especially those of wasps, from which it derives its specific appellation. This bird is very fat in winter, and good for eating, on which account snares are set for it.

There are a great number of other species described by naturalists as appertaining to this group. But not to mention that many of them do not appear to be well authenticated, there is nothing in the habits of any of them very different from what we have already detailed, or at all likely to amuse or instruct the reader. We shall, therefore, now proceed to that singular genus,

The Secretary. The single species which comprises this genus, is ranged by Gmelin in his falco, and by Latham, in his synopsis, among the vultures. Illiger isolated it under the name of *gypogeranus*, and the Baron under the name which heads this division, at the end of the diurnal birds of prey, founding this distinction on the legs entirely covered with feathers, the crooked and divided beak, projecting brows, and all the other details of its anatomy. Dr. Latham, in his last edition, has separated this bird, and placed it as a distinct genus at the end of the vultures; but M. Vieillot, in imitation of some other naturalists, has classed it with the grallæ, in consequence of its very long tarsi. We shall here extend a little the description of the text, from the important character of this species.

This bird, found at the Cape of Good Hope, is remarkable for very long legs, which seem to approximate it to the crane; for its robust beak, equal to that of a bird of prey; for its brows formed by a single rank of black hairs, placed very closely, and almost fifteen or sixteen lines in length; for its tuft, composed of a double rank of long feathers, hard, narrow at their origin, situated towards the base of the occiput; for its wings, armed with three osseous and rounded prominences; for the size of the mouth, whose commissure extends up to the eyes; for the skin of the neck, susceptible of very great extension; for the great amplitude of the crop; and, in fine, for the short and thick toes, armed with crooked and almost blunted talons. The assemblage of all these attributes constitutes a mixed, extraordinary being, not to be classed in any known group.

As M. Le Vaillant has seen several of these birds alive, we shall borrow our specific description from him. The secretary is rather more than three feet in height. The naked skin, surrounding the beak, is not red, as Buffon thought, but yellow, more or less partaking of orange; the bird can erect, at will, the sort of tuft which hangs like a mane on the back of the neck. The tail is much wedged; the two middle quills are double the length of the two following, and drag along the ground when held at all obliquely.

The male, in its full maturity, has the head, neck, chest, and entire mantle of a bluish gray; the wing coverts are also of the same colour, more or less shaded with a red brown; the quills are black; the throat and chest are shaded with white, and the lower coverts of the tail are of a very clear reddish; the abdomen is black, mingled, and as it were radiated, with red and white. A fine black, almost imperceptibly streaked with brown, is the prevailing colour of the limbs. The pen-feathers of the tail are partly black; they grow more gray, in proportion to their elongation, and are terminated by brown.

The female differs little from the male, except in having less deep colours in the tuft, which is not so long, and more mixed with gray, and in the two middle quills of the tail being shorter.

The osseous prominences of the wings are not observed in the young, nor indeed very apparent in the adult, except on examination. They are, in fact, nothing but the apophyses of the metacarpus.

In the season of reproduction, there are long and obstinate combats among the males. They strike with their wings, and the female always falls to the lot of the conqueror. These birds construct a nest, flat like the eagles', and lined internally with wool and feathers. They place it in the highest and thickest bush they can find, and sometimes even on large trees. The same nest serves for a long time for the same couple, who dwell alone, like the eagles, in a very extensive domain. The female lays two or three eggs, white, with reddish points, formed much like a goose-egg, but a little more elongated. The young are a long while before they take to flight. They cannot even run at the age of four or five months; but when full grown they run remarkably fast, and when pursued, prefer this motion to flying, and take uncommonly long steps. When nothing frightens them, their step is slow and grave. They are distrustful and cunning, and approached with considerable difficulty. The male and female seldom separate. They are found in all the dry plains in the neighbourhood of the Cape, particularly in Swartland. They are also very frequently seen on the eastern side, in the country of the Caffres, and in the interior. They are more rare on the western side, and especially in the country of the Namaquois.

When the secretary, says Quarhoënt, meets or discovers a serpent, he attacks it at first with his wings to tire it out. He then seizes it by the tail, raises it to a great height in the air, and then lets it fall. This operation is repeated, until the serpent is dead.

When the secretary is disturbed, he makes a hoarse kind of croaking sound. His natural disposition is mild. It is neither mischievous nor dangerous. The observer, above cited, has seen these birds living peaceably in a poultry-yard in the midst of the fowl. They were fed with meat, and were very greedy of intestines, which they kept under their feet in eating as they would have done a serpent. These birds, though armed like the carnivora, have nothing of their ferocity. They employ the beak, neither as an offensive nor defensive weapon. They fly instead of attacking, avoid approach, and, to escape even a feeble enemy, will make leaps of eight or nine feet in height. The secretary, when taken young, is easily tamed, soon grows accustomed to live with poultry, and will never do any harm unless suffered to fast too long. But if he suffers from hunger, he will make free with the chickens and young ducks. Naturally gentle and sportive, this singular bird seems to love peace: for if he sees a combat take place among the fowl, he will run to separate them. The inhabitants of the Cape, accordingly, rear these birds for the purpose of preserving peace in their farm-yards, and to destroy lizards, serpents, and rats, which often come to devour the eggs and fowls.

This African bird easily accommodates itself to our climates, and has been kept in menageries both here and in Holland. When desirous of repose and sleep, it lies on the ground, on its chest and belly. To eat at its ease, it gathers itself up on its talons, and thus, in a bended position, swallows its food. It kills its prey by striking it violently with the foot. It prefers living to dead animals, which distinguishes it from the vulture tribe, and flesh to fish, which characterizes it from water-birds. It will also eat small tortoises, which it swallows entire, after having broken the cranium. It destroys a great quantity of grass-hoppers and other insects. It has a cry analogous to that of the eagle, and usually walks with very long and wide steps, and for a long time without slackening its pace or stopping. From this it it probably derived the name of *messenger*. That of *Secretary* is given it from the tuft of feathers behind the head, bearing some fancied resemblance to a pen stuck behind a man's ear.

The endless aberrations of nature from given types; the unwillingness she seems to exhibit to be shackled by general universal rules; the excursive propensities, as it were, of her creative power, which defy the faculty of the zoological systematist, are equally observable, whether we regard her works in the mass or examine them in detail, whether we contemplate a class, a genus, or a subordinate group.

Thus, although we find, that the light and heat of the sun are agents of a most influential character in the developement of life in both the animal and vegetable kingdoms; although the rule is most extensively prevalent, that the day shall be the period for activity, and the display of all the ulterior objects of life, and the night for resuscitation and repose; yet this rule is by no means universal. A few beings are destined to an active existence only, while all other creatures sleep, and among these, in the present class, stand foremost the NOCTURNAL BIRDS OF PREY, THE OWLS.

The nocturnal habits of these birds, like, indeed, all the habits peculiar to any given animals, are decidedly predestinated by their physical characters. These habits are most evidently not the effect of accident, the caprice of the animal, or even of involuntary instinct, uncontroulled by physical causes. The owl is not made for the full light of day, and can live only, for all the active purposes of life, in partial darkness; the dusk of evening, or gray of the morning, is essential to the full exercise of her vision; the noonday sun, or even the presence of that luminary any-where above the horizon, dazzles and blinds her by the influx of too much light consequent on the unusual largeness of the disk of the eye-pupil: but this very circumstance, which is a source of so much inconvenience to the animal by day, is, in fact, an admirable contrivance for the perfection of vision during the comparative darkness of twilight or night. When the rays of light are diffused, and cannot find access in sufficient quantity to the ordinary pupils of diurnal animals, the capaciousness of those of the owl takes in enough for the perfect use of the eye: the shape of the pupil seems to be unimportant, but the capaciousness of its disk is certainly essential to nocturnal vision.

Although, however, the eyes of these birds will admit light enough for all purposes of vision during twilight, they will not enable them to see sufficiently during the darkness of night; and consequently, as they cannot see from redundancy of light during day, and from want of it during the greater part of many nights, they have very short space of time left them for procuring their food.

It is observable, from the quality of animal and vegetable food, that animals which feed on the former are capable of enduring abstinence much longer than these which subsist on the latter: if, therefore, this fact be considered in conjunction with the conditions of these birds just alluded to, we may fairly conclude, that if owls had been vegetable eaters, they would soon have all starved: for, without some special provision against such a consequence, the short spaces of time they could appropriate to procuring food would be insufficient to enable them to collect vegetable matter in sufficient quantity; but the owl, which is necessarily abstinent, is carnivorous:—so congruous are the works of nature.

The owl is enabled to make the most of the short time allowed for its predatory excursions, by the exposed situation of its prey, and by some other conditions of its own, which may deserve notice. Most of the small birds and quadrupeds pursued by the owl are the less able to guard themselves by flight, or concealment from the adversary, by the partial darkness, which, while it is advantageous to the owl, deprives them of the full advantages of sight. The quill feathers, moreover, of the owl are so light and downy, that it makes very little noise in flight, and gives, therefore, but little warning to its prey through the sense of hearing. With these advantages of its own, and disadvantages of its prey, therefore, the owl has little difficulty in redeeming its many hours of necessary inactivity; and the capacity of its throat, and undivided possession of its prey, consequent on its solitary habits, add still more to its facilities, and neutralize any apparent disadvantages incident to its condition in the pursuit of its food.

Some species of the owl are not so much nocturnal in their habits as others. The Great White owl, *S. nyctea*, and some others, will hunt occasionally by day; but they do so to considerable disadvantage, and the little birds may then be seen flying round about, though they will not venture to attack their too formidable adversary.

Ruined buildings and church towers are favourite places for nidification of these birds; a circumstance which, connected with their evening flight and melancholy cry, has doubtless assisted to inspire that ominous fear entertained so generally by the vulgar of these grotesque-looking birds. Some of the species, however, are found to build on tufts or grass, or even in little concavities on the bare earth, of their own making.

The popular notion that the owl is an harbinger of adverse fate is by no means confined to the superstitions of our own time or country. Virgil tells us that, on the death of Dido,—

> Solaque culminibus ferali carmine bubo
> Sæpe queri, et longas in fletum ducere voces.
> Multaque præterea vatum prædicta priorum
> Terribili monitu horrificant.

In Egypt, the fountain of European learning, as well as among the Roman fatalists, it was considered a bird of ill omen. In Greece, indeed, it was treated as emblematical of wisdom, and was therefore dedicated to Minerva. In America, New Holland, and in the islands of the Pacific, at the present day, it is both venerated and feared.

Among the double crested, or, as we must say in conformity with general usage, the eared owls, *The great-eared Owl (Strix Bubo)* of which there are probably some varieties, stands foremost. This species, which measures two feet or more from the extremity of the beak to that of the tail, is little inferior in size to the common eagle, but for its specific characters we must refer to the text. The tufts of feathers over the eyes, called the ears, are not always erect, and are prone, especially when the bird is unexcited.

It is most extensively located, being found generally, or occasionally, in most parts of the earth. In Europe, it is most common in Germany, in Russia, and the rock of Gibraltar; it is sometimes, though rarely, seen in England and Scotland, but has not been noticed in Ireland.

This species endures the light of day better than most of the others: it lives principally on the smaller rodentia, and even rabbits and hares; but, when pressed by hunger, will attack bats, snakes, and other reptiles and insects. Frisch, who kept some of these birds, states, that he sometimes gave them fish, and that they always broke the bones of fish and quadrupeds before swallowing them, which, together with the hair, were returned by the mouth in small pellets. These birds never drank; but, says M. Dumont, we are not, therefore, to conclude they never drink when in a state of freedom, for many of the the diurnal accipitres endeavour to conceal themselves when drinking.

The size of these birds does not hinder them from flying at a considerable height during twilight, when they are frequently attacked by a numerous body of crows, which they always beat off. They will even attack the buzzard, and sometimes carry off his prey. During day they fly very low.

Griffith sc

THE SUPERCILIOUS OWL,

S. GRISEATA.

London Published by G. B. Whittaker March 1828.

This species builds in caverns and the clefts of old walls; the nest is made of twigs of dry wood and pliant roots, and furnished within with leaves. It measures nearly three feet in diameter, though the bird lays but two or three grayish white eggs.

To pass here the several varieties of this species, we shall next notice the common long-eared or horned owl, *Strix otus*, which, as to its specific character seems to differ little or nothing from the *Strix bubo*, except as to size, this being considerably the smallest, and is about fourteen inches in length; the wings from tip to tip measure a little more than three feet. The tufts or ears are said generally to consist of six feathers; but Lewin and Dr. Latham have observed nine, and M. Temminck has mentioned ten. They are blackish brown, yellow on the edges, the eyes have the iris bright yellow, and are surrounded with a circle of whitish feathers, brown at the tips; the general plumage of the upper part of this species is brown, with different tints of rufous and whitish; the breast and belly are yellow with longitudinal brown spots, and transverse streaks of dark brown.

This species, which is rare in France, arrives in September and October in this country, and quits us again early in the spring, for the north. M. Temminck states, that she builds on the ground on some eminence, and in the marshes in the high grass. During the day, she remains concealed in the woods, which she quits in the evening to search for mice, small birds, and insects.

The Scops, or little-eared Owl, (*Strix scops*, Lin.) is varied all over with gray, reddish-brown, and black; lighter, as usual, underneath, but the tints of these colours vary considerably. The feet are feathered to the toes with rufous gray feathers, dotted with brown; the beak and toes are brown. The crests are composed of six or eight feathers, but Linnæus has erroneously stated, that they have each but one. This error of the great Zoologist is in all probability attributable to the bad state of the specimen under his observation, and, like all other errors of eminent men, has induced many more; for several have been named as distinct species with reference to the feathers of the crest, which seem to have no real pretensions to distinctive separation.

The Scops is extensively located, but seems rare every where. It has been said not to be British, but Dr. Latham denies that assertion. It builds on the branches of trees, and lays two or four round white eggs. It seems questionable whether this species be migratory.

The *red-eared Owl* of Pennant and Latham, or *Scops of Carolina*, (*Strix Asio*, Gm.) has the bill horn colour, and the irides saffron; the plumage, on the upper parts, bright ferruginous red; the feathers round the eyes are red, but the inner half is surrounded with white, meeting over the nostrils.

This species inhabits North America, from New York to the Carolinas. In summer, it remains in the woods, but in winter it frequents the houses in Pennsylvania and New York, and quickly clears the granaries of rats and mice; their eyes are so completely dazzled by the light of the sun, that they suffer themselves helplessly to be taken with the hand. They build in the clefts of trees, and are said to be monogamous.

We proceed to notice a few of the species with smooth heads, or destitute of the tufts, called ears.

The *Snowy Owl*, (*Strix nyctea*, Lin.,) is as big as the great horned species, but the head is smaller. The general plumage of this bird is of a dead white, varied with small brownish spots on the head, with transverse dorsal bars of the same tint under the wing and on the tail, but even the partial colouring is said to give place to an uniform white in winter. This species is an inhabitant of high northern latitudes, though Mr. Bullock states, that he saw one in the Orkney Islands. On the shores of Hudson's Bay, where this bird continues the whole year, it pursues in open day the ptarmigan, hares, and smaller rodentia It builds on elevated rocks, even in these inhospitable regions. Captain Parry met with it in Melville Island. The Calmucs have superstitious notions with regard to this bird, and predict futurity from its mode or direction of flight.

The *common White Owl* of this country, (*Strix flammea*, Lin.,) has the beak straight to near the tip, while it is arched from the base in the other species, from which circumstance some naturalists have separated it into a subgenus. It is full fourteen inches long; the eyes are encircled with a large circle of white plumes; the irides seem to vary from nearly black to yellow, the upper parts of the body, the wing coverts, and secondaries are pale yellow; on each side of the shafts two gray and two white spots are placed alternately; the outside of the quills are yellow, the inner white, marked on each side with four black spots; the upper sides of the tail feathers are marked with obscure dusky bars; the legs are feathered to the feet, which are covered with short hairs, and the edge of the middle claw is serrated. These characters will sufficiently distinguish it from the other species so common to this country, the brown or screech owl.

This species, so common in our own country, is perhaps equally so all over Europe. It is also found in Southern Africa, India, North and South America, and the West Indies, and seems indeed to be nearly cosmopolite.

The common white owl frequents barns, outhouses, and granaries, in search of those troublesome and destructive inmates, the rats and mice, on which, and on bats and beetles, it seems principally to feed. In winter they may be found in small parties of five or six in the clefts of old walls, particularly of churches and clock towers, in which, as well as in holes in trees, they build their nests about the month of April in rather a careless manner, in which the female lays two or four round eggs.

On quitting their perch, these birds seem at first rather to fall over than to fly, until they have gained their equilibrium after a few seconds. If taken young, they can easily be tamed, but they will not bear captivity if they have attained their full age in liberty.

The *Coquimbo Owl*, (*Strix cunicularia*, Gm.) This species which is called *Chouette à Terrier*, and *Chouette lapin* by the French, the *Uurcurea* of D'Azara, takes its name, in general, from its habits, and not, as might be supposed, by Gmelin's epithet, from its preying on rabbits : its English name, however, has reference to its locality.

It is nearly a foot long; the upper parts of the body are grayish, inclining to fulvous, or brown, covered with white spots which enlarge on the wings. It is found in St. Domingo, Chili, especially about Coquimbo, and various parts of America, and lives on small quadrupeds, reptiles, and insects. This species decidedly retires to burrow in the ground, a habit by no means singular ; but M. Feuillée has asserted that it makes these burrows itself. This assertion is repeated by M. Vieillot, who states that he himself saw one of the burrows, similar to that of a rabbit, and two feet deep, and that the freshness of the earth spread round the edge induced him to believe that it was recently formed, and therefore to open it, when at the bottom he found an egg lately laid on a bed of moss, grass and dry roots. He adds, that these birds usually lay two eggs of a brilliant white, and nearly spherical; and that the proprietor of the spot where this nest was found, stated that he had seen the young, when covered only with down, appear at the entrance of the burrow, into which they retreated as soon as they were approached.

Without questioning any of the facts here stated, we may nevertheless be permitted to doubt whether this burrow, which served for an asylum for the youn , were entirely formed by the parent bird. This species is not the only one which makes its nest in holes in the ground ready made for them by some of the digging mammalia; and when we consider that others of this genus do so, it seems the more improbable, unless the fact were stated by an eye-witness, to suppose that, in the case of this particular species, the excavation, as well as the nest formed therein, was made by the bird itself. That the bird, when she has selected a burrow in which to make her nest, may clear it of superfluous matter, or even in some degree enlarge it, seems not improbable; but it certainly demands proof of the fact, rather than presumption, to warrant the conclusion that she actually makes the hole.

The *Little Owl* of the English writers, (*Chevêche* or *Petite Chouette* of Buffon, pl. en. 439,) is about seven or eight inches in length; the head, back, and wings are of an olive brown colour; underneath it is white spotted with brown, and there is a circle of white feathers tipped with black round the face.

This species inhabits France, but is by no means common, and has been seen, though very rarely, in England. It is an inhabitant of deserted buildings, rather than of the woods, and is said to lay five yellow eggs, spotted with white. Its sight seems nearly perfect in the day time, as it is then seen to chase, but seldom to catch, small birds, preying principally, like its congeners, on mice and other small quadrupeds; in devouring its prey, it is observed not to swallow the animal whole, like others of the genus, but to tear off the flesh and reject the rest.

It is said, also, to inhabit Gibraltar, Russia, and India; but, so much uncertainty still prevails as to the specific identity of this with various other small owls that have been mentioned, that it is difficult to come to any conclusion on the subject of its habitat.

In conclusion of these brief observations on a few select species of this genus, we have to regret that, in no branch of zoology, does there appear to be more confusion and uncertainty than in this very limited, but well defined group of the nocturnal birds of prey. It would be no difficult task to present in detail the labours of practical ornithologists on the species of the owl; but these labours have been unfortunately almost confined to the nomenclature; and the result of them has by no means satisfactorily established the number of real species: the particulars of these labours here, therefore, would but little amuse or edify the general reader, for after all he would be obliged to confess that much uncertainty still prevails on the subject.

The owls are, in fact, very distinct from the diurnal rapacious birds. The former have obtuse sight, while the latter enjoy that sense to an exquisite degree of perfection. The owls have feathers immediately at the base of the bill, with the upper mandible in some degree moveable, as in the parrots; one of their anterior toes also is capable of being turned behind, and their flight is in general heavy and silent; while the diurnal acciptres, in general, have a denuded fleshy ridge at the base of the bill, with the upper mandible perfectly fixed, all the toes fixed, and a rapid, elevated, and noisy flight. In fact, there seems little else common to these divisions of the birds of prey than their carnivorous appetite, and consequent predacious habit.

PLATE III OF THE REGNE ANIMAL.

1, Wedge-tail Eagle. .... Vi. p. 36. 4, Java Honey Buzzard. .... Vi. p. 60.

2 Urubitinga .... Vi. p. 44. 5, Laughing Falcon. .... Vi. p. 52.

3, Great Harpy. .... Vi. p. 42. 6, White Striped Swallow Shrike. Ocyp. albovittatus Vi. p. 287.

London Published by G.B. Whittaker, Nov. 1828.

THE SECOND ORDER OF THE BIRDS,

OR

THE PASSERES,

Is the most numerous of the entire class. Its character appears at first purely negative, for it embraces all the birds which are neither swimmers, nor waders, nor climbers, nor rapacious, nor gallinaceous. Nevertheless, on a close comparison, we soon discover between the birds of this order a great resemblance of structure, and gradations so insensible from one genus to another, that subdivisions become difficult of establishment.

The Passeres have neither the violent character of the birds of prey, nor the fixed regimen of the gallinacea, or of the water-fowl. Their aliment consists in insects, fruits, and grains. It is more exclusively granivorous in proportion to the thickness of their bill, and more exclusively insectivorous, as the latter is more attenuated. Some, which possess a tolerably strong bill, are even found to pursue small birds *.

Their stomach is in the form of a muscular gizzard, and they have, in general, two very small cæcums. Among them we find the singing birds, and the most complicated conformations of the lower larynx.

* I have been unable to find, externally or internally, any proper character of separation between the passeres and the genera comprehended in the *picæ* of Linnæus, which are not climbers.

The proportional length of their wings, and the extent of their flight, are as variable as their mode of life.

Their sternum has usually but one slope on each side at its lower edge. There are, however, two in the rollers, the king-fishers, and the bee-eaters, and none in the martinets and the colibris.

Our first division shall be founded on the character of the feet, and our subsequent ones on the beak.

The first and most numerous division comprehends the genera in which the external toe is united to the internal, only by one or two phalanges.

The first family of this division is that of the

Dentirostres,

Whose beak is sloped on the sides of the point. In this family are found the greatest number of insectivorous birds. Still, they almost all of them also eat berries, and other tender fruits.

The genera are determined by the general form of the beak. It is strong and compressed in the shrikes, and in the thrushes; depressed in the fly-eaters; round and thick in the tanagers, slender and pointed in the fine-beaks, &c.

The Shrikes (Lanius, Linn.) Pie-Grieches, Cuv.

Have the beak conical or compressed, more or less crooked at the end.

The Shrikes properly so called, (Pie-Grieches,)

Have it triangular at the base, compressed at the

sides. Some have the upper crest arched: those in which its point is very strong and crooked, possess a degree of courage and cruelty which has caused many naturalists to associate them with the birds of prey. In fact, they do pursue small birds, and defend themselves with success against the larger; and they will even attack the latter when it is necessary to drive them from the nest.

The Shrikes live in families, fly unequally and precipitately, sending forth piercing cries. They nestle in trees, lay five or six eggs, and take great care of their young.

We have here four species of this subdivision.

The Great Cinereous Shrike. (*Lanius excubitor*, L.)
Enl. 445. Penn. B. Z. t. 73.

As large as a thrush, ash-colour above, white underneath; wings, tail, and a band round the eye, black. White on the scapulars, at the base of the quills of the wing, and at the external edge of the lateral quills of the tail. It remains the entire year in France.

The Lesser Gray Shrike. Lath. (*Lanius excubitor minor.* Gm.)
Enl. 32, 1.

Rather less than the preceding, wings and tail alike, ash colour above, reddish on the belly. The black bands of the eyes united on the forehead in a broad bandeau. This is a very distinct species; it learns extremely well to imitate the song of other birds.

The Red Shrike, Wood Chat. Lath. (*L. Collurio rufus* et

L. Pomeranus, Gm.) Enl. 9. 2. *L. Rutilus*, Lath. *L. ruficollis*, Sh. *L. rufus*, Briss. Vail. O. A. pl. 63. f. 1. 2.

The bandeau, wings, and tail of the preceding; the size a little less. The upper part of the head and neck a lively red; the back black, the belly and crupper white. It has also great powers of imitation.

The Red-backed Shrike, Lath. (*Lan. Collurio*, Gm.) Enl. 31. f. 1. 2. Penn. Br. Z. 1. Vail. O. A. t. 44. f. 12. *L. Spini Torquens*. Bechst.

Still smaller. The upper part of head and crupper, ash-colour; back and wings fawn; underneath whitish; a black band over the eye; the quills of the wings black, edged with fawn; those of the tail black, the lateral ones white at the base. It imitates naturally and immediately the voices of the best singing species. Too weak to attack birds, it destroys a great quantity of insects, which it sticks (according to report) on the thorns, to find them again when it wants them.

The three last species quit us during winter.

In foreign countries there are many more. The beaks diminish and grow weak in their points gradually, according to the species, so that it is impossible to establish a limit between this subgenus and the thrush.

Lanius Meridionalis. Temm.

Very like the great cinereous Shrike, but peculiar to middle Europe: the upper part is a deeper ash, and the lower part more reddish.

Lanius Ruficeps, Bechst.

Lanius Superciliosus, Lath. Pl. Enl. t. 477. f. 2.

Only differs from the former in the base of the bill being very red, in having no frontal band, in its white eyebrows and general ferrugineous tint. From Senegal.

Lanius Rufescens. Le Rousseau, Vail. Afric. t. 66. f. 2.

Differs from the former in being small. From India.

Lanius Nubicus, Licht. L. *personatus*, Temm. pl. coll. t. 256. f. 2.

Black; occiput, eye-brow, scapulars, central wing spot, and outer quills white; beneath ferrugineous; throat, middle of belly and vent white. Female gray above, duller. Length seven inches and a half. Nubia. Bill very short; tail wedge-shaped.

Collared Shrike, Lath. 10. *Lanius collaris*, Gm. pl. Enl. t. 477, f. 1. Vail. O. A. 61, 62.

Black, white beneath; primary quills white at the base; tail, middle feathers black, rest white. Length twelve inches. Cape of Good Hope.

Cape Shrike, Lath. *Lanius Brubru*, Lath. Suppl. Vail. O. A. t. 71. f. 1, 2. *Lan. Capensis*, Shaw.

Varied black and white above, beneath white; crown and nape black; eye-streak white: wing spot white; tail black; outer feathers white.

The Bou-bou, Vail. O. A. t. 60. *Lanius Bou-bou*, Lath. 49.

Black, chest and belly ashy; wings with two white bands; bill and feet yellow. Caffraria.

Blanchot Shrike, Lath. H. *Lanius. Le Blanchot*, Vail. O. A. t. 285.

Greenish olive; beneath brownish yellow; crown and nape slate-gray; forehead white. Size of a thrush. Senegal.

Madagascar Shrike, Lath. H. 46. *L. Madagascariensis*, Lin. pl. Enl. t. 299.

Ash, beneath white; eyebrows white; tail reddish; upper wing-coverts red; male, throat black; five inches long. Madagascar.

Blue Shrike, Lath. 26. *Loxia Madagascarina*, Lin. et *Lanius Bicolor*, pl. En. t. 298. f. 1. Vail. O. A. 73, 1, 2, 3. Nat. Misc. t. 521.

Tail nearly equal; above blue, beneath white; face black; six inches and a half long. Madagascar.

American Shrike, *Lanius Americanus*, Lath. 9. pl. Enl. 39.

Reddish-brown, beneath yellowish, crown gray, quills and tail black, throat and tail tips white. Eight inches long. N. America.

Blue striped Roller. *Coracias Pacifica*, Forst. Cor. *Striata*, Lath. *Philemon Sagittatus*, Vieill.

Blue black, streaked with bluish green; bill, tail, and feet black; length eight inches. New Caledonia.

Lanius Poliocephalus, Licht.

Above green, head gray, lores, beneath, and the lengthened thighs, bright yellow; quill and tail-feathers yellow-tipt; length ten and a half inches. Senegal. Not the *L. Policephalus* of Lord Stanley, in Salt, Voy. App. 1.

Hottuiqua Shrike, Lath. H. 26. *Lanius Cubla*, Lath. Vail. O. A. 72, 1. 2.

Black; loins white; scapulars half white: wing-coverts white; edged beneath whitish; quills all black, white fringed; female paler; length six and a half inches. Caffraria.

Lanius Gambensis, Licht.

Head above, ophthalmic region, and back of the neck, black; back and wing brown; scapular and loins lead-colour; wing-coverts white-edged; beneath white; tail-feathers entirely white: length seven and a half inches. Senegal.

Senegal Shrike, Lath. *Lanius Senegalensis*, Lin. pl. Enl. 97. 1. *Lanius Erythropterus*, Sh. from Vail. O. A. 70.

Grey, beneath white, crown and ocular streak, black; tail black, white-tipt; quills outer-edge reddish. Nine inches long. Senegal.

African Shrike, Lath. *Lanius Afer*, Lath. L. *Signatus*, Sh. Appears a doubtful species.

Corvine Shrike, Lath. *Lanius Corvinus*, Sh., et L. *Mellivorus*, Licht. Vail. O. A. 78.

Above, rufous-ash, streaked and waved with black; beneath white; chest streaked; bill brown; eye-streak black; eyebrow whitish, quill cinnamon brown tipt; tail long, wedge-shaped; length twelve inches. Senegal.

Ferruginous-bellied Shrike, Lath. *Lanius ferrugineus*, Gm. Freycinet, Voy. 17.

Blackish; crop and chest white, rump brown, belly and vent ferrugineous; length nine inches. Cape of Good Hope.

Cruel Shrike, Lath. H. *Lanius Pendens*, Lath. Suppl. 77. Vail. O. A. 66. 1.

Black; body above ash; belly and band above and below the eye, white. India?

Mustachio Shrike, Lath. H. *Lanius Mystaceus*, Lath. Vail. O. A. 65.

Above brown; neck, crest, and tail, red; chest-band, streak under the eye, white. South Sea Islands?

Silent Shrike, Lath. *Lanius Silens*, Shaw. Vail. O. A. 74, 1, 2.

Black, beneath white, longitudinal streak on middle of wing white; outer tail-feathers white-edged; female smaller, browner, gray beneath. Africa.

N. B. Consult Lath. Hist. gray-backed, 3. Bay-backed, 6. Keroula, 23. Indian, 31. White-cheeked, 53.

Lanius Scapulatus, Licht. *Geai noir à collier blanc*, Vail. par. 42.

Black; cross spot on side of the neck white; crown-feather very long, large, and flat; length eleven inches. East Indies.

? *Crested Red Shrike*, Lath. 17. *Lanius cristatus*, Lin.

Tail wedge-shaped; head crested; body reddish, beneath waved with fulvous and fuscous; behind the ear a black moon; length six and a half inches. Bengal.

? *Chinese Shrike*, Lath. 35. *Lanius Schah*, Lin.

Yellowish; forehead and wings black; head and neck gray above; beneath whitish; both primaries and tips of secondaries white. China.

? *Pacific Shrike*, Lath. 28. *Lanius pacificus*, Gm.

Black; head and neck greenish; belly, tail, and quills

blackish; feathers of head and neck narrow; eleven inches long. Pacific Islands.

Tabuan Shrike, Lath. 87. H. *Lanius Tabuensis*, Gm.

Olive-brown; crop and chest ash; belly yellowish brown; quills black; tail brown; nine inches long. Tabuan Island.

White Shrike, Lath. 87. H. *Lanius Albus*, Gm. Son. Voy. t. 72.

White; larger wing coverts and tail black; band on wing white. Panay.

Panayan Shrike, Lath. 40. *Lanius Panayensis*, Gm. Son. Voy. 70.

Brown; head, throat, crop, chest, and belly red. Panay.

Lanius Kirkocephalus, Lessron and Garnot, t. 11.

Bill long, pale; crested; reddish-brown; paler beneath; head and neck pale-brown; wing and tail fuscous brown. Tail-end rounded. New Guinea.

Lanius Karu, Lessron and Garnot, Voy. t. 12.

Bluish-black; beneath white, gray cross-streaked; band over eyes, tips of wing-coverts, outer tail-feathers and edge of secondaries, white; bill and feet black; nape bluish-white lunuled.

Black-headed Shrike, Lath. 29. *Lan. Melanocephalus*, Gm. Lath. t. 6. Hist. t. 19.

Olive; head black; tail with a broad black band; yellow tipt. Sandwich Islands. Six inches long.

Northern Shrike. *Lanius Septentrionalis*, Gm. *Lanius excubitor*, Wilson, A. O. *L. Borealis*, Vieillot.

Light-slate; beneath waved with brown; face whitish;

wings and tail black; tail-feathers, excepting the middle ones, partly white; third primary the longest, fourth equal to the second. North Europe and America.

Louisiana Shrike, Lath. No. 8. *Lanius Ludovicianus,* Lin. *Lan. Carolinensis,* Wilson, A. O. 22, 3. *Lan. Ardosiaceus,* Vieil.

Dark-slate; beneath white; face, wings, and tail black; tail-feathers, middle one excepted, partly white; second primary longest, first and fifth equal. N. America.

Natka Shrike. *Lanius Natka,* Gm. *L. Naotka,* Lath. 48.

Black; eyebrows, throat, collar, and larger wing-coverts, white; secondaries and four outer tail-feathers, black; seven inches long. Nootka Sound.

Cuvier has referred here *Tanagra Guianensis,* G. M. Vail. O. A. 76. It is a *Thamnophilus* of Temminck, and the type of the genus *Cyclarhis* of Swainson. The *Tan. Atricapilla,* Gm. Pl. Enl. is a Tanagra according to Temminck, and the type of the genus *Lanio* by Vieillot.

* * *Bill weak.*

Olive Shrike, Lath. H. 26. *Lanius Olivaceus,* Shaw. *L. Oleagineus,* Licht. Vail. O. A. 75, 76. 1.

Olive-green; forehead, and beneath, brownish-yellow; sides paler; orbit and neck-streak black, yellow-edged; tail, outer feathers, partly yellow. Cape of Good Hope. Size of Wood-chat.

Barbary Shrike, Lath. 43. *Lanius Barbarus,* Lin. Pl. Enl. 56. *Laniarius,* Vieillot. *Lanius* ***. Temm. Vail. O. A. 69.

Black; beneath red; crown, nape, thighs, and vent, fulvous yellow; nine inches long. Senegal.

Malimbic, and *Red-throated Shrike*, Lath. H. 13, 20. *Lanius Gutturalis*, Daud. Ann. Mus. iii. 15. Vail. O. A. 286. Shaw Nat. Misc. 637.

Deep green; forehead yellow; eye-streak going down the neck, and forming a broad crescent on the breast, black; throat and belly deep red; tail rather short. Malimba in Africa.

Ceylon Thrush, Lath. 80. *Turdus Ceylonus*, Lin. Pl. Enl. 272. Edw. 321. *Lanius Bacbakiri*, Shaw. *L. Ornatus*, Licht. Vail. O. A. 67.

Green; beneath yellow; eye-streak, forming a broad pectoral band, black; tail rather long. Cape of Good Hope.

Thick-billed Thrush, Lath. 30. *Turdus Crassirostris*, Gm. Lath. Syn. t. 37. *Tanagra Capensis*, Sparmann, Voy. 45.

Reddish-brown, beneath ash; reddish streaked; lateral tail-feathers dull-red; belly white; nine inches long. New Zealand.

Antiguan Strike, Lath. 16. *Lanius Antiguanus*. Gm. Sonn. Voy. t. 70.

Reddish-yellow; throat and chest white: head, quills, and tail black; lateral tail-feathers red-tipt. Antigua?

Some shrikes with straight beak have it very strong, and the lower mandible much enlarged.

Some are found in Africa, where they form the genus *Malaconotus* of Burchell.

Lanius Erythropterus, Sh. *Lan. rutilus*. Var. γ Lath. Pl. Enl. 479. 1. 297. 1. *Le Tchagra*. Vail. O.A. t. 70.

Rufous; beneath white; tail white-tipt; crown black; eyebrow white. Cape of Good Hope.

Lanius Atrococcineus. Burch.—Zoolog. Journal, jt. 18. Black; beneath scarlet; wings white streaked; tail, two outer feathers, red-tipt. Africa.

(Perhaps *L. Cubla* and *Bou-bou* should be placed in this group.)

Some are peculiar to America, especially the Southern part. The males are blackish and the females reddish; they have been divided into several minor groups. 1. The *Batara* of Azara and genus *Thamnophilus*, Vieillot.

The *Large Bush Shrike.* *Lanius Stagurus*, Licht. *Le grand Batara*, Az. 211.

Slightly crested; above black; beneath white; tips, wing-coverts, and sides of all the tail-feathers white; female, above, cinnamon; beneath, dirty-white; wing-coverts gray-tipt; eight and a half inches long; male varies; wing-spots larger, and more crowded; quills white-edged. Bahia.

Pied Shrike, Lath. H. 50. *Lanius doliatus*, Lin. Pl. En. 297. 2. Edw. 226. *Batara rayé*, Azara, 212. *Le Rousset*, Vail. O. A. 77. f. 2. *Lan. ferrugineus*, Act. Paris.

Tail rounded; body with crowded black and white bands; female, above chestnut; beneath ferrugineous; with a black and white varied collar; length six and a half inches. Cayenne.

Black-topped Shrike. Lath. H. 94. *Lanius Atricapillus*, Gm. Merrem. Icon. ii. t. 10. *Tyrannus Atricapillus*, Vieillot, 48?

Mouse-gray; beneath bluish-ash; crown, nape, shoulders, and wings black; secondaries and coverts white-edged; tail side-feathers white tipped; five inches long. Surinam.

Crested Shrike, Lath. 18. *Lanius Canadensis*, Lin. Pl. Enl. 479. 2.

Crested; reddish; beneath white; cheeks white-spotted; throat reddish-brown spotted; quills and tail white-edged; six inches long. Canada.

Spotted Shrike, Lath. 51. *Lanius nævius*, Gml. *Lanius Punctatus*, Sh. Vail. O. A. 77. 1. Zool. Misc. 17. ♂. *Batara noire et plombe*, Azara. 213. ♀ *B. Mordoré*, Az. 214.

Lead-colour; middle of nape black; wing and tail black, white-spotted; quill, outer edge, white; *female*, above olive-brown; crown chestnut; belly ashy, marked like the male; length five inches and a half. Brazil.

Muscicapa, Temm. Pl. Col. 17. 1. ♂ 2. Jun. *Lanius Cæsius*, Licht.

Lead-coloured, slender; *female*, olive-brown; wings reddish; throat white: chest fuscous; belly ferrugineous; vent cinnamon; length five inches and a half. Brazil.

Lanius guttulatus, Licht.

Olive-green; crown and nape lead-colour; sides of head and wing-coverts black, white sprinkled; throat white; crop with brown spots; middle of belly and vent yellowish; sides ash; *female*, nape brown; throat white; belly yellowish, scarcely spotted.

Rufous-winged Bush Shrike. *Thamnophilus Torquatus*, Swainson.

Grayish, beneath whitish; throat and breast black-banded; wings rufous, immaculate; tail black, rounded, white-spotted.

*

Rufous-crowned Bush Shrike. Thamnophilus Ferrugineus, Swain.

Ferruginеous brown, beneath pale fulvous; crown rufous; wings brown; spots on the back and wing-coverts white; tail rufous; length six inches; a female?

Vigors' Bush Shrike. Thamnophilus Vigorsii, Such. Zool. Jour. t. 7 & 8. *Thamnophilus Cinereus*, Vieil. *Vanga Striata*, Gaims. Frey. Voy. 19. 18.

Crested; above black, finely white-banded; cheeks and beneath slate-colour; crest black; *female*, crested, crest fulvous, black-tipt; above band black, and fulvous; beneath pale-fuscous; length thirteen inches. Brazil.

Leach's Bush Shrike. Tham. Leachii, Vigors. ♂. ♀

Black; head and back white-spotted; quills slightly pencilled with fulvous; throat, breast, and middle of belly, and tail black; sides of belly and rump white-banded; length ten inches.

Lineated Shrike, Lath. H.? *Thamnophilus Lineatus*, Leach. Zool. Mis. t. 6.

Black, finely white-banded; bill and feet black. Berbice.

Red-headed Bush Shrike. Thamnophilus ruficeps, Such.

Black-spotted; head lined, and the secondaries, rump, and tail, and abdomen banded with fulvous; length nine inches. Brazil.

Black Bush Shrike. Thamnophilus Niger, Vigors. Crested; black; quills obscurely banded with brownish; length eight inches. Brazils.

Lanius Severus, Licht. ♀ *Thamnophilus Swainsonii,* Vigors. Zool. Jour. t. 5. Suppl.

Crested black; wings sooty; tail graduated; female crested; crown chestnut; body, wings, and base of tail, with crowded ferrugineous and ash-coloured wavy bands.

? *Lanius Domicilla,* Licht.

Black; humerus snow white; wing-coverts white tipt; female, above brown; tail black; beneath ashy olive. Bahia. Seven inches long.

Lanius Luctuosus, Licht.

Crested entirely black; outer edge of the scapulars and tips of the tail white. Parag. Is this *Thamnophilus Albonotatus,* Spix?

** Tail rounded, long.

Black and White Shrike, Lath. H. 22. *Thamnophilus Albiventer,* Spix, Brazil, t. 32. f. 1. ♂ 20. *Thamnophilus Bicolor. T. Cinnamomeus,* Swainson.

Crested above deep black; beneath white; tips of wing covers, edge of quills, and interrupted bars on tail, white; body and tail three inches and a half long. Brazils, *female* above cinnamon brown; beneath white. Considered distinct by Swainson.

Barred Shrike, Lath. H. 8. *Lanius Palliatus,* Licht. *Thamnophilus Lineatus,* Spix, Braz. t. 33. f. 1. ♂ 29. *Thamnophilus Fasciatus,* Swainson.

Above chestnut; beneath black, with small white bands;

head black. *Female,* crown cinnamon; bill black; body five inches and three-quarters, tail two inches and a half.

Thamnophilus Radiatus, Spix, t. 25. f. 2. 5. t. 38. f. 1. ♀.

Above black, with white wavy bands: beneath white, black-banded; tail black, with speckled; head black-crested. *Female,* above cinnamon; wing and tail black-banded; beneath yellowish, black banded; neck streaked; body six inches, tail two inches and a half long. Brazil.

Lanius Meleager, Licht. *Thamnophilus guttatus,* Spix, Brazil, t. 35. f. 1. *Thamnophilus maculatus,* Swainson. Zool. Jour. t. 6. Suppl.

Above black, yellow-speckled; beneath yellowish-white; bill weak; chest black-spotted; wing and tail yellow-banded; body seven inches and one-third, tail three inches and a half long. Brazils.

Thamnophilus Strigilatus, Spix, Brazil. t. 36. f. 1.

Olivaceous brown; beneath yellowish-white; head and back yellowish; streaked wings, and tail cinnamon; body six inches and a half, tail three inches long. Brazil.

† *Sylvoïdes.*

Thamnophilus Agilis, Spix, Braz. t. 34. f. 1.

Olive green, spotless; beneath white; head ashy; superciliary streak white; lores fulvous, feet short; crown; wing-coverts green-edged; body four and a half, tail one inch and a half long. Brazil. Allied to *Lanius Guyanenis*, but smaller.

Thamnophilus Affinis, Spix, Brazil, t. 34. f. 2.

Head and above green; beneath greenish-ash; bill short slender; tarsi blood-red; no streak above the eye; quills brown, green edged; tail olive; body and tail one inch and three-quarters long. Brazils. Differs from Mus. Diope. Tem. pl. col. 44. 1. Bill compressed.

Thamnophilus Melanogaster, Spix, Braz. 43. 1.

Head and above lead-coloured; beneath deep black; tail very short, black; tarsi very short; sides white; wing coverts and scapulars white-tipt; bill very slender; body three inches three-quarters, tail one inch. *Female*, wing brownish. Brazils.

Short Tails,

Thamnophilus Stellaris, Spix, B. t. 16.f. 2.

Lead-coloured; paler beneath; head black, white tipt; wing-coverts black, white tipt; tail very short; bill very long; cheeks ashy; quills black brown; inner base red. Body four inches and a half, tail one inch long. Brazil.

Thamnophilus ruficollis, Spix, Braz. t. 37. f. 1 ♂.

Sooty-ash; head, neck, and beneath reddish; wing-coverts, and tail white-edged; bill above blackish, beneath whitish; body five and a half, tail two inches long.

Thamnophilus Albonotatus, Spix, Braz. 27, 2 ♂. 38. 2. ♀.

Lead-coloured; wing and tail black; wing-coverts white fringed; base of back feathers white; tail white-tipt; grey under the eyes. *Female*, cinnamon-brown; beneath fulvous; quills red-margined; tail yellow-tipt; body five, tail two inches long. Brazil.

Thamnophilus Gularis. Spix, Braz. 41. 2.

Above reddish; beneath ashy; throat black, white speckled; wings and tail, reddish; wing-coverts black, yellow tipt; bill slender; tail short; body four inches, tail one inch long. Brazil.

Thamnophilus Melanoceps. Spix. 39. 1.

Chestnut, head and neck black, tail rather short; bill rather strong; base of the soft dorsal feathers ashy; crown subcrested; legs yellowish; body six and a half, tail two and a half inches long. Brazils.

Thamnophilus Leuconotus. Spix, Braz. 39. 2.

Deep black; nape with a white collar; bill slender, rather long; frontal-plumes linear; bill black, feet reddish; body six, tail twenty-three and a quarter inches long; perhaps the male of the former. Brazil.

* * Outer tail-feathers short.

Thamnophilus Griseus. Spix, Braz. 41. 1♂. 40, 1 ♀. *Myothera Superciliaris*, Licht.?

Above brown; beneath black; bill slender; eyebrows white; wing-coverts, and tail black, white-tipt; wings short, black, sides whitish; female above chestnut; eye-streak, and beneath white; bill, sides of neck, wing and tail white-tipt; body four and a half, tail two inches long. Brazil.

Thamnophilus Striatus, Spix, Braz. 40. 2.

Reddish above, fulvous streaked; beneath whitish; varied fulvous and black; bill rather thick, short; sides rufous; throat white; body four and a half, tail two and a half inches long. Brazil.

Thamnophilus Myotherinus. Spix, Braz. 42. 1. 2. ♂. ♀.

Blackish, lead-colour; beneath ashy; black forehead; eyebrows, streak white; throat, lores and cheeks deep black; tail short, scarcely longer than the wings; body four and a half, tail one and a quarter inch long; female beneath darker; wing-covert, paler edged.

Thamnophilus Caudautus, Vieillot.

Greenish-brown; tail-feathers blackish-brown, acute; bill above fuscous, base beneath white; length seven and a half inches. America.

Thamnophilus Choloropterus. Vieill.

Above brownish rufous; beneath banded black and rufous; smaller wing-coverts pale rufous; quill outer green, inner brown; tail black, white and gray-banded; feet blue; length 8 inches. Guiana.

Turdus Alaspi. Lath. Pl. Enl. 701. 2.

Olive-brown throat, and chest black; belly ash; tail blackish; length six inches. South America.

Thamnophilus cœrulescens, Vieil. *Batara Negro y aplomado*, Azara. No. 213.

Above blackish; lead-colour; throat and chest bluish;

crown, wing and tail black; belly bluish-white; length five inches and three-quarters. Paraguay.

Thamnophilus Atricapillus. Vieil. Therrem. fasc. t. 10. ♀.

Crown black, body above gray; beneath bluish ash; wing-coverts and secondary quills white-edged; wings white-tipt; tail black; bill and feet black; length five inches.

Turdus Cinnamomeus, Lath. *Batara gola nigra,* Azara, n.—Pl. Enl. 560. 2.

Above cinnamon, beneath paler; temples, cheeks, chin, wing-coverts, throat and chest gray; torque clouded white; bill and feet black; length five inches. S. America.

Thamnophilus Auratus. Vieill. *Batara Pardo dorado,* Azara. 214.

Body above, golden brownish-lead colour; beneath rufous and golden red mixed; sides of head bluish, white-dotted; throat gray; bill bluish-black; feet lead-colour; length five inches, and three-quarters. Paraguay.

Turdus rufifrons, Lath. Pl. Enl. 614. 1.

Brown; beneath, forehead and temples rufous; vent whiter, tail, feet and bill ash. S. America.

Turdus rubiginosus, Lath.

Crested, body above reddish-brown; beneath yellowish-red.

Thamnophilus Albicollis. Vieil.

Above brown, throat white; cheeks and chest black; sides of neck with a black and white streak; wing-coverts varied black and white; bill black; feet brown; length five inches; allied to *T. Cinnamomeus.* S. America.

Sylvia Grisea, Lath. pl. Enl. 643. 1. 2.

Ashy-gray; crown, throat and chest black; eye-streak, tips of wing-coverts, belly and crest white; bill black; feet ash; length four and a half inches. Cayenne.

Turdus Cirrhatus, Lath. Vail. O. A. Sept. t. 48. ♂. t. 49. ♀. Vail. O. A. t. 77.

Ashy; tail white-edged and tipt; crown crested; throat varied, black and white; chest black; length six inches. S. America.

Thamnophilus Longicaudatus, Vieil.

Black; throat and tail white-spotted; bill and feet black; length eight inches. S. America. Mus. Paris.

Thamnophilus Guttatus, Vieil.

Above white; black spotted; beneath black, white spotted; bill yellow; feet brownish; length seven or eight inches. S. America. Mus. Paris.

Thamnophilus Radiatus, Vieil. *Batara Lisado,* Azara, 212.

Crest black; capistrum, head and neck above black and white marbled; cheeks and chest whitish, black streaked; body beneath white; bill blue; base blackish; feet pale lead-colour; length six and a half inches. S. America.

Thamnophilus Lineatus, Vieil.

Black, with reddish white cross stria; crown rufous; length six inches. Brazils. Mus. Paris.

Thamnophilus Rubicus, Vieil.

Reddish brown; beneath reddish; crown ash; cheeks white, brown-spotted; bill black; feet brown; length nine and a half inches. S. America. Mus. Paris.

Thamnophilus Rutilus, Vieil. *Batara Roxa*, Azara, 215.

Rufous; beneath yellowish-white; wing-coverts blackish; bill blackish; feet lead-colour; length seven inches. Paraguay.

Thamnophilus Cyanocephalus, Vieil. *Batara Obscuro y Negro*, Azara, n. 237.

Blackish; beneath dusky; crown shining blue; middle white streaked; nape and neck black; wing-coverts white-edged, and spotted; tail dull blue; feet blackish; female greenish. Paraguay.

Thamnophilus Ruficapillus, Vieil. *Batara Aconaledo*, Azara, 215.

Crown red; throat and chest black and white banded; belly whitish; back varied, blue and brown; middle tail feathers blackish; outer black; outermost white tipt; bill above black, beneath pale blue; feet lead-coloured; length six inches and a quarter. Paraguay.

Thamnophilus Viridis, Vieil.

Green; forehead, throat, hinder parts and tail above, black and white banded; length six inches and three quarters. S. America. Mus. Paris.

Thamnophilus Virescens, Vieil.

Crown greenish-gray, black spotted; quill black, white dotted; body above greenish; beneath reddish-gray; tail black, white tipt. S. America.

Thamnophilus Rufinus, Vieil. Pl. Enl. 711.

Above rufous; beneath ashy; bill black; feet yellowish. Cayenne.

Thamnophilus Cristalillus, Vieil.

Head reddish-brown; body above, reddish-brown, and yellowish-banded; beneath dull red; feet and bill brown; length ten inches. Brazil.

Thamnophilus cinereus, Vieil.

Crown black; body above black and white banded; beneath, cheeks, and throat, bluish-gray; bill brown white edged; feet brown; length ten inches. Brazil. Mus. Paris.

Another group, which appear to have habits intermediate between the thrushes and the warblers, which are peculiar to South America, with slender bills, rounded wings, and tail, long slender tarsi, have been separated by Mr. Swainson under the name of *Drymophila*.

White-legged Ant-Thrush, Drymophila Leucopus, Swainson.

Rufous brown; beneath whitish; vent, eye-streak, wing-cover spotted fulvous; breast with a concealed black collar; legs whitish; male chin black; *female* chin and throat fulvous; length five inches and a half. Bahia.

Long-legged Ant-Thrush, Drymophila Longipes, Swainson.

Rufous, beneath black; sides of the crown ash; belly white; legs long, pale; length six inches. Brazil.

White-shouldered Ant-Thrush, Drymophila trifasciata, Swainson.

Black; scapulars, interscapulars, and two bands on

the wing covers snowy white; length seven inches. Brazil.

Black Ant-Thrush, Drymophila Atra, Swain.

Black, base and edge of the interscapular feathers white; length seven inches.

Drymophila Variegata, Vigors. Zool. Jour. I. 559.

Above, olive-brown; head black, white-striped; eyebrows white; wings and tail-feathers black, white-tipt; breast, belly, and rump, red; length five inches.

Drymophila Velata, Temm. Pl. Col. 334

Blue-black; face, cheeks, and forehead black; throat chestnut; length seven inches. Timor.

Others of a small size peculiar to America, which have short, rounded wings, graduated tail, and moderate and slender tarsi, have been called *Formicivora* by Mr. Swainson.

White-spotted Ant-Wren, Formicivora Maculata, Swainson.

Above black, with many white spots; beneath ashy-white, varied with black; secondaries yellow-tipt; tail graduated; length five inches. Brazil.

Black-throated Ant-Wren, Formicivora Nigricollis, Swainson.

Above grayish, beneath black; sides and eye-streak snowy; tail graduated, black, white-tipt; length four inches three quarters. Brazil.

Short-tailed Ant-Wren, Formicivora Brevicauda, Swain.

Cinereous; throat and breast black; shoulders and

wing covers spotted white; tail very short; length three inches and a half. Brazil.

The second division of the genus *Thamnophilus* of Temminck is the genus *Cyclaris* of Swainson.

Tail square, and the bill slender and scarcely toothed; but strongly curved at the tip.

Gray-headed Tanager, Lath. 24. *Tanagra Guianensis*, Gm. Vail. O. A. 76. 2.

Green, head hoary-ash; forehead and double occipital band red; five inches and a half long. Guiana.

Other shrikes have the upper mandible straight in its length and crooked only at the end. They are all foreign, and their form passes by insensible degrees to that of the warblers and other slender beaks.

Others, the VANGA of Buffon, have the bill compressed in its whole length, the end very much hooked, and the lower jaw recurved.

They are all found in the ancient continent, and particularly the Indian and Oceanic Islands. They are the genus *Vanga* of Vieillot. Their tail is wedge-shaped.

Hooked-billed Shrike, Lath. 15. *Lanius Curvirostris*, Lin. Pl. Enl. 228. *Thamnophilus Leucocephalus*, Vieill.

Body white, back black; primary quills with five white spots; outer tail feathers black, white-tipt; occiput greenish-black; length ten inches. Madagascar. Brit. Mus.

* In putting the names to the species, I have only mentioned the name used by the first describer, without the specific name has been altered; otherwise I might have added three or four names and often more to each of the species; for almost every author thinks he gives a good reason to alter almost every generic name.—J. E. G.

Destroying Vanga, Vanga destructor, Temm.

Lastly, some have the bill straight and slender, and are remarkable for a crest of recurved feathers.

The genus *Prinops*, of Vieillot.

Geoffroy's Shrike, Lath. H. 22. *Lanius Plumatus*, Shaw. Vail. O. A. t. 80, 81. *Prinops Geoffroyii*, Vieillot, Gal. 142.

Blue-black, middle of the back, tips of quills, and beneath white; back of head and orbits dusky; two outer, and tips of other tail feathers white; seven inches long?

Cuvier has added here *Pipra Albifrons*, Gm. observing that it is not a pipra. But later authors have placed it as a myothera, or ant-eater.

Near the shrikes, properly so called, are grouped some foreign sub-genera, which differ from them more or less, and which we shall now point out.

THE LANGRAYEN or SWALLOW-SHRIKES. OCYPTERUS, CUV.

Have the beak conical, rounded everywhere, without crest; triflingly arched towards the end, with a very fine point; slightly sloped on each side; the feet rather short, and the wings of the same length, or rather longer than the tail. Their capacity of flight is the same as that of the swallows, but they have the courage of the shrikes, and do not fear to attack the raven.

The species are tolerably numerous on the coasts, and in the islands of the Indian Ocean, where they fly continually and rapidly in pursuit of insects.

The *Aratamus*, of Vieillot, and the *Leptopteryx* of Horsfield.

* Wing rather rounded.

Lanius Leucocephalus, Gm. Pl. Enl. 374.

Head, neck, and lower part of body white; back of neck, back, rump, scapulars, wings, and tail greenish-black; tail beneath black; tail rather long. Length eight inches. Madagascar.

Wings long.

Leptopteryx Leuchorhynchos, Horsf. *L. Leucogaster*, Mem. Mus. xi. t. 7. f. 2.

Head, wings, and tail grayish-black; back and rump fuscous; chest, abdomen, and upper tail coverts white. Timor and Manilla.

Aratamus Cinereus, Vieil. Mem. Mus. t. 9. f. 2.

Ashy-gray, rump and vent black; tail black, white-tipt; length seven inches and a quarter. Timor.

Ocypterus Albovittatus, Cuv. Reg. Anim. iv. *Turdus Sordidus*, Lath. Supp. *Aratamus Lineatus*, Vieillot, t. 3. f. 6. Mem. Mus. vi. t. 8. f. 1. 2.

Body brownish, wings slate-coloured; the 2d, 3d, and 4th quills white-edged outwardly; tail black, white-tipt. Timor.

Aratamus Minor, Vieil., *Ocypterus Fuscatus*, Valen. Mem. Mus. vi. t. 9. f. 1.

Body brownish, wings and tail slate-coloured; vent and rump black; tail white-tipt beneath. Pacific Islands.

Ocypterus Rufiventer, Valenc. Mem. Mus. vi. ♀. f. 1. *Arutamus Fuscus*, Vieil. Dict. H. N.

Head gray, back ashy brown; belly reddish; wings and tail slate-coloured; rump, vent, and tail tips white. Bengal.

Lanius Viridis, Gm. *Tschachert*, Buf. Pl. Enl. t. 32. f. 2. Briss. ij. t. 15, f. 2.

Head, wing and body above dull green; beneath white; tail black. Madagascar.

Loxia? Melanoleuca, Forst. Mss. *Lanius Manillensis*, Briss. ij. t. 18, f. 2. Pl. Enl. t. 9, f. 1. *Lanius leucorhynchus*, and var. β. Lath. and *L. Dominicanus*, Gm. Sonnervat, Voy. 1, 25.

Head, back of neck, wings, back, and tail black; lower wing coverts, rump, thigh, and body beneath white; length seven inches. New Caledonia.

THE CASSICANS, Buff. BARITA, Cuv.

Have a large straight conical beak, round at the base, beginning on the feathers of the forehead by a circular slope; rounded at the back, compressed at the sides, with a point crooked and sloped laterally.

These are large birds of New Guinea and New Holland, which naturalists have arbitrarily dispersed through many genera. The finest has been put among the birds of Paradise, *Paradisea Viridis*, Gm. Enl. 634. Its whole body is of a brilliant black, with the feathers of the head and neck goffered. It comes from New Guinea, as do the birds of Paradise.

The others are varied with white and black, and

inhabit New Holland, and the adjacent isles. Their habits are noisy and voices shrill. They pursue small birds.

* * Bill with a distinct ridge, which extends up the forehead; wings rounded. The sixth feather the largest.

Paradisea Viridis, Gm. *P. Chalybea*, Lath. Pl. Enl. t. 634. Vail. O. A. t.

Blue-green; head silky-black; woolly; back, rump, belly, and tail, shining steel black. New Guinea, Papua.

* * Bill without any ridge, surface nearly flat. Wing moderate. Fourth or fifth quill longest.

Barita Anaphonensis, Tem. Pl. Col.

Blackish ash; upper wing coverts, tips of quills, and tail-feathers, white; tail not graduated, as said by Cuvier. His bird was in moult. Oceania.

Coracias Strepera, Lath. White, Jour. Zool. Misc. t. 86. Vail. O. P. t. 24.

Wing with a white spot at the base of wing and tail white. Jun. reddish beneath. Oceania.

Coracias Tibicen, Lath.

Black; nape, wing-coverts, vent, and tail, white, latter black-tipt. Oceania.

Coracias Varia, Lath. Pl. Enl. t. 628.

Black above; loins, rump, and upper tail-coverts, white; tail equal black, white-tipt. Philippines.

Barita Destructor, Temm. Pl. Coll. t. 273.

Blackish-ash, beneath dirty white; lores, chin, throat, side of neck, upper and lower tail-coverts, edge of secondaries, white; crown, ears, quills, and tail, black, (latter white-tipt); length ten inches. Oceania. Also New Holland.

Doubtful species: *Corvus Pacificus* et *Tropicus* et *Cyanoleucus* et *C. Melanoleucus,* Lath. Not seen by Temminck. The genus *Craticus,* Vieil.

THE BECARDS, Buff. PSARIS, Cuv.

Have the beak conical, very thick and round at the base, but not sloping from the forehead. The point is slightly compressed and crooked.

There is but one species, of America, ash-coloured, with head, wings, and tail, black.

Lanius Cayanus, Gm. Enl. 304 and 377.

Its habits are those of our Shrikes.

The base of the bill reddish; lores and orbits naked; base of the quills ash when young; back and chest streaked. Is the *Titria cinereus,* Vieil. Gal. t. 134. *Pachyrhynchus Cayanus,* Spix, t. 44. f. 1. ♂.

Lanius Inquisitor, Olfers.

Differs from the preceding; the bill quite black; lores feathered; inner web of the quill with a basal white spot; head quite black; back and chest spotless. *Young;* front orbits and ears reddish; back black-marked; chest with narrow streaks. Brazil.

Pachyrhynchus Semifasciatus, Spix. Braz. t. 44. f.2. ♂.

Lead-coloured above; beneath ashy white; occiput

whitish; forehead, lore, inframaxillary streak, and chin deep black; tail whitish, with a large central black band; quill black, first short, slender, falcate. Variety of the former?

Psarius Cristatus, Swainson. *Lanius Atricilla*, Cuv. Mgs.

Brown, beneath pale fulvous; base of the wings with a concealed white spot; crown, black, slightly crested; length seven inches. Brazil.

Tityra Viridis, Vieil.? *Psaris Cuvierii*, Swainson. Zool. Ill. j. t. 32. *Pachyrhynchus Cuvierii*, Spix Braz. 45. 1.

Green beneath; yellowish head; above black; nape ash; throat white; chest yellow.

Psaris Erythrogenys, Selby Zool. Journ. ii. 483.

Above ash gray; cheeks red; crown, wings, and tail, black; beneath grayish white.

Pachyrhynchus Niger, Spix. Braz. t. 45. f. 2.

Dull black; head and wing-coverts shining violet; wing with two white bands; outer tail-feather white, tipt; bill black; feet rather short; black; body five, tail two inches long. Brazils.

Pachyrhynchus Cinerascens, Spix. Braz. t. 46. f. 1.

Above ashy, beneath reddish; wing and tail chestnut; bill thick, black; cheek and lores rather naked; sides of the neck reddish; feet black; body five and a half, tail two and a half inches long. Brazils.

Pachyrhynchus Rufescens, Spix. Braz. t. 46. f. 2.

Above chestnut, beneath reddish white; base of feather

black; bill beneath yellow; wings spotless; quills blackish, red-edged; tail chestnut; feet brown; body four and a half, tail two inches. Brazils.

? *Lanius Validus*, Licht. Jeune. *Distingue roux à tête noire*, Azara, 209.

Above sooty; head rather crested; black; rump olive; beneath ashy; throat whitish; crop and chest reddish; base of dorsal feathers and quills white; second quill shortest. *Young*, back olive; quill and tail red and black varied; beneath reddish ash; length seven inches and a half. Paraguay.

? *Lanius Mitratus*, Lich. ♂. *Lanius pileatus*, fem. Lath. 31? ♀ *Muscicapa Aurantia*. Lath.?

Ashy; beneath white; wing-coverts and secondaries, obsoletely white-lined; crown black; forehead and ears white; *female:* cinnamon; crown brown; forehead ashy; wing coverts and secondaries obsoletely ferrugineous edged; length 5—6 inches; allied to *Lanius Atricapillus?* Cayenne.

* *

Psaris niger, Swainson. l. c.

Black; beneath gray; tail slightly graduated, black, white-tipt. Length five and three quarters inches long. Brazil?

Pachyrhynchus Variegatus, Spix. Braz. 43. 2.

Ashy; wing and tail black; wing-coverts and wedge-shaped tail white-tipt; quills white-edged. Brazils.

The Choucaris, Buff. Graucalus, Cuv.

Have the beak less compressed than the Shrikes; the upper crest is sharp; arched equally in its whole length; the commissure is also a little arched; feathers, which sometimes cover the nostrils, have occasioned their approximation to the ravens, but the sloping of the beak will not allow this.

They come, like the Cassicans, from the remotest parts of the Indian seas.

United with the genus *Ceblephyris* of Cuvier, by Temminck and others, and with the *Coracina* of Vieillot.

Papuan Crow, Lath. 15. *C. Papuensis*, Gm. Pl. En. 6. 30. *Ceblephyris Javanesis*, Hors.

Grayish-ash; belly white; quills blackish-brown; eye-spot black; bill yellow; feet black, short; length eleven inches. India, Sumatra.

Corvus melanops, Lath. *Rollier à masque noir*, Vail. O. A. t. 20.

Above bluish-ash; beneath paler; face and throat black; quills black, gray-edged; bill black; feet dark; female, beneath brown; banded with only one black band. Oceania.

New Guinea Crow, Lath. 14. *Corvus Novæ, Guinea*, Gm. Pl. En. 629. *Coracina Fasciata*, Vielloit.

Above deep bluish-ash; face, eye-streak, and tail, black; loins, rump, belly, thighs, and vent, black and white-banded; quills white-edged; female grayer; length twelve inches. New Guinea.

Graucalus Mentalis, Vigors and Horsfield.

Ashy-brown; beneath paler; frontal band, gular spot, quills, and tail, black; chin, vent, and tips of tail, white; length nine inches. Young of C. Melanops?

Turdus Orientalis, Lath. Pl. En. 271. 2.
Ceblephyris, *Striga*, Horsf.

Differs from the former in the eye-brows being white; loins and rump lead-coloured; tail-feather black, base white-tipt, and beneath quite white; *female*, chest and belly with blackish wavy-lines; length six inches.

Others are peculiar to India and its islands.

Rufous Shrike, Lath. 35. *Lanius Rufus*, Lin. Pl. En. 298. 2.

Rufous; beneath whitish; head greenish-black; length eight inches. Madagascar.

Dial Gracle, Lath. 9. and *Mindanao Thrush*, Ib. 95. *Gracula Saularis*, Lin. *Turdus Mindanaensis*, Gm., and *Sturnus Salaris*, Daud. Pl. En. 627. 1. alb. iii. t. 17. 18. Edw. 181. Vail. O. A. 109.

Bluish-black; belly, broad band on rump, wings and outer tail-feather, white; size of a thrush. Bengal.

Cuvier speaks of a species of a bright violet-brown colour, and the female greenish, which is probably one of the following which form

The genus Ptilonorhynchus, Kuhl. (Beitr.) which appear to be allied to the Roller. Their bill is short, hard, and strong, not nicked; swollen out in the middle; nostrils basal and hid, and the tarsi strong and short; the toes united at the base, and the wings moderate, with fourth and fifth quill the longest.

Temminck has changed the name to Citta, which Waggler has used for Pitta.

Pyrrhocorax Violaceus, Vieillot.

Ptilonorhynchus Holosericeus, Kuhl. Pl. Coll. t. 396.

Ptilonorhynchus Macleayii, Lath. H.

Shining purple-black; tail and wings dull black; bill yellowish; length thirteen inches. New Holland.

Corvus Squamulosus, Illiger.

Ptilonorhynchus Holosericeus junior, Kuhl. *Fem.* Tem. Pl. Col. 422.

Above dull-green; quills pale olive; black internally; beneath whitish yellow, dull green edged; length thirteen inches New Holland.

Varied Roller, Lath. *Ptilonorhynchus Viridis*, Waggler. *Citta Virescens*, Temm. Pl. Col. 896. *Ptilonorhynchus Smithii*, Lath. Mss.

Above dull parrot-green; beneath paler; shaft and tips of feathers with a square white spot; quills white-tipt; length twelve inches.

THE BETHYLES. BETHYLUS, *Cuv.*

With thick short beak, inflated on every side, slightly compressed toward the end.

But one species is known, the forms and colours of which resemble those of our common pies.

The *Lanius Leverianus* of Shaw; the *L. Picatus* of Latham, the *Corvus collurio* of Daudin. It has been considered as a *Tanagra* by Illiger and Temminck. Mus. Lever. t. 59. Vail. O. A. t. 60. Gal. Ois. t. 140.

To these genera must be added:

The FALCUNCULUS of Vieillot, which is peculiar for the lower and upper jaw, both being incurved.

Peculiar to the Oceanic Islands.

Frontal Shrike, Lath. H. N. 86. t. 20. *Lanius Frontalis*, Lath. Pl. Col. t. 77. Vieil. Gal. 137.

Crested; brown; beneath yellow; head and neck black; sides with two white bands. N. Holland.

Falcunculus Gutturalis, Vigors and Horsf.

Fuscous brown; beneath paler; forehead and throat white; crest and throat black; vent fulvous. New Holland.

At the end of this family may be placed the genus PSOPHODES, Vigors and Horsfield; which have a strong, short, compressed, keeled, but unnotched bill, furnished with strong incumbent bristles; short, rounded wings, and long graduated tail. They have been placed with the Honey-eaters. They are only found in New Holland.

Coach-whip Fly-catcher, Lath. *Muscicapa Crepitans*, Lath. Supp.

Olive-brown; greenish head: crested; throat and chest black; broad band under the eyes, and tips of tail white; belly varied with white; thighs reddish. New Holland. Length?

The genus COLLURICINCLA of Vigors and Horsfield, which agree with the American *Thamnophili* and African *Melanoti* in bill and wings, but differ in the tail being quite even, and the skin of the shins. They are allied to the thrushes, and appear to be peculiar to Oceanic Islands.

Colluricinca Cinerea, Vigors. l. c.

Above ash; beneath paler: throat and space before the eye white; quills internally brown; female, ashy beneath; throat black-streaked; length eight inches and a half. New Holland.

The TRICOPHORUS, or CRINIGER of Temminck: which has a short, strong, conical bill, furnished at the base with very long bristles, the wing moderate, and the sixth quill longest. And many are furnished with a bunch of hair on the nape. They are confined to the east part of Africa; five species are known, but only one has been described.

Tricophorus Barbatus, Temminck. Pl. Col. t. 88.

The genus *Sphecotheres*, Vieillot, which he has placed with the thrushes, has been placed in this genus by Quoy,

Sphecotheres viridis, Vieil. Gal. Ois. t. 147. Frey. Voy. t. 21.

Olive-green; throat, chest, and nape, ash; head, cheeks, quills, and tail, black; outer tail-feathers white-tipt; fem. ? nape green, and tail-tips black.

The TANAGERS. (TANAGRA, Lin.)

Strong conical beak, triangular at the base, lightly arched on the crest, sloped towards the end; wings and flight short; resembling our sparrows in their habits, and seeking grain, as well as berries and insects. The majority are remarkable for lively colours. We subdivide them as follows:

Consult for this genus Desmarest's work, with plates, by Th. Pauline de Courcelles.

The BULL-FINCH, or EUPHONIAN TANAGERS.

With the bill short, and, when viewed vertically, exhibiting an enlargement on each side of the base. The tail short.

Golden Tanager, Lath. 33. *Tanagra Violacea*, Lin. Pl. En. t. 114. 1. 2. Desm. t. 21. 23. β. 24. 25.

Violet; occiput, and beneath deep yellow; middle quills and lateral tail-feathers internally white; three inches and a half long. Brazil.

Negro Tanager, Lath. 34. *Tanagra Cayana*, Lin. *T. Cayenensis*, Gm. Pl. Enl. t. 114. f. 3. *Pipra Serena*, Desm. t. 26. 27.

Shining black; sides of the chest, and beneath the wings yellow. Brazil.

Tuneful Manakin, Lath. 28. *Emberiza Flavifrons*, Sparmann. *Pipra Musica*, Mus. Carls. iv. t. 92. Desm. t. 19. 20.

Dusky-black; lower part of the back, rump, and beneath orange; crown and nape blue; chin, cheeks, and throat black; forehead yellow or black. St. Domingo.

Cayenne Tanager, *Tanagra Chlorotica*, L. *T. Violacea*, Var. Lath. *E. Pusilla*, Kuhl.

Shining black; belly, chest, and forehead yellow; outer tail-feather with a white spot internally; three and a half inches long. Brazil.

Red-bellied Tanager, *Euphone Rufiventris*, Licht.

Steel-black; sides of the chest bright yellow; belly chestnut; female green; middle of the belly and occiput ash; rump chestnut; length four inches and a quarter. Brazil.

* *

Jacarine Tanager, Lath. 32. *Tanagra Jacarina*, Lin. Edw. t. 306. Pl. En. t. 224. 3. *Emberiza*, Vieil.

Violet-black; wings beneath whitish; tail divaricated, forked. Brazils.

Tanagra Viridis, Vieil. Pl. Coll. 36. 1. 3.

Green; back bluish; chest and belly yellow; crown yellowish-green. Brazils.

* *Variable Tanager*, Lath *T. Variabilis*, Lath.

Shining-green; rump greenish; tail blackish; bill horn-colour; feet brownish.

Black-necked Tanager. T. Nigricollis, Vieil. Not. Lath.

Forehead, sides of head and chin, throat, back, wing, and tail black; crown and back of neck blue; chest and body beneath yellow; bill black; feet chestnut.

Golden Tanager, T. Aurata, Vieil. *Lindo Azul*, n. 99. Azara.

Forehead, cheeks, chin, quill, and tail-feathers black; top of head sky-blue; top of neck and small wing-coverts blue; back and body beneath blue; bill black; feet chestnut. Paraguay.

Tanagra Olivacea, Desm. is the young of *T. Rubra*.

The Gross-beak Tanagers.

Bill conical, large, swollen out, higher than broad, back of the upper-jaw rounded.

The genus *Saltator* of Vieillot.

Grand Tanager, Tanagra Magna, Gm. *Saltator Olivaceus*, Vieil. Gal. t. 77. Pl. En. t. 205. Desm. t. 43.

Olive-brown; forehead and cheeks blue; maxillary streak black; throat and vent red; eyebrow and throat-spot white. Brazil.

Black-faced Tanager, Lath. 10. *Tanagra Atra*, Gm.

Tanagra Melanopis, Lath. Pl. En. t. 714. 2. Desm. t. 42.

Ash; front of head and neck, the whole of the lower part black; seven inches long; female brown beneath. Guiana.

Cayenne Roller, Lath. 14. *Coracias Cayenensis*, Gm. Pl. En. t. 616. *Fringilla Coracina*, Kuhl. *Saltator Viriscens et Cærulescens*, Vieil. *Tanagra Decumana*, Licht. *Habia à sourcil blanc*, Azara, 81. *Tanagra Superciliaris*, Spix. Braz. t. 57. f. 2.

Brownish-green; eyebrows white; a black streak on the side of the throat; lower part of neck and breast ashy; tail wedge-shaped; length nine inches. Cayenne.

Orange-billed Tanager, Lath. H. *Saltator Aurantiarostris*, Vieil. *Habia à bec orangé*, Azara, ni. 83, 84?

Lead-coloured; crown dusky; beneath, brown mixed with rufous; superciliary streak white beneath; outer tail-feather white-tipt; bill orange; length eight inches. Paraguay.

Blue-winged Tanager, Saltator Cyanopterus, Vieil. D.

Bluish-ash; front of wings blue; quills black, green-edged; tail bluish; bill and feet blackish. South America.

Saltator Flavus, Vieil. *Habia*, n. 87. Azara.

Eyebrows and body beneath deep yellow; above yellowish brown; bill blackish, beneath bluish; feet blackish brown. S. America.

Saltator Melanoleucus, Vieil.

Body above, throat and front of chest black; beneath

white; bill above black, beneath yellow, feet black. S. America. Mus. Paris.

Saltator Rubicus, Vieil. from *Habia Roxisa*, Azara, p. 8.

Crest fine red; forehead, side of head, and nape reddish-brown; throat and body beneath dirty red; body above dull red; bill blackish; feet reddish. South America.

Saltator Albicollis, Vieil.

Above brownish gray; beneath pale gray, brown-spotted; eyebrows and throat white; bill and feet brown. S. America. Mus. Par.

Tanagra Psittacina, Spix, Braz. 57. 2. ♂.

Ashy-black; bill very large, thick, high, convex, hooked and ferrugineous; wings above brownish, beneath white; body eight, tail four inches long. Brazil. Allied to T. Magna, Lath. Is it a *Pitylus*, Cuv.?

Saltator Cærulescens, Vieil. *Habia Cega blanca*, Azara. n. 81.

Head and body above bluish; beneath reddish white; bill blackish. South America.

Saltator ruficapillus, Vieil.

Head and body beneath red; throat, chin, and tail bluish-gray; forehead, cheeks, and belly black; bill bluish-black; base yellowish; feet black. S. America. Mus. Par.

Saltator Viridis, Vieil. from *Habia Verde*, Azara. No. 9.

Crown brown; eyebrows reddish; sides of head and occiput pale lead-colour; throat and upper part of body yellowish-green, beneath white; bill red, beneath blue; feet brown and blue. S. America.

Saltator Niger, Vieil.

Shining violet-black; bill and feet dull black. Brazils.

Saltator Ruber, from *Habia Purizo*, Azara, 88.

Eyebrows and body beneath reddish; above, varies red and brown; upper wing-coverts and quills deep brown; bill dull blue; feet lead-coloured. South America.

Saltator Validus, Vieil. from *Habia Robustana*, Az. 84.

Crown black, body above brown, beneath reddish-white; wing coverts gray; bill orange; base black; feet yellow. South America.

Saltator Maculatus, from *Habia Cobigas Pintadas*, Azara, 86.

Back brown, quills blackish, white-spotted; body beneath reddish; throat brown-streaked; bill above blackish, beneath blue; feet blackish. South America.

Tanagra Capistrata, Spix, Braz. t. 54, 1.

Ashy, beneath ferrugineous; head brownish; band at the base of the bill black; middle of the belly white; feet black; bill thick; quills black, pale-edged.

Tanagra Diademata, Natter. Pl. Coll. t. 243. *Pyrula Azurea*, Vieil. Gal. Ois.

Azure-blue; face, wing, and tail-feather black, latter blue-edged; crown red; nape white.

Tanagra Flammiceps, Pr. Max. Pl. Coll. 177.

Bright red-brown; beneath paler; head duller; crown crimson; bill black; quills and tail red-edged. Brazils.

Saltator Atricollis, Spix, Braz. t. 56, f. 2, ♀. *Tanagra Gugularis*, Licht. *Habia Gola Negra*, Azara, n. 42, Vieil.

Above brownish, beneath ferrugineous; tail ferrugineous beneath, obscurely banded; cheeks, throat, and crop black; quills whitish; bill thick, yellowish; body seven inches and three quarters; tail three and three quarters long. Brazil.

The True Tanagers.

The bill conical, shorter than the head; broader than high; upper jaw acute, arched.

Paradise Tanager, Tanagra Talao, Lin. Edw. t. 349. Pl. Enl. t. 7, 1, and t. 127, t 2. Desm. t. 1.

Violet, back black; rump fulvous; head green; chest and wings violet; quills and tail black; female, rump orange. Brazil.

Green-headed Tanager, Lath. 29. *Tanagra Tricolor*, Gm. Pl. Enl. t. 33, f. 1. Desm. t. 3. t. 4.

Shining green, beneath yellowish-green; wing coverts violet; face and upper part of the back black; rump fulvous,—Var. Crown and throat violet; and an orange band from each eye over the nape. Cayenne.

Black and Blue Tanager, Lath. 23. *Tanagra Mexicana*, Lin. *T. Flaviventris*, Vieil. Pl. Enl. t. 290, t. 2. 155, 1. Edw. t. 350. Desm. t. 5.

Black, beneath yellowish; crown, chin, chest, and rump blue; sides, black and blue, spotted; length five inches. South America. *Blue-shouldered Tanager*, Lath. *T. Virens*, Lin. is perhaps the young.

Red-headed Tanager, Lath. 26. *Tanagra Gyrola*, Lin. Pl. En. t. 133. t. 2. Edw. t. 23. Desm. t. 16, t. 17.

Green ; red head ; collar yellow ; chest blue ; spot on the wing-coverts reddish-yellow ; length four and three quarters inches. S. America.

Rufous-headed Tanager, Lath. 25. *Tanagra Cayana*, Lin. Pl. Enl. t. 201. 2. ♂. 290. 1. ♀. Desm. t. 10. t. 11.

Fulvous ; back green ; crown rufous ; cheeks black. *Female*, green ; beneath yellow-green ; crown red.

Bishop Tanager, Penn. Lath. 17. *Tanagra Episcopus*, Lin. Pl. Enl. t. 178. 1. 2. Edw. t. 351. 1. Desm. t. 15.

Ash ; wings and tail blue externally. *Female*, brown ; head, neck, and chest, bluish ; belly grayish ; wing and tail black ; length seven inches. Cayenne.

Archbishop Tanager, Lath. H. *Showy Tanager*, Lath. *Tanagra Ornata*, Lath. *Tanagra Archiepiscopus*, Desm. t. 17. 18. Mus. Carl. iv. t. 95. Spix, Braz. t. 55. 1. ♂.

Lazuline blue ; wings and tail greenish ; smaller wing-coverts silky yellow ; back blackish azure ; body seven, tail three inches long. Brazil.

Sayacu Tanager, Lath 18. *Tanagra Sayaca*, Lin. Pl. Enl. 301. 1. *Gracula Glauca*, Sparm. Mus. III. 14. *T. Episcopus*, Lath. *T. Cœlistis*, Spix. t. 56. 1. ○.

Glaucous ; wings and tail bluish ; smaller wing-coverts silky milk-white. Brazil.

Tanagra Olivascens, Licht. *Tanagra Sayaca*, Fem. Lath.

Shining olive; crown greenish; wings and tail brown; base of the quills and coverts greenish-gray. Brazil.

Red-bellied Warbler, Lath. 146. *Motacilla Velia*, Gm. *Tanagra Varia*, Desm. t. 2. Pl. t. Enl. 669. 3. Edw. t. 22.

Fine black, mixed with brown; belly and breast rufous; greater wing-coverts, quills and tail black, blue-edged; forehead bluish-green; rump gilded-green. S. America.

Spotted Green Tanager, Lath. 19. *Tanagra Punctata*, Lin. Pl. Enl. t. 133. 1. Desm. t. 8. 9. Edw. t. 262.

Green, black-spotted; beneath yellowish-white. The young or female of *Tanagra Sayaca?* Var. *Spotted Emerald Tanager*, Lath. H.

Palm Tanager, *Tanagra Episcopus*, Desm. ♀, *T. Palmarum*. Pr. Max. Pl. Col. t. 178. 2. Desm. t. 16.

Tanagra Chloroptera, Vieil.

Above, pale yellow; quills and tail green; throat and front of neck black; bill and feet pale brown. Brazils.

Tanagra Melanotha, Vieil.

Forehead, sides of head, and back black; crown and back of neck blue; quills and tail black, blue-edged; body, beneath red; bill and feet black. Brazils.

Tanagra Formosa, Vieil. *Lindo Bello*, Azara, 94.

Yellow cheeks, throat, neck, and chest black; larger

lower wing-coverts silvery; bill black, beneath blue; feet violet-black. South America.

Tanagra Canora, Vieil. from *Xiuhtototl,* Fernandez.

Blue varied with fulvous; tail black, white-tipt; wings partly blue and partly fulvous; bill reddish-white; feet gray. New Spain.

Tanagra Leucocephala, Vieil. *Lindo Azul Cabeza blanea,* Azara, 93.

Dull violet, crown bluish-white; bill black; feet blackish. South America.

Tanagra Desmarestii, Vieil.

Forehead black, crown blue; nape, chin, and body beneath yellow; above yellow and black varied; bill brown; feet flesh-coloured. Brazil.

Orange Finch, Lath. *Fringilla Zena,* Lath. *Tanagra Multicolor,* Vieil. Gall. Ois. t. 76. Catesby, Carol. 1. t. 42.

Head, body above, quills and tail black; eyebrows white; throat yellow; chest and rump reddish; belly yellow; bill and feet black.

Tanagra? Melalictera, Guldest. Nov. Act. Petrop. xix. t. 13, 14.

Above ferrugineous; beneath deep yellow; head and nape, wings and tail brown, white streaked; bill and feet livid; very doubtful if a Tanager. Russia.

Tanagra Striata, Vieil. *Lindo Celeste Oro y Negro,* Azara, 94.

Beneath yellow, head blue and black; back above

black; quills, and larger upper wing-coverts and tail black, blue-edged; bill blackish, white beneath; feet bluish. S. America.

Tanagra Peruviana, Vieil. *T. Cayana,* mas Desm.

Crown and back of neck foxy-red; throat and chest green; lower part of back pale-yellow; primary quills and tail brown, greenish-blue edged; feet and bill brown. South America.

? *Tanagra Rudis,* Lath.

Throat brownish-ash; chest, belly, and rump ferrugineous red; bill and feet brownish-ash. Coromandel.

? *Tanagra Ornata,* Lath.

Beneath shining ash, above brownish-green; head, mask, and shoulders violet; wing spot golden yellow; bill dull gray. East Indies.

Tanagra Fasciata, Licht.

Olive-gray; beneath ash; cheeks and wing-coverts black; throat and basal wing-band white; length six inches. St. Paul, Brazil.

Tanagra Leucophæa, Licht.

Gray; face bluish-black; crown, throat, chest, and rump pale ferrugineous; middle of belly white; length six inches. Brazil.

Blue-headed Tanager, T. Tricolor, B. Lath. *Tanagra Cyanocephala,* Vieil. *T. Rubricollis,* Tem. Kuhl. Pl. Enl. 33. 2. ♂. Pl. Coll. 245. 2. ♀.

Green; face, upper part of back, quills, and tail black,

two latter green-edged; top of head, upper part, and throat blue; side of face, ears, and lower part of neck red.

Tanagra Thoracita, Tem. Pl. Coll. 42. 1.

Greenish; beneath paler; face, bill, and spot on throat black; throat, side of neck, small wing-coverts, and vent orange. Brazils.

Tanagra Citrinella, Temm. Pl. Coll. 42. 2.

Yellow; face, spots on throat, and spots on back black; wings and tail green; beneath bluish. Brazils.

Tanagra Vittata, Temm. Pl. Coll. 48. 1. 2.

Blue; beneath pale-brown; forehead, bill, side of face, ears, and upper part of back black; female, back, wings, and tail greenish.

Tanagra Canicapella, Swain. Zool. Illust. iii. t. 174.

Tanagra Tephrocephala, Vieil.

Crown and neck above ash; back, wings, and tail olive; forehead, cheeks, and throat yellow; body beneath bluish-gray in the centre; bill and feet black. S. America.

Tanagra Cyanoventris, Vieil.

Face black, crown, nape, and chin yellow-green; back and upper wing-coverts yellow and black; chest blue; belly blue with a yellow centre; bill black; feet flesh-coloured. Brazils.

Tanagra Rufiventris, Vieil.

Blue-black, side of chest yellow; centre of belly reddish; bill black; feet brown. Brazils.

Green Tanager, Lath. *Tanagra Virens*, Lin.

Above green, beneath yellowish; lores, cheeks, and throat black; bill dull black; feet brown. Brazil.

Tanagra Chlorocyanea, Vieil.

Olive green, throat, neck, and middle of belly blue; bill and feet brown. South America.

Tanagra Graminea, Spix, Braz. t. 53. 2.

Green, beneath pale yellow; quills and tail black; crown green-edged; bill finch-like; throat yellow-green; feet yellowish; body four and three quarters; tail two inches long. Brazils.

Schrauk's Tanager, Tanagra Schrankii, Spix, Braz. t. 51. 1. ♂. 2. ♀.

Varied black and green, beneath golden; forehead and lores black; rump and crown orange; throat and sides greenish; bill short, compressed, black; body four, tail one inch and a half long. *Female* not yellow-crested; back duller. Allied to *T. Citrinella,* Pl. Col. Brazils.

Tanagra Axillaris, Spix, Braz. ii. t. 54. f. 2 ☿.

Dull ash; beneath white; quill and tail blackish, partly banded.

Tanagra Viridis, Spix, Braz. 48. 2. ♂.

Green, beneath yellowish; bill compressed, scarcely thick; feet yellowish-white; body six and three quarters; tail three inches long; whiskers black; tarsi long, strong; quills blackish, brown tipt.

Some are peculiar for their bills being very slender, like the Warblers.

The genus NEMOSIA, of Vieillot.

Red-throated Tanager, Lath. *Tanagra gularis*, Lin. Pl. Enl. t. 155. 2. Desm. t. 12. 14.

Black, beneath white; head red; throat crimson; length seven inches. Brazil.

Hooded Tanager, Lath. 11. *Tanagra Pileata*, Gm. Pl. Enl. t. 72. 2. Desm. t. 41.

Bluish-ash, beneath white; crown and streak on side of neck black; a white spot between eyes and beak; female bluish-ash, beneath white; length seven inches. Brazil.

Cuvier observes, that the *Black-throated Tanager*, Lath. *T. Nigricollis*, Gm. Pl. Enl. t. 720. 1. is a Warbler *(Sylvia)*. Vieillot places it in this genus.

Yellow-throated Tanager, *Nemosia Flavicollis*, Vieil. Gal. t. 75.

Black, throat, back, rump, and vent yellow; chest and belly white; bill brown, beneath white; feet black. Brazils

Tanagra Speculifera, Temm. Pl. Coll. 36. 1. 2.

Black; throat, sides of neck, spot on wings, back, and rump, and vent yellow; beneath yellowish-white; *female* olive-green; edge of quills, tail-feathers, and beneath yellow. Brazils.

Red-headed Tanager, *Nemoria Ruficapillu*, Vieil. D.

Head and throat red; sides of neck, and rump pale-yellow; body above green-olive; bill brown; beneath yellow. South America.

The Oriole Tanagers,

With the bill conical, arched, acute, and nicked at the end.

The genus Tachyphonus of Vieillot.

Crested Tanager, Lath. 9. *Tanagra Cristata*, Lin. Pl. Enl. t. 7. 2. Desm. t. 47—49.

Blackish, crest orange; throat and rump fulvous; *female*, brown beneath; rump and margin of the quills bright ferrugineous; bill brown; base of the lower jaw white.

Tanagra Martialis, Temm. Pl. Enl. 301. 2. *Tachyphonus Desmarestii*, Swainson.

Glossy black; crest and rump fulvous; vent rufous; under wing-coverts snowy.

Red-necked Tanager, *Tanagra Ruficollis*, Licht.

Black, throat chestnut, paler near the chest; belly whitish; double spot on the wing; and loins white; bill and feet black; six inches long. Is it of this section? Brazil.

White-winged Oriole, Lath. 31. *Tanagra Nigerrima*, Gm. Pl. Enl. t. 179. 2 ♂. 711. ♀. *Oriolus Leucopterus*, Gm. *O. Melaleucus*, Sparmann. Mus. Carls. ii. 31. ♂. G. Ois. 82.

Black; wing-spot white; *female*, cinnamon-brown; beneath ashy; length eight inches. S. America.

Cuvier has referred *Tanagra Olivacea* to this section, but it has since been proved to be the young female of *T. Rubra*.

Tanagra Speculifera, Temm. Pl. Col. 36. 1. 2.

T. Cirrhomelas, Vieil. D. Desm. t.

Head, back, belly, wings, and above the tail, black; three outer tail-feathers yellow; shoulder-spot white; tail, beneath fulvous; bill black, beneath yellow; feet black. South America.

Palm Thrush, Lath. 108. *Turdus Palmarum*, Lin. Pl. Enl. t. 539. 1.

Olive green; beneath ashy; occiput and cheeks black, with three white spots on each side; length six inches. Brazil.

Tanagra Quadricolor, Vieil. D.

Forehead, cheeks, wings, and tail, black; inner webs of the quills white; crown and body beneath yellow; cheeks and back of neck dull ash; feet and bill, brown.

Tachyphonus Olivaceus, Swain.

Olive; beneath fulvous-white; crown ash; orbits yellow.

Tachyphonus Vigorsii, Swain.

Violet-black; crest red; grapular and under wing-coverts snowy.

Black-faced Finch, Lath. 4. *Fringilla Cristata*, Gm. Pl. Enl. t. 181. 1. *T. Flammea*, Lath. not Lin. Vieil. Ois. Chant. t. 29. *Tachyphonus Rubescens*, Swain.

A. Loxia of Temminck?

Tachyphonus Fringiloides, Swain.

Ash; beneath whitish; crest crimson; sides black; bill short conic.

Tachyphonus Suchii, Swain.

Olive beneath; pale fulvous; crest yellow; wings black; inner base of quills white.

Tachyphonus Tenuirostris, Swain.

Glossy, olive-black; scapulars white; under tail-covers niforus; bill slender.

Tachyphonus Ruber, Vieil.

Dull red; crown purple-red; body beneath, rosy red; bill and feet reddish. South America.

Tachyphonus Chloristerus, Vieil.

Above, wings and tail green; beneath yellow; bill brown; feet reddish. Brazils.

Tanagra Saira, Spix. Braz. t. 48. 1. ♂.

Above, yellowish-green; beneath lemon-yellow; bill thick, black; forehead yellowish; crown not crested; throat cinnamon colour; body seven, tail three inches long. Brazil.

Tanagra Penicillata, Spix, Braz. t. 49. 1.

Yellowish-white above; beneath orange; head ashy; occipital crest drooping, olive and white; throat, crown white; bill, short, subulate; tarsi slender; quills yellow-green; tail olive-green; body six and a half, tail three inches long. Brazils.

Tanagra Brunnea, Spix, Braz. t. 49. 2 ♂.

Above brown, beneath ferrugineous; occipital feathers long, reddish; bill short, compressed, and arched; rump reddish; feet blackish; body four and a half inches, tail two and three quarters. Brazils.

Spix has placed here *Muscicapa Rubricollis*, Pl. Enl. t. 381, which Cuvier calls an *Ampelis*, and Temminck a *Coracina*.

Golden-crested Tanager, *Tanagra Auricapilla*, Pr. Max. Spix, Braz. t. 52. 1. ♂. 2. ♀. Azara. N. 101.

Above olive, beneath fulvous ferrugineous; crown lemon-colour; wings and tail black; base of the secondary quills white; body six and three quarters, tail three inches; *female* darker beneath; crest very small. Brazils.

Red-bellied Tanager, *Tanagra Rufiventer*, Spix. Braz. t. 50. 1.

Black, beneath and hinder part of back reddish-yellow; crown yellow; wing-coverts white; bill arched, compressed; sides reddish; body six, tail three inches long; differs from *T. Cristata*, Lin. crest not purple; belly not black. Brazil.

Golden-forehead Tanager, *Tanagra Aurifrons*, Spix, Braz. t. 50. 2.

Above brown; feathers gray-edged; beneath whitish; forehead of the males, especially over the eye and shoulders, yellow; bill conical, compressed, keeled; body four, tail one and a half inches long. Brazil.

Red-necked Tanager, *Tanagra Ruficollis*, Spix, Braz. t. 54. 3. ♂.

Fuscous brown; beneath white; head and spot over the ears black; streak over the eyes and on the crown white; napal collar reddish; body five and a half, tail two inches and a quarter. (*Emberiza cap.* N. S. Pl. Enl. 386?) Allied to *Palm Thrush*. Brazil.

Tanagra Cristatella, Spix, Braz. 53. 1.

Black; brown above; dirty white beneath; base of the feathers ashy black; crown crested, black, purple centred; bill finch-like; wings spotless; tail black; body five, tail two inches and a half long. *Tanagra Pileata*, Lath.? *Fringilla*, Newied?

Cardinal Tanagers.

With the bill conical, slightly swollen, and with a blunt prominent tooth at the sides.

The genus *Pyranga* of Vieillot.

Mississippi Tanager, Lath. *Muscicapa Rubra*, Lin. *Tanagra Mississipensis*, Gm. *T. Variegata*, Gm. *Loxia Virginica*, Gm. *Tanagra Æstiva*, Gm. *The Summer Red Bird* of Catesby. Car. 1. 54. Edw. t. 239. Pl. Enl. t. 741. Lath. H. t. 93. Desm. t. 32. 33. Wils. A. O. i. t. 6. f. 3. 4.

Red Tanager, Lath. 45. *Loxia Mexicana*, Lin. and *T. Olivacea*, Gm. *The Scarlet Sparrow*, Edw. 343. t. 44. Pl. Enl. t. 156. 1. Brisson, iv. t. 2. 5. Desm. Tang. t. 34. 37. *Pyranga Erythromelas*, Vieil.

Red; wings and tail black; tail white-tipt; female and young olive-green; beneath white; throat and chest yellow; quills and tail brown.

Black-headed Tanager, Lath. 13. *Tanagra Atricapilla*, Gm. Pl. Enl. 809. 2.

Red; head, wings, and tail black; streak on wing white; length seven inches. Guiana.

Cuvier has proposed to place this with *Lanius*, and Vieillot makes it a genus called *Lanio*. But Temminck places it here.

Pyranga Cyanictenis, Vieil. Gal. Ois. t. 81.

Above blue; body beneath yellow; bill black; feet pale-yellowish. South America.

Pyranga Cinera, Vieil.

Dark gray; wing and tail-coverts white-spotted; tail white-tipt; feet and bill black. South America.

Tanagra Ludoviciana, Wilson. A. O. t. 20. 1. *Pyranga Erythropis*, Vieil.

Back, wings, and tail black; large wing-coverts yellow-tipt; smaller, yellow; body beneath yellowish-green; face and chin pale-red; bill yellowish; feet blue. S. America.

P. Icteropus, Vieil. D.

Head, back of neck, and back greenish; quills and side tail-feathers brown, blue-edged; chin, throat, and beneath yellow; bill brown; feet yellow. South America.

P. Icteromelas, Vieil. D.

Above black; beneath and middle of the throat yellow; bill blackish; beneath horn-colour; feet reddish. S. America.

Green-headed Tanager. *P. Chlorocephalus*, Vieil. D.

Head greenish; body above very pale-blue; beneath yellow; bill brown; feet reddish. South America.

The Ramphoceline Tanagers.

The bill conical, and the branches of the lower jaw swollen behind.

The genus *Rhamphocelus* and *Rhamphopis* of Vieil.

Brazilian Tanager, Lath. 2. *Tanagra Brazilia*, Lin.

T. Rudis, Lath. Mus. Carl. iv. t. 92. ♀. Pl. Enl. t. 126. 1. 127. 1. Desm. Tang. t. 30. 31. *Rhamphocelus Coccineus*, Vieil.

Scarlet ; wings and tail black ; bill black ; middle of the base of the lower jaw white ; length six inches and a half. Brazil.

Red-breasted Tanager, Lath. 1. *Tanagra Jacapa*, Lin. *Lanius Carbo*, Pallas. Pl. Enl. t. 128. 1. 2. Edw. t. 267. Desm. t. 28. 29. Nat. Misc. t. 234. *Rhamphocelus Purpureus*, Vieil.

Black ; forehead, crop, and chest crimson ; six inches and a half long ; *female* duller. South America.

Black-throated Tanager, Tanagra Nigrogularis, Spix, 47. 1.

Crown, neck, sides, and rump scarlet ; face, orbits, cheeks, chin, middle of the belly, back, wings, and tail black ; body six, tail three inches long.

Latham notices the *Poppy Tanager, Habia Ponceau*, and the *Paraguayan Tanager, Habia Jaune* of Azara, all from Paraguay. These, with some other South American birds, form the genus *Saltator*, of Vieillot.

Tanagra Virens of Linnæus, *T. Variabilis*, and *T. Canora* of Gmelin, have not been figured. *T. Albifrons*, and *T. Amboinensis* are taken from Seba.

Tanagra Sinensis, Gm. from Sonn. 114, is perhaps a Finch, and *T. Capitalis*, Lath. 112, is perhaps a *Ploceus*.

Tanagra Cœrulea, Pl. Enl. t. 203. 2. is a *Finch* ; *T. Atrata* is a *Lamprotornis*. *T. Militaris*, an *Icterus*, and *T. Albirostris*, an Oriole. *T. Guianenis* is a *Lanius*.

Vieillot has here placed the genus *Dulus*.

St. Domingo Tanager, Lath. 16. *Tanagra Dominica*,

Lin. Pl. Enl. t. 156. f. 2. *Dulus Palmarum*, Vieil. Gal. Ois. t. 147.

Above olive-brown; beneath whitish; black spotted; tail slightly forked. St. Domingo.

And also the genus *Arremono*.

Silent Tanager. Tanagra Silens, Lath. from Sonn. t. 117. Pl. Enl. t. 742. Shaw Zool. x. t. 42. Misc. t. 761. Desm. t. 38—40. *Arremon Torquatus*, Vieil. Gal. t. 78.

Green; head and beneath horny; sides of head, eye-brows, throat, and shoulder, yellow; throat-bar black.

THE FLY-CATCHERS, GOBEMOUCHES, CUV. MUSCICAPA, Lin.

Have the beak depressed horizontally, furnished with hairs at the base, and the point more or less bent and sloped. Their manners are, in general, those of the shrikes, and they live on small birds or insects, according to their size. The weakest of them pass insensibly into the form of the slender beaks. We divide them as follows:

THE TYRANTS, TYRANNUS, CUV.

With a very long, strong, and straight bill. The upper ridge straight and blunt; the point is suddenly bent. These are American birds, of the size of our shrikes, and equally brave. They defend their young even against the eagles, and are able to drive from their nests all the predacious birds. The largest species prey on small birds, and will even sometimes attack carcasses.

The first section has the tail very longly-forked; wings long; inner web of quills nicked. *Gubernetes*, Such. (Vigors.)

Muscicapa Yiperu, Licht. 1823. *Gubernetes Cunninghami*, Such. Zool. Journ. 1825. ii. t. 4. *Muscicapa Longicauda*, Spix, Braz. ii. t. 17. *Yiperu*, Azara, 75. *Tyrannus Bellulus*, Vieil.

Ash-coloured; black streaked; red streak on middle of wing; tail and wing blackish, white-edged; throat white, with broad chestnut crescent; body fourteen, tail ten inches long. Brazil.

Muscicapa Vetula, Licht. (1823.) Spix, Braz. t. 18.

Ash; wings and forked tail sooty; body eight inches, outer tail-feather four and a half, inner three inches long.

Muscicapa Tyrannus, Lin. *Tyrannus Savanna*, Vieil. O. A. S. t. 43. Briss. t. 39. 3.

Body black above, white beneath; bill and feet white. South America.

Muscicapa Fucata, Spix, Braz. t. 19.

Olive; lemon-yellow beneath; crown orange; throat ashy-white. Brazil.

Tyrannus Longipennis, Swain.

Cinereous; chin whitish; tail brown. Brazils.

Muscicapa Despotes, Licht.

Head gray; base of crown-feathers scarlet; back olive; quills and tail black; throat, chest, and belly bright-yellow. Bahia.

Muscicapa Forficata, Lath.

Pale-gray; white beneath; quills and tail-feathers black, gray edged; bill and feet black. South America.

Tyrannus Melancholicus, Vieil. *Suiriri Guau*, Azara, n. 138.

Crown-feathers long, yellow or red; body above

blackish-brown; deep-yellow beneath; outer tail-feathers very long. South America.

Muscicapa Yetapa, Vieil. *Tyrannus Violentus*, Vieil. *Yetapa*, Azara. 190.

Crown-feathers yellow, black-tipt; bluish ash above, white beneath; quills, tail, bill, and feet black. South America.

In the second section, the tail is square. Some have the wings moderate, inner beard of the quills entire. *Platyrhynchos*, Temm.

Lanius Sulphuratus, Lin. *Corvus Flavus*, Gm. Pl. Enl. t. 296. 249. Vieil. O. A. S. t. 47. *Tyrannus Magnanimus*, Vieil. *Corvus Flavigaster*, Lath.

Brown; yellow beneath; head blackish, with a whitish ring; bill and feet black. South America.

Lanius Pitungua, Lin. *Tyrannus Pentaveo*, Vieil. Pl. Enl. 212. Vieil. O. A. S. 16. *Tyrannus Carnivorus*, Vieil.

Black; beneath yellow; crown-streak fulvous; ocular band, white; bill black.

Muscicapa Audax, Lath. Pl. Enl. 453. 2.

Black; yellowish beneath; crown yellow; face, &c., white. Brazils.

Muscicapa erincta, Lin. *Muscicapa Ludoviciana*, Lath. Wils. A. O. t. *Tyrannus Irritabilis*, Vieil. *Tyr. Cayanensis*, Briss. *Suiriri Brun et Rouge*, Azara, 95.

Head and neck bluish; belly yellowish; back greenish; quills and tail reddish; bill and feet brown. N. America.

Spiny-footed Tyrant, Tyrannus Calcaratus, Swain.

Cinereous brown; knees armed with small acute spines. Brazils.

Muscicapa Legatus, Licht.

Olive brown; white beneath; length five inches and a half. Bahia.

Tyrannus Bellicosus, Vieil. From *Suiriri Roxo Obscuro,* Azara, n. 189.

Crown and neck above reddish-brown; back blackish; red beneath. S. America.

Red-brown Tyrant, Tyrannus Pyrrhophaius, Vieil.

Greenish-brown above; red beneath. S. America.

Tyrannus Rixosus, Vieil. *Suiriri,* n. Azara, 137.

Pale brown; yellow beneath. South America.

Tyrannus Solitarius, Vieil. *Suiriri Chorreado todo,* Azara.

Crown-feathers, inside yellow, outside blackish; body beneath blackish; belly white and brown spotted. South America.

In others the wings are moderate, and the tarsi long.

Muscicapa Cinereus, Gm. Briss. Suppl. t. 3. f. 2. *Tyrannus Rufus,* Vieil.

Ash-rump; tail and body beneath reddish. S. America.

Yellow-rumped Flycatcher, Lath. *Tyrannus Rufescens,* Swain.

Ferrugineous; paler beneath; tail rufous. S. America.

Tyrannus Ambulans, Swain.

Olive brown; yellow beneath; crest orange. Pernambuco.

Black and White-Winged Tyrant, Lanius Nengeta, Lath. (not Syn.)

Gray; white beneath; wings and tail black. Brazils.

Tyrannus Albicollis, Vieil. *Suiriri Chorreado sin Roxo,* Azara, n. 186.

Body above blackish-green; beneath yellow, with blackish cross-bands. S. America.

Wings and tail equal; are unknown?

Muscicapa Ioazeiro, Spix. Brazils, ii. t. 23.

Olivaceous yellow above; sulphureous beneath; crown crimson, erectile; body six inches and one-third; tail two inches and three-quarters long. Brazil.

Muscicapa Polyglotta, Lich. Spix. Brazil. t. 24. *Pepoaza,* Azara, 201. *Tyrannus Pepoaza,* Vieil.

Ash above; ashy-white beneath; body eight inches, tail three inches and three quarters long. Brazil.

Muscicapa Similis, Spix, Brazil. ii. t. 25.

Olive-brown above, pale yellow beneath; crown orange, quill and tail black-brown; body six inches; tail two inches and three-quarters long.

Muscicapa Thamnophiloides, Spix. Brazil. 26.

Chestnut; paler beneath; body seven inches and a half; tail three inches and a quarter long. Brazil.

Muscicapa Cinerea, Lath.? Spix. Brazil. t. 26. 2.

Reddish chestnut; head and nape lead-coloured; throat whitish-ash; body seven inches and three-quarters; tail three inches long. Brazil.

Tyrannus Rufiventris, Vieil. Az. 205.

Throat, crop, and body, above lead-coloured; beneath reddish; bill and feet black. S. America.

Tyrannus Atricapillus, Vieil. Az. 204.

White; head, tail, quills, and wing-coverts, black. S. America.

In others, the beak is moderate; the wings long; inner beard of the quill nicked; and the tarsi short; tail various.

Thick-billed Tyrant, *Tyrannus Crassirostris*, Swain.

Gray-brown; pale-yellow beneath; bill strong, Mexico.

Noisy Tyrant, *Tyrannus Vociferus*, Swain.

Olive-gray; yellow beneath; crest red; primaries pointed. Mexico.

Lanius Tyrannus, var. Lath. *Tyrannus Pipiri*, Vieil. *Tyrannus Intrepidus*, Vieil. Wils. A. O. ii. t. 13. 1. Vieil. Gal. Ois. t. 133. O. A. S. t. 44. *Muscicapa Animosa*, Licht.

Ash; beneath white; crown blackish, with yellow streak. S. America.

St. Domingo Tyrant, Lath. *Tyrannus Griseus*, Vieil. O. A. t. 46. Briss. ii. t. 32. 2.

Cinereous; white beneath; tail forked. Mexico.

Tyrannus Crudelis, Swain.

Olive; yellow beneath; crest orange; tail forked. Brazils.

Tyannnus Verticalis, Say.

Rocky Mountains.

Tyrannus Leucotes, Swain, Pl. Enl. t. 820. 2? *M. Barbata*, ♀. Buffon?

Gray-brown; whitish beneath; crest yellow; quills pointed; tail even.

Tyrant Flycatcher, Lath. *Muscicapa Ferox*, Gm. Pl. Enl. t. 571. 1. *M. Fusca*, Lin. *M. Nunciola*, Wilson, A. O. t. 13. 4. *Suiri Pardo Aplomado*, Azara.

Brown; chin, throat, chest, ash; belly yellowish; feet blackish; bill brown. America.

Muscicapa Atra, Gm. *Muscicapa Phœbe*, Lath.

Ashy-olive; beneath yellowish; chest ashy; tail-feathers white-tipt.

Lanius Tyrannus, Var. Lath. *Tyrannus Matutinus*, Vieil. Pl. Enl. t. 537.

Body gray above; crown-feather orange, ashy-tipt; chest ashy-white; belly dull white. North America.

Tyrannus Vorax, Vieil.

Gray; paler beneath; bill very broad; first quill very deeply nicked.

Tyrannus Coronatus, Vieil. *Muscicapa Vittigera*, Licht. *Pepoaza Couronné*, Azara. n. 202.

Forehead, band above the orbit, and all beneath, white; crown black; tail black, ashy-fringed. Paraguay.

Muscicapa Velata, Licht. Spix, ii. t. 22.

Ashy; forehead whitish; belly, rump, and lower half of tail white; end of tail white; length eight inches. Paraguay.

Muscicapa Cinerascens, Spix. Braz. t. 21. f. 1. ♂. 2. ♀.

Ashy; paler beneath; bill strong; length seven inches. Brazil.

Querula Cinerea, Vieil.? *Muscicapa Plumbea*, Licht.

Olive-gray; paler beneath; length ten inches, tarsi one inch.

Muscicapa Pagana, Licht.

Olive above; throat whitish; chest grayish, sprinkled with yellowish-green; wing-coverts yellow-tipt, forming two bands; bill attenuated; length five inches. Bahia.

Muscicapa Strangulata, Licht.

Olive-green; white beneath; bill rather long, black; length five inches, tarsi one inch. St. Paulo.

Muscicapa Oleaginea, Licht.

Olive-green; pale ferrugineous beneath; length four inches and a half, tarsi seven lines; bill, base broad, depressed and nearly subulate.

Muscicapa Sulfurea. Spix, Braz. t. 20.

Greenish-olive above; lemon-colour beneath; crown orange-yellow; tail square; length eight inches, tail three and a quarter.

Muscicapa Sibilans, Licht. Le Siffleur d'Azara, n. 191.

Back olive; crown and tail black; belly amianthus; length seven inches and a half. St. Paulo.

Muscicapa Galeata, Licht. *Lindo Bruna-huppe Guane*, Azara, 101.

Forehead and orbits black; crest and occiput orange; back olive; entirely fulvous beneath; tail rounded and black; length seven inches. St. Paulo.

Muscicapa Nigriceps, Licht.

Green; chest yellow; throat and belly white; crown black; length six inches. Rahia.

M. Comata, Licht. *Muscicapa Galeata*, Spix. Braz. t. 27. ♂ 28. 1. (not Licht.)

Violet-black; crown crested; body seven, tail three and three quarters inches; *female?* crown not crested. Brazil.

Muscicapa Flavicauda, Spix. Braz. t. 28. 2.

Above, sooty-olive; yellow, white beneath; tail reddish-yellow; body five, tail two inches long. Brazil.

White Fly-catcher, *Muscicapa Mœsta*, Licht. *Tyrannus Irupero*, Vieil. *Muscicapa Nivea*, Spix, Braz. t. 29. 1. *Irupero*, Azara. 204.

Snow-white; primaries and tips of tail black: body

five inches and three quarters, tail two inches and three quarters long. Brazil.

Muscicapa Albiventer, Spix, Braz. t. 30. 1. ♂. 2. ♀.

Blackish above; forehead, and beneath, snow-white; tail deep-black; body four inches and a half, tail one inch and a half.

Muscicapa Dominicana, Licht. Spix, Braz. t. 29. 2. *Viuva Brazilians*, *Pepoaza Dominicain*, Azara, n. 203. *Tyrannus Dominianus*, Vieil.

Black; head, neck, and beneath, white; body five, tail two inches. Brazil.

Muscicapa Rufina, Spix, Braz. t. 31. 1. ♂. 2.· ♀.

Brown; yellowish-white beneath; tail and bill long; male, crown yellow; body five and a half, tail three inches long.

Muscicapa Mystacea, Spix, Braz. t. 31. *. 1. ♂ 2. ♀.

White; streak above the ears, wing, and white-tipt tail, black; middle of the back dirty-white; tail somewhat wedge-shaped; body five, tail two inches one-fifth long. Brazil.

Muscicapa Varia, Vieil. *Suiriri Chorradeo Debazo*, Azara, 178.

Blackish, pale-yellow beneath. South America.

Muscicapa Flava, Gm. Pl. Enl. t. 569. 2. Vieil. Ois. Am. Sept. 41.

Brown, yellow beneath; bill and feet brown.

Others, which inhabit New Holland, have the bill

very broad and strong, furnished with strong bristles; nostrils round, and the tail equal. They form the genus *Monarcha* of Vigors and Horsfield.

Muscicapa Carinata, Swain. Zool. Ill. t. 147.

Lead-coloured; cheek and side of neck paler; forehead and throat black; belly, lower wing-coverts, and vent ferruginous; length New Holland.

THE FLYEATERS, MUSCIPETA, CUV.

HAVE the beak long and very depressed; twice the width of its height even at the base. The crest is very obtuse, and yet mobile. The edges have a slight oval bend. The point and the notching are weak, and there are long threads at the base of the beak.

Their weakness allows them to take only insects; and many of them are adorned with long plumes at the tail, or fine crests on the head, or at least with plumage of brilliant colours.

The majority of them belong to Africa or India. Some species allied to these are remarkable for a beak still more enlarged and depressed than the preceding.

Others, which have the beak large and depressed, are distinguished by high legs and a short tail. There are but two or three known, of America, which live on ants; whence they have been united to the little tribe of Thrushes called Anteaters.

Muscicapa Cristatus, Pl. Enl. 373. 2. Vail. O. A. iii. 142. 1.

Azure Flycatcher, Lath. 36. *Muscicapa Cœrulea*, Pl. Enl. t. 666. 1.

Shining blue; bluish-white beneath; length five inches. Philippines.

Muscicapa Cyanea, Vieil.

Deep blue; belly and vent red; length six inches. East Indies.

Collared Flycatcher, Lath. 11. *Muscicapa Melanoptera*, Gm. Pl. Enl. t. 567. 3. *Muscicapa Collaris*, Lath. *Platyrhynchus Collaris*, Vieil.

Ashy lead-colour; wing-band and beneath white; length four inches and three quarters.

Le Mantele, Vail. O. A. iv. t. 151. 1.

Le Molenar, Vail. O. A. t. 160. 1. 2.

Platyrynchus Perspicillatus, Vieil. *Le Gobe-mouche à lunettes*, Vail. O. A. t. 152. 1.

Deep brown above; white beneath; eyebrows and orbits white. Africa.

Yellow-necked Flycatcher, *Muscicupa Flavicollis*, Lath.

Green, throat yellow. China.

Desert Flycatcher, *Muscicapa Deserti*, Sparmann. Mus. Car. ii. t. 47.

Body ferrugineous and sooty black; belly yellowish. Africa.

Muscicapa Melanoleuca, Guld. Nov. Comm. Petrop. xx. t. 15.

White; chest yellowish; wings and tip of tail-feathers black; length six inches. Georgia.

Muscicapa Fuscesens, Lath.

Brownish, whitish beneath. China.

Muscicapa Afra, Lath.

Dull yellowish, black-spotted; length seven inches. Cape of Cood Hope.

Muscicapa Luzoniensis, Lath. Sonnerat, Voy. t. 27. f. 2.

Violet, black-gray beneath. East Indies.

Muscicapa Philippensis, Lath.

Gray brown, whitish beneath; eyebrows white.

Muscicapa Madagascariensis, Gm. Brisson. ij. t. 24. f. 5.

Olive; throat yellow; crop and chest yellowish. S. Africa.

Crested Promerops, Lath. *Upupa Paradisea.*
Muscicapa Paradisii, Lath. *Todus Paradisiacus*, Gm. Pl. Enl. t. 234. ♀. 2. Vail. O. A. t. 144.

Head black; body white. Cape of Good Hope, Madagascar.

Muscicapa Cristatus, Lath. Pl. Enl. t. 573. 2. Vail. O. A. t. 142. 1. 2.

Head crested; bay above; ash beneath. Africa.

Mutable Flycatcher, Lath. *Muscicapa Mutata*, Lin. Pl. Enl. t. 248. 2. Vail. O. A. t. 148.

Crested; varies in colour; length eleven inches.

Muscicapa Borbonica, Lath. Pl. Enl. t. 573. 1. *Platyrhynchus Borbonicus*, Vieil.

Ash; head greenish-black. Africa.

Muscicapa Labrosa, Swain. Zool. Ill. t. 179.

Muscicapa Carinata, Swain. Zool. Ill. t. 147. Genus *Monachra*, Vigors.

Muscicapa Senegalensis, Pl. Enl. t. 567. 1, 2. *Le Birit*, Vail. O. A. t. 161. *L'Agurous*, Vail. O. A. t. 158. 2.

Muscicapa Cingalensis, Brisson, Pl. En. t. 567. 1. 2? *Platyrhynchus Velatus*, Vieil.

Variegated. Africa.

Platrhynchus Melanoleucus, Vieil.

Black above; white beneath. Senegal.

Platyrhynchus Cyanoleucus, Vieil.

Deep blue; body white beneath. East Indies.

Platyrhynchus Albicollis, Vieil.

Brown; eyebrows and throat white; tail wedge-shaped. East Indies.

Platyrhynchus Polychopterus, Vieil.

Black; gray beneath. Australasia.

Todus rubecula, Lath. *Platyrynchus Rubecula*, Vieil.

Ash; throat and chest red; belly white. New Holland.

Todus Flavigaster, Lath. *Platyrhynchus Flavigaster*, Vieil.

Ashy-brown, beneath yellow. New Holland.

Platyrhynchus Rufiventris, Vieil.

Brown-black, belly reddish. New Holland.

Platyrhynchus Ruficollis, Vieil.

Blue above, throat and front of neck reddish. N. Holland.

Todus Plumbeus.

Head black; beneath white; crown-quills and tail blackish. Surinam.

Todus Maculatus, Desm.

Todus Regius, Gm. Pl. Enl. t. 289.

The genus *Onychorhynchus*, Fischer.

Muscicapa Barbata, Gm. Swain. Zool. Ill. t. 116. Pl. Enl. t. 830, 1.

Platyrhynchus Barbatus, Vieil.

Olive-brown, greenish-yellow beneath. S. America.

Round-crested Flycatcher, Lath. 8. *Muscicapa Coronata*, Gm. Pl. Enl. t. 675, 2.

Crest rounded and scarlet, brown above; beneath scarlet; length five inches and a half.

Yellow-rumped Flycatcher, *Muscicapa Spadicea*, Lath.

Rump, belly, and vent yellowish; length six inches. S. America.

M. Cinnamomea, Lath.

Yellowish-brown, length six inches. S. America.

Muscicapa Obscura, Vieil.

Above brownish-gray; belly reddish; length seven inches. S. America.

Muscicapa Albicapilla, Pl. Enl. t. 568. 1.

Body above greenish-gray; head crested; middle of chest white. Martinique?

Muscicapa? Melanops, Vieil. *Lindo Pardo Corpo Amarillo,* Azara, 101.

Brown above, whitish-red beneath; forehead black. Paraguay.

Muscicapa Nigerrima, Vieil.

Black quills, inner edge and base white. South America.

Black-headed Warbler, Lath. 22. *Muscicapa Ruticolla,* Lein. ♀. *Motacilla Flavicauda,* Gm. Edw. t. 257. ♂. Pl. Enl. t. 566. 1. 2. Edw. t. 80. Cates. Carl. 1. t. 67.

Black, white beneath; length ten inches and a half; *female* ashy-brown above. S. America.

Crested Warbler, Lath. 125. *Motacilla Crisata,* Gm. Pl. Enl. t. 391, 1.

Crest brown above; gray beneath; length four inches. Guiana.

Muscicapa Cyanorostris, Vieil. *Suiri Negro peco celesto,* Az. n. 181.

Black, bill blue, black-tipt.

Muscicapa Armillata, Vieil. O. Am. Sept. t. 42.

Above bluish-ash; beneath brown-red; bracelet yellow; length six inches. Martinique.

Muscicapa Phœnoleuca, Vieil. *Suiri Pardo y Blanco,* Az. n. 92.

Brown above, white beneath. S. America.

Muscicapa Nigricans, Vieil. *Suiriri Chorreado*, Az. n. 182.

Body above blackish, streaked with black, and reddish beneath. S. America.

Muscicapa Fusca, Vieil. O. A. Sept. t. 40.

Brown, ochraceous beneath. N. America.

Muscicapa Punctata, *Suiriri Puteado*, Az. n. 184.

Greenish-brown, white spotted; beneath yellow; length six inches. S. America.

Muscicapa Caudacuta, *Cola de Agudas*, Az. n. 227.

Blackish varied with reddish-white; beneath yellowish-white, varied with red; tail-feather acute. S. America.

Muscicapa Rubra, Vieil. *Suiriri Roxo*, Az. n. 188.

Red, chest and belly yellowish-white. S. America.

Muscicapa Sibilator, Vieil. *Suiriri Pitador*, Az. n. 191.

Brown varied with deep green, white beneath, shaded with greenish-gray. S. America.

Muscicapa Icteropterys, Vieil. *Suiriri Obscuro y Amarillo*, Az. n. 183.

Deep green, eyebrows and body beneath yellow. S. America.

Muscicapa Ruficapilla, Vieil. *Suiriri Cabeza y Rabadilla de Canela*, Azara, n. 178.

Head deep-red, reddish-brown above, beneath varied black and white. S. America.

Muscicapa Flaviventris, Vieil.

Reddish-gray above, beneath yellow.

Platyrhynchus Xanthopygus, Spix, ij. t. 9. 1.

Ashy-brown, red beneath; rump yellowish; body four inches and one-third; tail two inches. Brazil.

Platyrhynchus Ruficauda, Spix, Braz. ij. t. 9. 1.

Olive-brown, tail rufous, yellow-green beneath; body five and a half, tail two inches and three quarters. Brazil.

Platyrhynchus Chrysoceps, Spix, Braz. ij. t. 11. 2.

Brown, yellowish-white beneath; crown orange; body four and a half, tail two inches and three quarters. Brazil.

Platyrhynchus Sulphurescens, Spix, Braz. ij. t. 12. 1. ♂. 2. ♀.

Yellow-green, beneath greenish-yellow; body five and three quarters, tail two inches and a half long. Braz.

Platyrhynchus Hirundinaceus, Spix, Braz. ij. t. 13. 1.

Brownish-black, chestnut beneath; body seven, tail three inches long. Braz.

Platyrhynchus Cinereus, Spix, Braz. ij. t. 13. 2.

Black-brown, beneath lead-coloured; body five, tail two inches and a half. Braz.

Platyrhynchus Flavigaster, Spix.

Olive-green above, yellow beneath; body four, tail two inches and three quarters long. Brazil.

Platyrhynchus Brevirostris, Spix, ij. t. 15. 2.

Olive-green, pale yellow beneath; body four, tail two inches long. Braz.

Platyrhynchus Paganus, Spix, Braz. ij. t. 16. 1.

Olive-ash colour above, beneath pale yellowish; body five and a half, tail two inches and a half long. Brazil.

Platyrhynchus Murinus, Spix, Braz. ij. t. 16. 2.

Dull brown, sulphureous yellow beneath; body three and three quarters, tail one inch and three quarters.

Muscicapa Petechia, Pl. Enl. t. 568. 2.

Brown, ashy beneath, reddish spotted; length six inches. Martinique.

Platyrhynchus Leucophaius, Vieil.

Body above brown, yellow beneath, white streaked. S. America.

In others, the tail is uneven, and the two central feathers are much the longest. The subgenus, *Colonia*, gray, peculiar to South America.

Peruvian Flycatcher, Lath. H. t. 102. *Muscicapa Colonus*, Vieil. *M. Monacha*, Freyr. Licht. *Platyrhynchus Filicauda*, Spix, Braz. ij. t. 14. *Le Colon*, Azara, n. 180. *Platyrhynchus Platurus*, Vieil.

Black, crown gray; forehead and rump white; length nine inches; *young* middle feathers shorter. Bahia.

The genus *Platyrhynchus*, Tem. and *Platyrhynchus*, Swain.

Todus Rostratus, Gm. *Todus Platyrhynchus*, Gm. Pallas, Spix, t. 3. Desm. Tod. 1. Gal. Ois. t. 126.

Yellowish-brown, yellow beneath; bill very large.

Todus Regius, Lath. Pl. Enl. t. 289.

Platyrhynchus Regius, Vieil.

Black-brown, beneath reddish. S. America.

Great-billed Tody, Todus Nasutus, Gm. Lath. Syn. 20. *Todus Macrorhynchos*, Gm.

Black, belly and rump red ; bill very large. S. America.

Platyrhynchus Mystaceus, Vieil. *Bigotillos*, Az. 173.

Body above brown, beneath deep yellow. S. America.

Platyrhynchus Olivaceus, Temm.

Todus Obscurus, Lath. *Muscicapa Arcadica*, Wils. A. O. t. 13. 3. *Platyrhynchus Virescens*, Vieil.

Olive-green, greenish-yellow beneath. N. America.

Platyrhynchus Cancromus, Swain. Zool. Ill. t. 115, Pl. Col. t. 115.

Platyrhynchus Musicus, Vieil.

Crested ; black ; belly and behind white. Africa.

Platyrhynchus Ceylonensis, Swain. Zool. Ill. t. 13.

Muscicapa Aurantia, pl. Enl. t. 831. 1. *Platyrhynchus Aurantius*, Vieil.

Red with a greenish tint, white beneath ; chest orange. S. America.

Muscicapa bicolor, Lath. Pl. Enl. t. 675. 1.

Black, white beneath. S. America.

Platyrhynchus Melanops.

Body above reddish-gray ; beneath reddish-white ; Cheeks black. South America.

Platyrhynchus Ruficaudatus, Vieil.

Olive-green, belly green, olive spotted; tail reddish. South America.

Muscicapa Rufescens, Lath. *Platyrhynchus Rufescens*, Vieil.

Shining reddish; white beneath. South America.

Todus brachyurus, Lath. *Platyrhynchus brachyurus*, Vieil.

Black, white beneath; tail short. South America.

Pipra Nævia, Lath. *Platyrhynchus Nævius*, Vieil. Pl. Enl. t. 823. 2.

Brown, belly white; vent orange. S. America.

Turdus Auritus, Lath. *Platyrhynchus Leucotes*, Vieil. Pl. Enl. t. 822. 1, 2.

Olive and red varied, reddish beneath; long white feathers on each side of the neck. S. America.

Platyrhynchus Coronatus, Vieil. *Muscicapa Coronatus*, Lath. Pl. Enl. t. 453. 1. Cop. E. M. t. 192. 2.

Brown, crest rounded; temples and body beneath red. S. America.

Platyrhynchus Maculatus, Vieil. Desm. Tanag. t.

Deep olive-gray, pale-yellow beneath; throat white, brown-spotted. S. America.

White-headed Tody, *Todus Leucocephalus*, Pal. Spix, ij. t. 3. 2. *Platyrhynchus Leucocephalus*, Vieil.

Black, head slightly crested, and throat white. S. America.

Platyrhynchus Dupontii.

Nape pale bluish-ash; crown black; chest above yellow; back and rump olive-green. America.

THE FLYCATCHERS PROPERLY SO CALLED, (*Muscicapa*, Cuv.)

Have the mustaches shorter and the beak narrower than the Flyeaters. It is, nevertheless, depressed with a strongly marked ridge above, straight edges, and the point a little bent.

Two species of this subgenus inhabit France during summer; they live retired on the elevated branches of trees. The most common is

Spotted-Flycatcher, *Muscicapa Grisola*, Gm. Pl. Enl. t. 565. f. 1.

Is gray above, whitish underneath, with some grayish spots on the breast. In some countries they are kept in rooms to destroy the flies.

The other is

Pied Flycatcher, *Musc. Atricapilla*, Gm. Enl. 565. 2. f. and 3.

Is very remarkable for the change of plumage of the male, similar to the female in winter, that is gray, with a white band upon the wing; they assume, during the season of their loves, a pleasing distribution of pure black and white; the back, wings, and tail black; the front collar, underpart of the body, and a band on the wing and exterior edge of the tail, white. It builds in the trunks of trees.

The ancients were well acquainted with this bird under the name of *Sycalis* and *Ficedula* in its best plumage; but as the name *Beque-figue*, which answers to *Ficedula,* is applied in the south and in Italy, several naturalists have united the specific characters of these birds under one certain state of this Flycatcher, and have formed of it the imaginary species presented under the name of *Bec-figue* in Buffon and his followers.

'The *M. Collaris* of Bechstein and Temminck, and *M. Streptophora* of Vieil.

Europe also contains two other species.

English Flycatcher, *Emberiza Luctuosa,* Scopoli, *Muscicapa Atricapilla,* Gmelin ; *Motacilla Ficedula,* Gmel. *Muscicapa Muscipeta,* Bechst. *Motacilla, Atricapilla,* Gmel. Pl. Enl. t. 668. f. 1. Edw. t. 30. f. 1. 2.

Body above and tail-feather deep black; forehead, and beneath white; wings black; middle and large coverts white. The former has a white collar (which is wanting in this species).

Muscicapa Parva, Bechstein.

Above reddish-ash ; behind the ears bluish; quills ashy-brown ; four middle, and tips, side tail-feathers blackish ; throat and chest bright red, beneath whitish ; length four inches and a half.

The exotic species are those found on the old continent.

Muscicapa Senegalensis, Gm. *Le Pririt.* Vail. O. A. t. 161. Pl. Enl. t. 567. f. 1. 2. *Muscicapa Pririt,* Vieil.

Chest band and eye-streak black ; body beneath white, above bluish-ash ; crown bluish. Africa.

Muscicapa Azurea, Vieil. *L'Azuroux,* Vail. t. 153. f. 2.

Shining blue, beneath orange-red. South Africa.

Muscicapa Cœrulea, Gmel. Pl. Enl. t. 666. f. 1. Vail. O. A. t. 153. f. 2.

Blue, nape and chest black-spotted; belly and vent bluish-white; tail and quills blue-black. Africa and India.

Muscicapa Erythropis, Lath.

Spotted, white beneath; forehead red; country unknown.

Muscicapa Nitida, Lath.

Pale green; wing-coverts white-edged; quills and tail blackish, yellow-edged. India.

Muscicapa Cochinsiensis, Lath.

Olive-brown, beneath reddish; tips of three outer tail-feathers black and white spotted. India.

Muscicapa Torquata, Gmel. Pl. Enl. t. 572. f. 1. 2. *M. Capensis,* Kuhl. Vail. O. A. t. 150.

Black, beneath white; chest red; quills white-tipt. Cape of Good Hope.

Muscicapa Meloxantha, Sparman, iv. t. 96.

Ash, beneath yellow; crown, wings, and tail black; tail-feathers white-tipt. Country unknown.

Muscicapa Comata, Lath.

Black beneath, rump and tips of middle tail-feathers white; vent yellow; head crested. India.

Muscicapa Albifrons, Sparman, Mus. t. 24.

Black-brown, chest whitish; belly pale ferrugineous; forehead whitish. Southern Africa.

Muscicapa Manillensis, Gm. Sonnerat, Voy. t. 26. f. 2.

Occiput and back gray; head and nape black; loins bay; throat yellow; tail-feathers, middle white and black, side ones white. India.

Muscicapa Psidii, Gm. Sonnerat, Voy. t. 38. Cop. E. M. t. 192. f. 5.

Brown, beneath white; vent yellowish; eyebrows white; Crown lores black; length six inches. Manilla.

Muscicapa Pondicheria, Sonnerat, Voy.

Ash-gray beneath; white eyebrows; spot on wing-coverts, and half tail-feathers white; length seven inches.

Muscicapa Leucura, Lath.

Ashy-gray, beneath white; middle tail-feathers black, rest half white. South Africa.

Muscicapa Rosea, Vieil.

Head and back gray; chin white; body beneath rosy; the three first quills internally red-spotted, the rest partly red. India.

Muscicapa Cyanocephala, Gm. Sonnerat, t. 26. f. 1. Cop. E .M. t. 191. f. 4.

Red, beneath yellowish; head blue; tail-feather black-tipt; length six inches. India.

Muscicapa Cœruleo-Capilla, Gm. Sonnerat, Voy. t. 27. f. 2.

Head, neck, and throat blue; back, chest, and belly bluish-gray; quill and tail-feathers black. India.

Muscicapa Tectec, Brisson, Orn. ii. t. 39. f. 1.

Brown, dotted with red, beneath reddish; throat whitish; quill and tail-feather brown edged; latter red-tipt. South Africa.

Muscicapa Griseo-Capilla, Vieil.

Crown, neck above, and cheeks gray; back and rump olive-green; chin white; body beneath yellow. India.

Muscicapa Atricapilla, Gm.

Head black, back deep gray; throat whitish; quill and tail brown; vent red; rump dull white; length ten inches. China.

Muscicapa Superciliosa, Vieil.

Eyebrows, chest, and belly white; wings brown; head throat, neck, back, and tail black. India.

Muscicapa Variegata, Vieil.

Brown beneath; forehead and rump white. India.

Muscicapa Sinensis.

Greenish-gray, throat white; crop and chest gray; belly and vent yellowish; quills yellowish-green; head black. India.

Muscicapa Nigrifrons, Gmel.

Brown; beneath olive; forehead and temples black; chin and throat yellow; quills, two middle tail-feathers brown. Country unknown.

Muscicapa Grisea, Lath.

Black; beneath reddish; throat gray; wing-coverts forming a white band; tail slightly wedge-shaped. China.

M. Rufiventris, Gmel. Pl. Enl. t. 572. f. 3.

Black, vent red. South Africa.

M. Undulata, Vieil. Vail. O. A. t. 159?

Waved, brown and white; head blackish; wings reddish brown. Africa.

Red-vented Flycatcher, Muscicapa Hæmorhousa, Brown, Illust. t. 31.

Clouded crown, beneath white; vent red; tail, and slightly crested head, black. India.

Dun Flycatcher, Muscicapa Sebrica, Penn.

Brown; throat and vent spotted. Siberia.

Muscicapa Javanica, Sparmann, Mus. t. 75.

Black, and ferrugineous variegated beneath; eyebrows white; crop, bill, and feet black. Java.

Muscicapa Cyanomelas, Vieil. Vail. O. A. t. 151.

Head shining blue-black; body above bluish ash, beneath bluish white, varied with gray; wing-spot white. South Africa.

Muscicapa Scita, Vieil. Vail. O. A. t. 154.

Eye-streak black; middle of the throat and chest reddish; tail black and white. India.

Muscicapa Pristrinaria, Vieil. Vail. O. A. t. 160.

Ferrugineous brown, varied with olive; eye streak, throat, and bands on chest, black; crop and belly white. South Africa.

Muscicapa Ædon, Pallas.

Beneath yellowish-white; tail ashy-brown; long; wedge-shaped. Tartary.

Muscicapa Nitens, Gm.

Golden-green; wings black; throat and chest reddish; rump and belly yellow; tail long; tail-feather and quill green, black edged. India.

Muscicapa Melanictera, Lath. Brown. Illust. t. 82.

Cheeks black; back and wing-coverts ash, brown and yellow; chest yellow; tail-feathers and quills black, India.

Sitta Chloris, Lath. Mus. Carls. III. t. 53.

Green; beneath snow-white; tail-feather black, outermost yellow-tipt. Cape of Good Hope.

Papuan Manakin, *Pipra Papuensis*, Gm. Pl. Enl. 707. f. 2.

Greenish-black, beneath white; chest with an oval fulvous spot; two middle tail-feathers shortest; three inches and a half long. New Guinea.

Obscure Flycatcher, Lath *Musc. Obscura*, Horsf. Z. I. t. f. 2. *M. Hirundinea*, Reim. Pl. Col. t. 119. f. 1. 2.

Bluish-black; beneath and rump white; length five inches. Java.

Indigo Flycatcher, Lath. *Musc. Indigo*, Horsf. 3. R. t.

Dusky sky-blue; quills and tail black; tail base of axillaries, belly and vent whitish; length four inches and three quarters. Java.

Banyumas Flycatcher, Lath. *Musc. Banyumas*, Horsf. *M. Cantatrix*, Temm. Pl. Col. t. 226.

Above deep-azure blue; beneath chestnut; belly paler; quills and upper side of tail black; length five inches and a half. Java.

Javan Flycatcher, Lath. *Musc. Javanica*, Mus. Carls. iij. t. 75.

Dusky, varied with ferrugineous; forehead and half collar blackish; belly and vent yellow; throat and tips of outer tail-feathers white; length six inches. Java.

Muscicapa Hyacinthina, Temm. Pl. Col. t. 30. f. 1. 2.

Blue; front of chest and beneath reddish; female chin and throat reddish. Length seven inches.

Muscicapa Flammea, Forster Zool. Ind. Pl. Col. t. 263. f. 1. 2.

Black; chest and beneath rump, three spots of the wing and sides, feather of tail below orange; *female*, lead-coloured; wings black; the forehead, sides of face, throat, and where orange in male, yellow.

Muscicapa Miniata, Temm. Pl. Col. t. 156.

Black; chest, side of neck, and below, back, rump, large spot on wing, and outer feather of tail below, bright red; female like male; face and throat red, and back dull red; seven inches long. Java.

Malabar Titmouse, *M. Subflava*, Vieil. *Parus Malabaricus*, Lath. *Parus Peregrinus*, Lath. Forst. Zool. Ind. t. 15. male. Mus. Carls. t. 48. 49. Vieill. O. A. t. 155. *Sylvia Peregrina*, Vieil. Sonnerat, Ind. t. 114. f. 1.

Ashy; beneath white; rump scarlet. Malabar.

Australasian Flycatcher. *Muscicapa Rhodoptera*, Lath.

Slightly crested; brown; beneath white; lower half of the quills and tail-feathers rosy.

Muscicapa Australis, Lath. White Voy. t. at p. 239.

Ash; body beneath, and eyebrows, yellow. New Holland.

Muscicapa Obscura, Lath.

Brown; beneath ash; belly reddish; *tail-feathers long, equal, sharp-pointed.* Sandwich Islands.

Muscicapa Cambaiensis, Lath.

Shining-black; back yellowish-green; body beneath fulvous; wing-coverts with a double white band.

Muscicapa Cucullata, Lath.

Black; beneath white; quill and smaller wing-coverts white-edged. New Holland.

Muscicapa Melanocephala, Lath.

Head and neck black; back fulvous; body beneath black and white-spotted; wings and tail black. New Holland. A Stone Chat?

Yellow-fronted Flycatcher. Muscicapa Flavifrons, Lath.

Yellow-olive; forehead, eyes, and beneath yellow; quill brown; tail-feathers blackish, both yellow-edged; eyebrows white.

Muscicapa Sandwichensis, Lath.

Brown; beneath ochraceous; forehead yellow; eyebrows white; chin black, streaked; quills and tail-feathers white-tipt. Sandwich Islands.

Muscicapa Barbata, Lath.

Brown; beneath white; crown and gular spot black. New Holland.

Muscicapa Maculata, Lath.

Ferrugineous; beneath pale-bay; quills black; wing-coverts whitish-tipt; tail-feathers brown, outermost white-tipt. Polynesia.

Muscicapa Passerina, Gm

Blackish; beneath white; tail black. Polynesia.

Muscicapa Rhodogastra, Lath.

Brown; beneath pale; chest rose red; wing-coverts white-edged.

Muscicapa Coccinigastra, Lath.

Olive; throat white; chest and belly scarlet; forehead black; quill and tail half black and half white. New South Wales.

Red-bellied Flycatcher, Lath. *Muscicapa Multicolor*, Gmel. *M. Erythrogaster*, Lath. n. 50. Hist. t. 100.

Black; forehead, wing-coverts, spot, band of quills, streak on side, tail-feathers and vent white; chest and belly scarlet—*var.?* no white on wing or tail. Lath. New Holland.

Muscicapa Lathami, Vig. Zool. Jour. 1. t. 13. Jardine Illust. Orn. t. 8. f. 1.

Black; chest and belly rosy-purple; frontal spot and vent white—*var.?* three outer tail-feathers internally white-edged; length four inches and a quarter. New Holland.

Muscicapa Goodenovii, Vig. and Horsf. Jardine Illust. Orn. t. 8. f. 2.

Black; belly longitudinal, wing-streak, and edge of

two outer quill-feathers white; forehead and chest vivid scarlet; length four inches.

Tyrannula Affinis, Swain.

Olive; beneath pale-fulvous; wing-coverts and quills with pale margins; base of the lesser quills with a blackish band; bill small; under mandible yellow; tail divaricated; length six inches and a half. Mexico.

Tyrannula Obscura, Swain.

Above olive-gray: beneath yellowish-white; wings short; brown, with two whitish bands; tail brown, even, the outer feather with a pale-yellow edge; length five inches and a quarter. Mexico. Perhaps *Muscicapa Querula*, Vieil. O. Amer. t. 39?

Tyrannula Barbirostris, Swain.

Beneath pale-yellow; crown blackish; chin and throat white; bill large, and strongly bearded; tail even; length six inches. Mexico.

Tyrannula Nigricans, Swain.

Blackish-brown; head and throat darker; vent, under tail-coverts, and margin of the outer tail-feathers, white; length seven inches. Mexico.

Muscicapa Coronata, Gmel.

Mexico.

Muscicapa Cayenensis, Gmel.

Mexico.

Tyrannula Pallida, Swain.

Pale gray; beneath ferrugineous; throat hoary; tail black; length seven inches. Mexico.

Tyrannula Musica, Swain.

Cinereous crown; beneath dirty-yellow; tail forked; wings lengthened, brown; bill strong, hooked; length seven inches and a half. Allied to Tyrannus. Mexico.

Muscicapa Saya, Bonaparte, Amer. Orn. t. 2. f. 3.

Dull cinnamon-brown; belly rufescent; tail nearly even; first primary longer than the sixth. North America.

Pewit Flycatcher, *Muscicapa Nunciola*, Wilson, Amer. O. II. t. 13. f. 4. *Muscicapa Fusca*, Gm. *M. Phœbe*, Lath. *M. Atra*, Gmel.

Dark olive-brown; head blackish; beneath pale-ochreous; bill quite black; tail nicked; outer feather whitish on the outer web. North America.

Wood-Pewee Flycatcher, *Muscicapa Virens*, Lin. *M. Rapax*, Wilson, Amer. O. II. t. 13. f. 5. *Todus Obscurus*, Gmel. *Mus. Querula*, Vieil. O. A. t. 39.

Brownish-olive; beneath pale-ochreous; bill black, beneath yellow; tail nicked, second primary the longest. North America.

American Red-start, *Muscicapa Ruticilla*, Lin. Cates. Car. Wilson, A. O. I. t. 6. f. 6. V. t. 45. f. 2. *Setophaga*, Swain. *Muscicapa Flavicauda*, Gm.

Black; belly white; sides of the breast, base of primaries, and tail-feathers, the two side ones excepted, orange, becoming greenish-olive in autumn. North America.

Muscicapa Bicolor, Gm. Pl. Enl. 566. f. 3. Edw. t. 348. f. 1.

Black; body beneath, forehead, orbits, rump, wing-band, and tips of tail-feathers, white. South America.

Muscicapa Melanoptera, Kuhl. Pl. Enl. 675. f. 1. *M. Bicolor*, β. Lath.

White; nape, back of neck, wings, rump, and tail black; female gray. South America.

Setophaga Ruticella, Swain. Ann. Phil.

Cinereous; breast and body beneath vermilion; tail black; side feathers of tail partly white. Mexico.

Setophaga Rubra, Swain. Ann. Phil.

Entirely red; ear-feathers of a silky-whiteness. Mexico.

Muscicapa Fuliginosa, Gmel.

Black brown; feathers yellow-edged, beneath whitish; quills and tail-feathers white-edged. South America.

Muscicapa Rufifrons, Lath.

Brown; forehead, back, and base of tail, red; quills black; ears and chest black; spotted tail; long wedge-shaped. Brazils.

Muscicapa Canadensis, Gmel. Brisson, ii. t. 39. f. 4.

Ash; body beneath and lores yellow; crown black-spotted. North America.

Muscicapa Ferruginea, Merrem. t. 6.

Reddish-brown; beneath reddish-white; throat white; wings black, brown-edged; tail-feathers beneath glaucous, above brown-edged; outermost very short, white. North America.

Muscicapa Minuta, Gmel. Enl. t. 192. f. 4.

Olive-gray; wings blackish; body and wings streaked with ochraceous. South America.

Muscicapa Cristata, Lath. Pl. Enl. t. 391. f. 1.

Brownish; beneath greenish-gray; crest blackish-brown, white-edged. South America.

Muscicapa Ochroleuca, Lath.

Dull olive; beneath ochraceous; throat and edge of wings yellow; primary quills and tail olive. North America.

Muscicapa Agilis, Gm. Pl. Enl. t. 573. f. 3.

Olive-brown; beneath whitish; quills and tail-feathers black, olive-edged. South America.

Muscicapa Pygmea, Lath.

Head red and black spotted; body above deep ash; beneath pale yellow. South America.

Muscicapa Surinama, Lath.

Olive-black; beneath white; tail white-tipt. South America.

Muscicapa Suiriri, Vieil. *Suiriri Ordinario,* Azara. n. 179.

Head and neck pale lead-colour; back and rump brown, varied with green; throat and body beneath bluish-white. South America.

Muscicapa Virgata, Gm. Pl. Enl. t. 574. f. 3.

Brown; beneath brownish-white brown-streaked; crown slightly crested; varied ashy and yellow; edge of quills and two band of wing-coverts red.

Muscicapa Obsoleta, Natter. Pl. Col. t. 275. f. 1.

Greenish-ash; beneath whitish; crown and nape gray;

wings brown, with two bands of reddish spots; quills reddish, gray-edged; bristles very short; between *Parus* and *Muscicapa*. Brazils.

Muscicapa Ventralis, Natter. Pl. Col. t. 275. f. 2.

Greenish; beneath dirty yellow; face and orbits streaked greenish and white; wings green, edged with two bands of yellow spots; three last secondaries yellow-tipt. Brazils.

Muscicapa Virescens, Natter. Pl. Col. t. 275. f. 3.

Greenish; beneath dirty yellow; face and orbits streaked green and white; wings brown, green-edged, with two bands of yellow spots; secondaries not tipt.

Muscicapa Cæsia, Pr. Max. Pl. Col. t. 27.

Muscicapa Diops, Temm. Pl. Col. t. 144. f. 1.

Ash, beneath paler; spot before each eye, white.

Muscicapa Eximia, Temm. Pl. Col. t. 144. f. 2.

Blueish ash; beneath yellowish; nape, ear, quills, and tail, black; latter pale-edged; side of face and over eye, white.

Muscicapa Gularis, Natter. Pl. Col. t. 167. f. 1.

Blue green; beneath blueish-white; side of face and of throat reddish; two wing-bands yellow.

Muscicapa Straminea, Natterer, Pl. Col. t. 167. f. 2.

Blue; throat blueish-white; crest and eye-streak, and two wing-bands, white; tail and quills black; belly and chest yellowish.

Muscicapa Stenura, Temm. Pl. Col. t. 167. f. 3.

Tail long, wedge-shaped; throat, belly, and eye-streak, white; crown, wing, and tail, black.

Muscicapa Flamiceps, Temm. Pl. Col. t. 144. f. 3.

Brown; beneath white; quills and tail blackish; secondaries and wing-coverts white-tipt; crest scarlet.

Prince Maximilian in his travels mentions—1. *Muscicapa Vociferans*, 1. p. 38. *M. Ampelina* of Illiger. 2. *M. Rupestris*, ii. 151. *M. Rivularis*, ii. 167, and *M. Mastacalis*, iii. 50. All from Brazils.

In others the tail is compressed; the side feathers oblique; the middle ones longer, vertical; the central rib of all ending in a point. The genus *Alecturus* of Vieillot. It is peculiar to South America.

Alecturus Tricolor, Vieil. *Muscicapa Alector*, B. Max. *M. Alectura*, Temm. *Le Petit Coq*, Azara, n. 225. Plate made up. Pl. Col. t. 156. Gal. Ois. t. 132.

Above black; beneath white; broad interrupted chest-band, capistrum, rump and humerus white; quills black; white-edged secondaries; inner web white; tail black. *Female and young* sooty; where male black; tail flat; feathers square. Length five inches and a half.

In some, from South America, the tail is also peculiar for having the two outer feathers very long, and only feathered at their tips.

Yetapa Flycatcher, Muscicapa Psalura, Temm. Pl. Col. t. 286. ♂. 296. ♀.

An American bird has been formed into the genus *Icteria* by Vieillot; it appears to be intermediate between Flycatchers and the Tanagers.

Chattering Flycatcher, Ampelis Lutea, Sparman. *Muscicapa Viridis*, Gm. Catesby, Car. 1. t. 50. *Icteria Dumicola*, Vieil. Gal. Ois. t. 85. *Pipra Polyglotta*, Wilson, A. O. I. t. 6. f. 2. *Tanagra Olivacea*, Desm. *Gamulus Australis.*

Greenish-olive; throat and breast yellow; belly and line encircling the eye white.

In other American birds the bill is rather compressed and arched. Vieillot has formed them into a genus under the name of *Vireo.*

Yellow-throated Flycatcher, Muscicapa Sylvicola, Wils. A. O. II. t. 7. f. 3. *Vireo-flavifrons*, Vieil.

Yellow-olive; throat, breast, frontlet, and line round the eye yellow; belly white; wing two-banded with white; tail blackish. North America.

Solitary Flycatcher, Muscicapa Solitaria, Wilson, A. O. II. t. 17. f. 6.

Olive-green; head bluish-gray; line round the eye and belly, and two wing-bands white; breast pale ash; sides yellowish; tail blackish. North America.

White-eyed Flycatcher, Muscicapa Noveboracensis, Gmel. *Vireo Musicus,* Vieil. *Muscicapa Cantatrix,* Wilson, A. O. II. t. 18. f. 6.

Yellow-olive ; beneath white ; sides, line round the eye and spot near nostrils, and two wing-bands yellow ; tail blackish, under white. North America.

Warbling Flycatcher, Muscicapa Silva, Vieil. *Muscicapa Melodia,* Wilson, A. O. V. t. 42. f. 2.

Pale olive-green : head inclining to ash ; line over eye and all beneath white ; wings dusky ; bandless ; bill short ; irides brown. North America.

Red-eyed Flycatcher, Muscicapa Olivacea, Lin. Wilson, A. O. II. t. 12. f. 3.

Yellow-olive ; crown ash, with a black side line ; line over eyes and all beneath white ; wings bandless, bill long ; irides red. North America. Somewhat allied to Sylvia.

? *Vireo Virescens,* Vieil.

Crown blackish ; eyebrows white ; body above greenish ; beneath grayish-white ; bill above brown ; beneath horny. North America.

Muscicapa Longipes, Lesson and Garnot. Voy. t. 19. f. 6.

Tarsus very long ; feather pale-edged ; belly and vent white.

Some species, which have the ridge rather more elevated, and bending into an arch toward the point, approach the form of the stone-chats.

Black and Scarlet Thrush, T. Speciosus, Lath.

Black; belly, loins, middle wing-coverts, edges of quill and tail-feathers, scarlet. India.

Muscicapa Stelluta, Vieil. *Gobemouche Etoilé*, Vail. t. 157. f. 2.

Olive-green; beneath yellow; head and throat bluish-gray; collar white; a black star above the eyes. Africa.

Collared Platyrhynchus, Platyrhynchus Collaris, Jardine, Illust. Orn. 1. t. 9. f. 1.

Above shining blue-black; beneath white; pectoral band black; eyes caruncled; length five inches and a quarter.

Desmarest's Platyrhynchus, Plat. Desmarestii, Jardine, Illust. Orn. 1. t. 9. f. 2.

Above gray; throat white; neck and chest chestnut; tail and quills black: eyes caruncled; length four inches and a half.

Platyrhynchus Pusillus, Gm.

Olive-brown; beneath yellowish-white; wings with two pale bands; tail moderate, even; bill small; head crested; length five inches and a half; bill six-tenths of an inch. Marine parts of Mexico.

There are several genera, or sub-genera, which approach certain links of the series of Flycatchers, as the

GYMNOCEPHALES, OR BALD TYRANTS,

which have nearly the beak of the Tyrants, except that the keel is rather more arched, and a considerable part of their face is denuded of feathers. There is but one species known, which is of Cayenne, as large as a crow, and of the colour of Spanish snuff.

Corvus Calvus, Gmelin, Vail. O. Amer. et Ind. t. 49. *Coracina Gymnocephala*, Vieil. Pl. Enl. t. 521.

Ferrugineous brown; forehead and nape, bald, or scarcely feathered; bill black.

The CEPHALOPTERES, Geoff.,

on the contrary, have the base of the beak furnished with inclining feathers, which, spreading at their upper parts, produce a large panacle in the form of a parasol. But one species is known of America of the size of the jay. It is black, and the plumes of the lower part of the breast form a sort of hanging. The *cephalopterus ornatus* of Geoff. Ann. du Mus. xii. t. 15.

Umbelled Chatterer, *Ampelis Umbellata*, Shaw, N. M. Also the *Coracina Ornata* of Spix, Braz. t. 49, and *Coracina Cephaloptera* of Vieillot. Pl. Col. t. 255.

Referred to *Coracina* by Spix, Vieillot, and Temminck.

The COTINGAS (AMPELIS,) Lin.

have the bill depressed, like the Flycatchers in general, but in a shorter proportion; broad and slightly arched.

Those with the bill stronger, and more pointed,

living chiefly on insects, are called *Peauhace,* from their cries. They are peculiar to America, and fly in troops in the woods in the pursuit of insects. They are

Muscicapa Rubricollis, Gmel. Pl. Enl. 381.

Black; throat with a large red spot.—See *M. Rubricollis,* Spix.

Coracias Militaris, Shaw. Vail. O. A. t. 25, 26. *C. Rubra,* Vieil.

Red; quill, tail, and beneath, blackish; bill red. South America.

Red-breasted Roller. Coracias Scutata, Lath. Pl. 40. Mus. Lever.

Black; throat crimson. Brazils.

Cuvier places here *Ampelis Cinerea,* as being more allied to this than to the following genus.

The Common Cotingas,

whose beak is rather weak, besides insects, seek also berries and tender fruits. They reside in the humid places of America, and are remarkable by the purple and azure colours of the plumage of the males during the breeding season. During the rest of the year both sexes are tinted gray or brown.

The *Ampelis Carnifex.* L. Pl. Enl. t. 378,

has the hood, the crupper, and the belly scarlet;

the rest reddish-brown; the fourth quill feather of the wing is narrowed, shortened, and as if hardened.

The *A. Cuprea.* Merrem, Icon. t. 1. 2.

Le Pompadour, A. Pompadora, L. Enl. 279,
is of a fine bright purple colour, with the wing-quills white; the large coverts have the barbs red, and disposed on two planes in an acute angle, like a roof.

The *Cordon bleu, A. Cotinga,* Z. Enl. 186 and 188,
is of the finest ultramarine, with the breast violaceous, often traversed by a large blue stripe, and marked with rosy spots.

Vieillot makes Pl. 186 a new species under the name of *A. Cœrulea.*

Ampelis Cayana, Pl. Enl. t. 624. *A. tersa* and *A. variegata,* Gmel.

Shining blue; neck beneath violet; quills and tail black, blue-edged.

Ampelis Cristata, Vieil.

Crested; wings and tail black; belly and cheeks white; back red.

Ampelis Maynana, Pl. Enl. t. 229.

Shining blue; throat violet, silky.

Ampelis Fusca, Vieil.

Body above black; brown beneath; crown, chest, and middle, white; streaked sides, with violet-brown. Brazils.

Ampelis Cinerascens, Vieil. Vail. Ois. Rar. t. 144.

Ash; beneath paler; quills and tail brownish. South America.

Ampelis Aureola, Vieil.

Purple; crown, front of wings, chest, and sides, orange-yellow. South America. Perhaps a Var. of *A. Pompadora.*

Ampelis Hypopyrra, Vieil.

Deep gray; back greenish; sides orange-red.

Ampelis Purpurea, Licht. *Ampelis Astro-Purpurea*, Pr. Max.?

Shining black-purple; quills white; primaries black-tipt; side tail-feathers externally red, internally white; when young, purplish-ash; wings black. Bahia.

Ampelis Cuprea, Merrem, Ic. Av. I. t. 2. is a *Carnifex.*

M. Le Vaillant properly separates from the Cotingas

The Echenilleurs, Ceblephyris, Cuv.,

whose singular character consists in the slightly elongated stalks of the feathers of their croup. They live, in Africa and India, on caterpillars, which they gather from the highest trees, and have little of the character of the true cotingas. The tail, rather forked in the middle, is wedged on the sides.

The name of the genus is taken from the Greek name of an unknown bird.

The genus *Campephaga* of Vieillot.

Muscicapa Cana, Lath. Pl. Enl. t. 541. *Ceblephyris Madagascariensis*, Vieil.

Slate-gray; head black; quills blackish, gray-edged; tail-feathers, except the middle ones, black, gray-tipt. Madagascar.

Ceblephyris Levaillantii, Temm. Vail. O. A. t. 162, 163. *Ceblephyris Cana*, Vieil.

Slate-gray; beneath paler; face, cheeks, and forehead black; first quill brownish, white-edged externally; bill and feet black; *female*, face, slate-gray.

Campephaga Niger, Vieil. Vail. O. A. t. 165.

Shining metallic-black; lower wing-coverts greenish; bill and feet black; length seven inches.

Campephaga Flava, Vieil. Vail. O. A. t. 164. β *Muscicapa Bicolor*, Mus. Carls. t. 45 ?

Above greenish-gray; black banded; crown and back of neck gray, varied with olive; scapulars yellow; rump gray; throat, and beneath, brownish, black and yellow spotted; outer tail-feathers blackish, rest olive; all yellow-edged; length seven inches.

Ceblephyris Lobatus, Temm. Pl. Col. t. 279.

Base of bill with a red wattle, head and neck black; rump and beneath red; back greenish; vent yellow; *female*, beneath yellow. Congo and Sierra Leone.

Ceblephyris Bicolor, Temm. Pl. Col. t. 270.

Black; rump, chin, and beneath white.

Ceblephyris Fimbriatus, Temm. Pl. Col. t. 249, 250.

Black-gray; wing and tail shining black; outer tail-feathers gray-tipt; bill and feet black; *female*, gray; beneath banded; feathers white-edged; seven inches and a half long. Java.

Temminck places in this genus *Corvus Melanops*, *C. Papuensis*, and *C. Novæ Guineæ*, of Lath., but Cuvier forms them into a sub-genus of Lanius.

Dr. Horsfield has placed here *Turdus Orientalis*, Lath., as *Ceblephyris Striga*, but Cuvier calls it a *Lanius*.

Campephaga Leucomela, Vigors and Horsfield.

Above black; beneath white; finely black-banded throat; tips of wing and tail-feather, and edge of quills, white; vent fulvous; length of body three inches and a half. New Holland.

African Flycatcher, Lath. 17. *Muscicapa Ochracea*, Sparman, Mus. Carls. t. 22.

Neck and chest ashy ferrugineous; feathers lanceolate; wing and tail ashy-black; head and back brown; ears ciliated with long feathers; belly yellow-brown.

Tanagra Capensis, Sparman, Mus. t. 45. *Campephaga Ferruginea*, Vieil.

Above ferrugineous-brown; beneath varied ferrugineous and white; tail blackish; side-feather reddish-brown; bill yellow; feet black. Cape of Good Hope.

Ceblephyris Lineatus, Swain. Zool. Jour.

Ash; breast and body, and lower wing-coverts beneath, white, banded by narrow black lines; tail-feathers and lores black; quill black, white-edged; length ten inches. New Holland.

Ceblephyris Tricolor, Swain. Zool. Jour.

Glossy-black; beneath white; rump and upper tail-coverts cinereous; wing-coverts and tips of tail-feathers white; bill rather slender; nostrils partly exposed; length six inches, tarsi 5-8ths. New Holland

We may also separate from the Cotingas,

The Chatterers, Bombycivora, Temm.,

which have another singular character in the secondary quills of the wings, of which the end of the stalk enlarges into an oval disk, pliant and red.

There is said to be one in Europe, but without much authority*.

European Chatterer. *Ampelis Garrulus*, L. Enl. 261.

Rather larger than a sparrow, with the head crested, the plumage of a vinous gray, the throat black, the tail black, bordered with yellow at the tip; the wings black varied with white. This bird arrives in Europe in flocks at long and irregular intervals, whence it was long considered ominous. It is stupid, is easily taken, eats a great deal, and of every thing. It is presumed to build in high northern latitudes.

* Except as a bird of passage.—Ed.

Cedar Bird. *Ampelis Garrulus Var.* Lath. *Ampelis Americana,* Wils. A. O. I. t. 7. f. 1. *Bombyciphora Zanthocœlia,* Meyer. *Bombycilla Cedrorum,* Vieil. A. A. S. t. 37. *Bombycilla Canadensis,* Brisson.

Drab frontlet, and line over the eyes black; belly yellow; vent white; wings and tail blackish, latter yellow-tipt. North America.

Bombicivora Japonica, Seibold, Bull. Sci. Nat. 1827. 87. Japan.

The genus *Bombycilla* of Brisson, and *Bombyciphora,* Meyer.

MM. Hofmansegg and Illiger separate, with still more reason, from the Cotingas,

The Procnias, Hofm.,

whose beak, very weak and depressed, is cleft as far as under the eye. They are American, and feed on insects.

One species,

Hirundo Viridis, Temm.,

is distinguished by a naked throat.

Procnias Ventralis, Illiger, Pl. Col. t. 5. the male, of which *Hirundo Viridis* is the female. *Procnias Hirundinacea,* Swain. Zool. Ill. t. 21. *Tersina Cœrulea,* Vieil. Gal. t. 119. The *Azure Chatterer* of Lath. H. and *Procnias Cyanotropeus,* Pr. Max., are all of this species, but not *Ampelis Tersa,* Lath., which is a Tanagra.

Procnias Cucullata, Swain. Zool. Ill. t. 37.
Ampelis Cucullata, Temm. Pl. Col. t. 363.

Head, neck, and chest black; collar, and beneath, yellow; back and scapulars brown; wing-coverts black, yellow-edged; quills and tail blackish, green-edged. Brazils.

Prince Maximilian also describes a bird of this name.

*Ampelis Caruncula*ta, Gm. Pl. Enl. t. 793.,

is distinguished by a long soft caruncle, which it carries at the base of the beak. Both this and *Hirundo Viridis* are white in their perfect state, greenish the rest of the year, and come from South America.

Araponga, Ampelis Nudicollis, Vieil. *Cas. carunculatus,* Spix, B. ii. t. 4. *Casmarhynchos Nudicollis,* Temm. Pl. Col. t. 368. ♂ 383. ♀.

White cere; region of the eye and throat naked, green, with black hairs; bill black; feet red; ten inches long; *female,* ash-green, white-spotted beneath. Brazils.

The genus *Casmarhynchos* of Temminck, and *Procnias,* Swain., *Ampelis* and *Tersina,* Vieil.

Variegated Chatterer, Lath. 10. *Ampelis Variegata,* Gmel. Pl. Enl. t. 793. Pl. Col. t. 51.

Ash, varied greenish and black; head dull-brown; quill blackish; under the throat two long fleshy caruncles; female without any caruncle; when young, caruncle and throat naked. Brazils.

? *** *Australasian.*

Muscicapa Melanopis, Vieil. N. D.

Face black; body above deep ash; beneath red; bill base bluish and greenish. Australasia.

Muscicapa Mystacea, Lath.

Brown; beneath white; crown and gular spot black; tail long; bill and feet black. New Holland.

Muscicapa Caledonica, Lath.

Olive; beneath ochraceous; chin and vent yellow; quills ferrugineous. New Holland.

Muscicapa Novæ Hollandiæ, Lath.

Brown; beneath whitish; streak under eye to the ears yellow; tail slightly forked, long.

Muscicapa Pectoralis.

Greenish-yellow; beneath yellow; head, sides of neck, and band on chest black; throat and crop whitish; length seven inches. New Holland.

Muscicapa Nævia, Lath.

Black; middle of back and shoulders white-spotted; length eight inches. New Holland.

Muscicapa Ochrocephala, Lath.

Head, neck, and chest, golden; body above yellowish-green; beneath white; bill and feet black; length five inches. New Holland.

Muscicapa Lutea, Lath.

Ochraceous yellow; tail feathers black and tipt; length five inches. New Holland.

Muscicapa Flavigaster, Lath.

Ashy; beneath yellow; quills and tail-feathers dull. New Holland*.

Other Flycatchers have a short broad bill, furnished with strong bristles, and a moderately broad, equal or slightly forked tail.

Some are found in New Holland; they form the genus *Myiagra* of Vigors and Horsfield.

Myiagra Rubecoloides, V. and H.

Head gray; throat and chest red; belly whitish; wings and tail brown; length five inches and a half. New Holland.

Myiagra Plumbea, V. and H.

Above brown, lead-colour; head, nape, and throat, shining lead-blue; belly and vent white; length four inches and a half. New Holland.

Myiagra Macroptera, V. and H.

Olive-brown above; beneath whitish; quill and tail brown; outer tail-feather, throat, and vent, white; length five inches. New Holland.

Some of the *Muscicapæ* belonging to New Holland have a long patulous, rounded tail, whence called fan-tails. They form the genus *Riphidura* of Vigors and Horsfield. The bristles of the mouth exceed the length of the tail.

Fan-tailed Flycatcher, Lath. Hist. t. 9. Cop. Gm. t. 193. f. 3. *Muscicapa Flabellifera*, Gmel.

Brown-black; superciliary and postocular spot, throat

* Latham and Vieillot described these birds as *Muscicapa*. Many of them will probably be found to be *Meliphaga*.—J. E. G.

and wing-covert tips, shaft and tips of tail-feathers white; belly ferrugineous. New Holland.

Riphidura Motacilloides, Vig. and Horsf.

Black; superciliary spot, middle of chest, belly, and vent white; quills black-brown; length seven inches and a quarter.

Black-tipped Flycatcher, Lath. *Motacilla Atricapilla.*
Rufous-fronted Flycatcher, Lath. *Muscicapa Rufifrons.*

Fuscous brown; eyebrows, lower part of back, base of tail, lower part of belly, red; crop black; throat and chest white, black-spotted; quills and tail brown; latter white-tipt.

Other New Holland Flycatchers agree with the last in the length of the tail, but it is nearly even; the bill is longer and more depressed, and is only furnished with short bristles. It forms the genus *Seïsura* of Vigors and Horsfield.

Volatile Thrush, Lath. *Turdus Volitans*, Lath. H. 151.

Black above; beneath white; head metallic black; quills brown. New Holland. The *Dishwasher* of the Colonist.

The genus *Pachycephala* of Swainson, peculiar for its large head, has been arranged among the *Pipridæ.* It may remain near the Chatterers.

Pachycephala Fusca, Vigors and Horsfield.

Olive-brown; beneath paler; throat and belly white quills and tail brown; ferrugineous edged; length five inches. New Holland.

Pachycephala Olivacea, Vigors and Horsfield.

Above olive-green; beneath yellowish; head grayish; throat white, marked; quill and tail brown, olive-edged; length seven inches and a half. New Holland.

Pachycephala Fuliginosa, Vigors and Horsf.

Testaceous gray, beneath paler, rather yellowish; throat whitish; length six inches. New Holland.

Southern Motacilla, White, Voy. t. at p. 239. *Muscicapa Australis.*

Above gray; lower part of back yellowish; beneath yellow; quill and tail brown.
young? throat whitish, called *Yellow Robin.* New Holland.

Finally, should be placed immediately at the end of the Cotingas

The GYMNODERES, Geoff.,

with the beak only a little stronger, but with the neck naked, and the head covered with downy plumes. The species known is also of South America, principally frugivorous, about the size of a pigeon; black with bluish wings. This is the *Gracula Nudicollis* of Shaw; the *Corvus Nudus*, and the *Gracula Fœtida* of Gm. Enl. 609.

The *Coracina Gymnoderma*, Vieillot, Vaillant, Ois. Amer. et Ind. t. 45, 46.

Placed in the genus *Coracina* by Vieillot and Temminck, and in *Ampelis* by Lichtenstein.

The Drongos—Edolius, Cuv.

belong also to the grand series of Flycatchers; the beak is also depressed and sloped at the end; the upper crest is lively; but they are principally distinguishable by the two mandibles being slightly bent the whole length; the nostrils are covered with feathers, and they have long hairs which form mustachios.

The species of this genus are numerous in the countries which border on the Indian Seas. They are generally coloured black, with the tail forked, and live on insects; some are said to have a song like that of the nightingale.

The genus *Dicrurus* of Vieillot.

Forked-tail Crested Shrike, Lath. *Lanius Forficatus*, Lin. *Dicrurus Longus*, Vail. Pl. Enl. t. 189. O. A. t. 66. *D. Cristatus*, Vieil.

Greenish; black frontal; crest erect; length ten inches. *L. Drongo*, Shaw.

Cineraceous Shrike, Lath. *Edolius Cineraceus*, Horsf.

Dark uniform ash-coloured; tips of quills and outer side of the outer tail-feathers black; length eleven inches. Java.

Malabar Shrike, Lath. *Lanius Malabaricus*, Lath. *Dicrurus Platurus*, Vieil. Vail. O. A. t. 175. *Edol. Retifer*, Temm. Pl. Col. t.

Bluish-black; quills and tail black; outer tail-feathers longer, naked, and inside feathered. Java. Also *Cuculus Paradiseus*, Gmel.

Fork-tailed Shrike, Lath. *Lan. Cœrulescens*, Lin. Vieil. O. A. t. 172. Edw. t. 56.

Bluish glossy black; abdomen white; breast dark-ash; tail forked; outer feathers white-tipt.

Corvus Balicassius, Gmel. Pl. Enl. t. 163.

Greenish black, bill and feet black.

Dicrurus Macrocercus, Vieil. *Le Drongolon*, Vail. O. A. t. 174. *Muscicapa Biloba*, Licht.

Black; tail deeply nicked, longer than the body; tail feather slender near the end; length ten inches; tail five inches and a half; habit slender. East Indies.

Dicrurus Æneus, Vieil. *Le Drongo Bronzé*, Vail. O. A. t. 176.

Shining black, reflecting violet and golden green. Bengal.

Dicrurus Lophorinus, Vieil. N. Dict. H. N. ix. t. d. 2. f. 2.

Iridescent black; forehead with a small crest of free and erectile feathers, perhaps a var. of *Corvus Balicassius*, Gmel.

Dicrurus Leucophæus, Vieil. *Drongi*, Vail. O. A. t. 170.

Gray, lead-colour; tips of quills blackish-brown; outer web of quills black; tail long, forked; length nine inches. Ceylon and Java. Perhaps young?

Dicrurus Leucogaster, Vieil. Vail. O. A. t. 174.

Above gray, beneath white; bill and feet lead-colour; a var. of former?

Edolius Azureus, Temm. Pl. Col. t. 225. ♀.

Fine blue; bill, quill, tail, and legs black; tail nearly even.

Edolius Remifer, Temm. Pl. Col. t. 178.

Above shining-black blown, beneath dull black; tail square; two outer tail-feathers very long, middle beardless, filiform, and dilated; length nine inches; *female*, outer tail-feathers like the rest. Java.

Dicrurus Mystaceus, Vieil. *D. à Moustaches*, Vail. O. A. t. 169? *Muscicapa Divaricata*, Licht.

Black; tail slightly nicked, as long as the body; tips of the tail-feathers dilated, divaricated; length nine inches, tail four inches and a half; habit stout. Senegal.

Dicrurus Musicus, Vieil. *Muscicapa Emarginata*, Licht. *Drongear*, Vail. O. A. t. 167, 168.

Black; tail slightly nicked, shorter than the body, divaricated; length nine to ten inches, tail four to four and a half. Africa.

Lichtenstein very justly remarks that the distinction of the species is very difficult; the young have the belly grayish; the length of the tail and wings varies; the adult are quite black, and the jaws and the bill are the same in all the species; they all have mustachios at the base.

Coracias Puella, Lath. Pl. Col. t. 70. 255. *Irena Puella*, Horsf. Java, t.

Nape, neck, and lesser wing-coverts splendid blue;

tail dusky-blue; middle of back, head, front of neck, and beneath, black.

The BLACKBIRDS, MERLES, (TURDUS, Lin.) have the beak compressed and bent, but the point does not make a hook, and its notches do not produce a denticulation so strong as in the shrikes. Nevertheless, there are, as we have said, gentle gradations from one genus to the other.

The regimen of this genus is more frugivorous. They live pretty generally on berries. Their habits are solitary.

The name of blackbird is more especially applied to the species whose colours are uniform, or distributed in large masses.

The most extended is,

The *Common Blackbird.* *T. Merula.* Lin.

The male, Enl. 2., is black, with a yellow beak; the female, Enl. 555, is brown; above reddish-brown; underneath spotted, with brown upon the breast. It is a bold bird, though easily tamed, and taught to sing, or even to speak. It remains here the whole year.

This bird is sometimes found entirely white, or partially varied with that colour, when it is the *Merula Leucocephala*, *Varia* and *Candida* of Brisson.

An allied species, but a bird of passage, which likes mountainous situations best, is the

Ring Ouzel. *T. Torquatus,* L. Enl. 168, and 182.

whose black feathers are in part edged with whitish, and the breast marked with a patch of the same colour.

In the south of France there is also at times seen,

The *White-tailed Ouzel.* *T. Leucurus,* Lath. Syn. ii. Pl. 38.

Smaller, black, the croup, and tail (the extremity excepted,) white.

In the high mountains of the south of Europe are. found

The *Rock-Crow.* *T. Saxatilis,* Enl. 512, and the *Solitary Thrush.* *T. Cyaneus,* Enl. 250,

from which the *T. Solitarius,* according to M. Bonelli, does not differ.

The first, which lives more commonly in the north, is best known; it builds in steep rocks and old ruins; sings well. The male has the head and neck ashy-blue, the back brown, the croup white, the under part and the tail orange colour.

We may conclude, with Shaw, that it is by confounding this species with the Jay of Siberia, that Linnæus has attributed to it the habits of the harpy, and has named it at one time *Corvus,* and at another *Lanius infaustus.*

These two birds form the section Saxicola of the genus Turdus of Temminck, to which Mr. Vigors has given the generic name of *Petrocincla.* The *Solitary Thrush* of Montague is a young Starling.

There are also two other blackbirds found in Europe.

Black-necked Thrush. *T. Atrigularis*, Temm. *T. Dubius*, Bechst. iii. t. 5. f. 1. 2. Young.

Olive-ash; beneath whitish, brown spotted; face, cheeks, throat, black. Austria and Russia.

Brown-eared Thrush. *T. Naumanni*, Temm. *T. Dubius*, Naum. Voy. t. 4. f. 8. not Bechst.

Reddish-brown; crown and ears deep-brown; beneath, brown spotted. Russia and Hungary.

Allied to the *Rock-Crow* are the

T. Rupestris, Vieil. *T. Rupicola*, Lath. *Rocar*, Vail. O. A. t. 101 and 102.

Blackish, varied with red and bluish; head and neck bluish-black; rump and body beneath red. South Africa.

T. Explorator, Vieil. *L'Espionneur*, Vail. O. A. t. 103.

Bluish-ash; wing-coverts and quills blackish-brown, white-edged; chest foxy; rump red. Cape of Good Hope.

The species allied to the Solitary Thrush, from the beauty of their plumage, are

T. Manillensis, Gml. Pl. Enl. t. 564. f. 2. and 626. *T. Violaceus*, Sonnerat, Voy. t. 108?

Blue-ash; rump blue; wing and tail blackish, red edged; throat and chest yellow spotted; belly orange-blue and white waved. India.

Hermit Thrush. *T. Eremita*, Gml. Pl. Enl. t. 364.

Orbits white; crown olive; higher occipital feathers brown tipt, black and white banded; lower ones pale-red, brown-edged; rump ash. India.

The foreign species of blackbirds are numerous. Belonging to the Old World may be noted,

T. Senegalensis, Gml. Pl. Enl. t. 563. f. 3.

Fuscous-gray; belly whitish; quills and tail brown. Africa.

T. Ornatus, Vieil. Vail. O. A. t. 86.

Black; golden-green gloss; tail short, nearly equal.

T. Nigricapillus, Vieil. Vail. O. A. t. 108.

Olive-brown; beneath bluish-ash; crown black. Africa.

T. Perspicillatus, Gml. Pl. Enl. t. 604.

Greenish-brown, beneath yellowish; head and neck ash; forehead and band on each side of the eye black. India.

T. Dominicanus, Gml. Pl. Enl. t. 627. f. 2.

Brown, glossed with violet and blue; beneath brownish-white; tail-base bluish, end greenish.

T. Squammeus, Vieil. Vail. O. A. t. 116.

Head, neck, and chest black; feathers of belly and beneath dirty-white, black tipt; of wing-coverts and back black, yellow-edged; tail subcuneate. Africa.

T. Tibicen, Vieil. Vail. O. A. t. 112. f. 2.

Brown-spotted, beneath pale-gray; tail wedge-shaped, pointed. Africa.

T. Phœnicurus, Vieil. Vail. O. A. t. 111.

Olive; eyebrows white; eye-streak black; quill and two middle tail-feathers bay; sides, throat, and chest red. Africa.

T. Importunus, Vieil. Vail. O. A. t. 106.

Olive-green ; quills, side-feathers, and tail yellowish edged. Africa.

T. Melanicherus, Vieil. Vail. O. A. t. 117.

Crested yellow ; quills and tail black ; tail wedge-shaped. Africa.

T. Macronnus, Gml. Lath. Syn. t. 93. Vail. O. A. t. 114. *T. Tricolor*, Vieil.

Shining purplish-black, beneath dull foxy ; rump and three outer tail-feathers half white. India.

T. Australis, Lath. Sparm. Mus. Carls. t. 59.

Blackish-brown ; chest and belly white, New Holland.

T. Chrysogaster, Gml. Pl. Enl. t. 221.

Green, above bluish, beneath orange ; bill and feet brown. Africa.

T. Ouravang, Gml. Pl. Enl. t. 557. f. 2.

Ash ; crown greenish-black ; head, chest, and body above olivaceous ; belly and vent yellowish ; bill yellow. Africa.

T. Miniatus, Sparm. Mus. Carls. t. 68.

Ferrugineous-brown, beneath ferrugineous ash ; throat whitish ; wing and tail black, and ferrugineous varied.

T. Erythropterus, Gml. Pl. Enl. t. 356.

Black ; wings red ; vent and tail-feathers (except the middle ones) white tipt ; tail wedge-shaped. Senegal.

T. Reclamator, Vieil. Vail. O. A. t. 104.

Brown, varied with blue, ash, and olive, beneath orange. Africa.

T. Atricollis, Vieil. Vail. O. A. t. 113.

Bluish; wing-coverts red, spotted and edged; quill black; throat and crop ochraceous; body beneath yellowish-red; collar blackish. South Sea Islands.

T. Hispaniolensis, Gmelin. Pl. Enl. t. 273. f. 1. t. 558. f. 9.

Olive; beneath varied olive and green; tail brown; inner edge white, outer olive; middle feathers olive. America.

T. Pratensis, Vieil. *T. Braziliensis, T. Atricapillus,* and *Gracula Longirostris,* Gml. *Batara Agallaspeladus,* Azara Pl. Enl. t. 292.

Black, beneath ferrugineous-yellowish; rump ferrugineous; tail slightly wedge-shaped; outer tail-feathers entirely, and rest white-tipt. South America.

T. Senegalensis, Pl. Enl. t. 539.

Shining-black; feathers yellow-edged; throat quills, and tail black. South Africa.

T. Madagascariensis, Gml. Pl. Enl. t. 557. f. 1.

Brown; belly and vent white; tail, two middle feathers entirely, and margin of rest bright golden-green; outermost white-edged. Africa.

T. Carbonarius, Licht.

Black; wings sooty; back, rump, sides, and vent slate;

Female, olive-brown; wing reddish; belly slate; bill brown. Brazils.

Merula Flavirostris, Swain.

Gray; back and wings tinged with ferrugineous; beneath white; breast and flanks ferrugineous; chin spotted; bill yellow; length nine inches and a half. Mexico.

Merula Tristis, Swain.

Olive-brown; beneath whitish; chin with black spots; under wing-covert pale ferrugineous; bill and legs brown; length nine inches. Mexico.

Turdus Pectoralis, Gml. Pl. En. t. 644. f. 2.

A *Thamnophilus* of Temminck.

Black-crested Thrush. Cinnamomeus, Gml. Pl. Enl. t. 560.

Reddish-brown; beek black, white-edged; wing-coverts black; small, white, middle and longer red-tipt. Cayenne.

Rufous Thrush. T. Rufifrons, Gm. Pl. En. t. 544. f. 1.

Brown; nape, sides of head and body beneath red; wing-coverts black, yellow-edged; tail ash; vent white. Cayenne.

T. Plumbeus, Gml. Pl. Enl. t. 560.
T. Ardosiaceus, Vieil. ?

Bluish; cheeks black; tail wedge-shaped; bill and feet red. North America.

Indian Thrush *T. Indicus,* Gml. Pl. Enl. t. 564. f. 1.

Olive-green; quill, inner web brown, outer yellow. India.

Black-headed Thrush. *T. Atricapillus,* Gm.Pl.En.392.

Blackish; head black; belly and rump rufous; wing-spot white. Cape of Good Hope.

Palm Thrush. *T. Palmarum,* Gml. Pl. Enl. t. 539. f. 1.

Olive-green, beneath ashy; nape and cheeks black, with three white spots on each side. Cayenne.

Gracula Athis, Gml.

Green; belly yellow; feet red-brown.

Muscicapa Hæmorrhousa, β Gml.
T. Hæmorrhousa, Hors.

Grayish-brown; head black; cheeks, throat, and belly white; rump yellow. Java.

Emerald Thrush, Lath. *T. Viridis.* Hors.

Emerald-green, uniform; above slightly olivaceous; chin yellowish; inner webs of quills and tail beneath pale brown; length eight inches. Java.

T. Javanicus, Hors. *T. Concolor,*
Temm. Pl. Col.

Body brown; gular-streak and abdominal spots dull ferrugineous; length eight inches and a half. Java.

Varied Thrush, Lath. *T. Varius,* Hors. Java.

Testaceous chestnut; tips of feathers deep brown; quills brown, edged externally with chestnut: belly whitish;

sides varied with chestnut and black; vent banded with white and black; tail beneath brownish; length eleven inches. Java.

Gular Thrush, Lath. *T. Gularis*, Hors.

Brownish olive; wings and tail ferrugineous; chin white; belly yellow; crown ferrugineous gray; length seven inches Java.

T. Arsinæ, Licht.

Ash brown; head black; belly dull-white; vent snow-white; length seven or eight inches. Egypt.

T. Falcklandii, Quoy and Gaim. Voy. t. Falkland Islands.

The following unfigured species of the Old World have been referred to this genus.

T. Arcuatus, Lath., China. *T. Canorus*, Asia. *T. Africanus*, Gml., Africa. *T. Splendidus*, Gml., *T. Abyssinicus*, Lath. *T. Obscurus*, Lath., North Asia. *T. Albicapillus*, Vieil., Africa. *T. Columbinus*, Gml., and *T. Nigricollis*, Gml., India. *T. Ruficollis*, Pallas, North Asia. *T. Leucocephalus*, Sonn., India. *T. Griseus*, Sonnerat, India. *T. Suratenes*, Sonnerat, India. *T. Borbonicus*, Gml., Africa. *T. Flavus*, Sonneret, India, an *Oriole?* *T. Kamtsckensis*, Penn, North Asia. *T. Leschenhaulti*, Vieil., Java. *T. Monachra*, *T. Asiaticus*, Lath., and *T. Speciosus*, Lath., India. *T. Sibiricus*, Lath., Siberia. *T. Tripolitanus*, Gml., an *Oriole?* West Africa. *T. Viridi-Olivaceus*, India. *T. Oonalaschkæ*, Penn., Siberia. *T. Validus*, Lath. *T. Persicus*, Vieil.

T. Ruficaudus, Lath., Africa. *T. Shannu*, Lath., China. *T. Tricolor*, Africa. *T. Virescens*, Lath., China. *T. Barbaricus*, Gml., Africa. *T. Daoma*, Lath. *T. Olivaceus*, Africa. *T. Phillippensis*, Gml., India.

Of the New World:—*T. Rufiventris*, Vieil., Brazils. *T. Chochi*, Vieil., Azara, N. 79, Paragua. *T. Albicollis*, Vieil. *T. Minor*, Gml., Carolina. *T. Dentirostris*, Vieil., Martinique. *T. Leucomelas*, Vieil., Azara. *T. Brevicaudatus*, Vieil., Brazils, a *Myothera.?* *T. Melanocephalus*, Vieil., Brazils. *T. Americanus*, perhaps an *Icterus*. *T. Leucogenus*, Gml., and *T. Fuscus*, Penn., North America. *T. Brachypus*, Vieil., Martinique. *T. Curæus*, Molina, Chili. *T. Cinereus*, Martinique. *T. Leucopterus*, Vieil., Brazils. *T. Triurus*, Vieil., Azara, N. South America. *T. Flavipes*, Vieil., Brazils. *T. Nævius*, Penn., North America. *T. Stratus*, and *T. Variegatus*, Gml., Polynesia, South America.

From Polynesia:—*T. Brachypterus*, Lath. *T. Musicola*, *T. Tenebrosus*, Lath. *T. Melanophrys*, Lath. *T. Dubius*, Lath., a *Meliphaga?* *T. Frivolus*, and *T. Albifrons*, Lath. *T. Crassirostris*, Lath. *T. Dilutus*, Lath. *T. Varius*, Lath. *T. Pacificus*, Lath. *T. Sanwichensis*, Lath. *T. Cyaneus*, Lath. *T. Maxillaris*, Lath. *Longirostris*, Lath. *T. Suerii*, Vieil. *T. Leucophrys*, Lath. *T. Macei*, Vieil. *T. T. Melanops*, Lath. *T. Poliocephalus*, Lath. *T. Leucotis*, Lath. *T. Peronnii*, Vieil. *T. Badiuus*, Lath. *T. Inquietus*. *T. Ulietensis*. *T. Novæ Hollandiæ*. *T. Gutturalis*. *T. Praursus*, Lath.

T. Fuliginosus, Lath. *T. Lunulatus*, Lath. *T. Harmonicus*, Lath. *T. Cyanocephalus. T. Punctatus. T. Ardosia- ceus*, and *T. Melinus*, Lath.

The name Thrush (*Grive*) is given to the species marked with black or brown spots. We have few of them in Europe altogether brown on the back and the breast spotted. They are singing birds and live on insects, and are gregarious in large flocks. They are good eating.

The *Missel Thrush, T. Viscivorus*, Enl. 489. Frisch 25.

is the largest of them, the under part of the wings is black; this species feeds much on the mistletoe, and contributes to spread this parasitical plant.

The *Fieldfare, T. Pilaris*, Frisch 26.

is distinguishable from the last by the ashy tint of the upper part of the head and of the neck.

The *Thrush*, properly so called, *T. Musicus*, Enl. 406. Frisch 27.

has the under part of the wings yellow. This is the best singer and is the most eaten.

The *Red Wing Thrush, T. Iliacus*, Enl. 51. Frisch 28.

is the smallest, and has the under part of the wings and the flanks red.

The foreign species of this genus are very numerous. We shall cite here only

The Mocking Bird, Moqueur, T. Polyglottus, Catesby 27. South America.

Ashy above; pale brown underneath; with a white band on the wing. It is famous for its astonishing power of imitating immediately the song of other birds and even all the voices it hears.

T. Orpheus, Lin. Edw. t. 78. Spix. t. 71. f. 12 Brazils.

Cinereous, spotted with brown and white; breast and belly pale gray; quills and tail white at the end.

T. Dominicus, Gml. Pl. Enl. t. 558. f. 1.

According to Prince *Masignano*, these are varieties of the former.

T. Lividus, Licht.

Ash; beneath white; sides brown-spotted. Brazils.

T. Saturninus, Licht.

Brown-ash; beneath ashy, sides streaked. Brazils.

Cat Bird, T. Lividus, Wils. A. O. II. t. 20. f. 3. *Muscicapa Carolinensis*, Gml. *T. Felivox*, Vieil.

Deep slate; beneath paler; vent rufous; crown and tail black; latter rounded. North America.

American Robin, T. Migratorius, Lin. Wils. A. O. 1. t. 2. f. 2. Pl. Enl. t. 586. f. 1.

Dark ash; beneath rufous; head and tail black; two outer feathers white at the inner tip. North America.

Red Thrush, T. Rufus, Gml. Wils. A. O. 2. t. 14. f. 1. Pl. Enl. t. 645.

Reddish brown; beneath whitish, black-spotted; tail

very long; rounded wing with two white bands; bill long, entire. North America.

Wood Thrush, T. Mustelinus, Gml. *T. Melodus*, Wils. A. O. 1. t. 2. f. 1. Vieil. O. A. S. 2. t. 62.

Brown-fulvous; head reddish; rump and tail greenish; beneath white, black-spotted; tail short, slightly nicked; bill moderate. North America.

Hermit Thrush, T. Minor Gml. *T. Solitarius*, Wils. A. O. 5. t. 43. f. 2.

Olive-brown; tail reddish; beneath white; sides and breast dusky; tail short, nicked; bill short. North America.

Tawny Thrush, T. Mustelinus, Wils. A. O. 5. t. 43. f. 3. *T. Wilsonii*, Pr. Masignano. *T. Silens*, Vieil.

Tawny-brown; beneath white; throat brown-spotted; tail short, nearly even; feathers pointed; bill short. North America.

T. Fuscatus, Vieil. O. A. Sept. 2. t. 57.

Brown; beneath ash, brown-spotted; side tail-feathers white-tipt; bill deep yellow. North America.

T. Olivaceus, Licht. *Le Griveron*, Vieil. O. A. t. 98, 99.

Bill and feet yellowish; olive-gray; throat white-brown streaked; belly ferrugineous; vent white. Cape of Good Hope.

T. Rufiventris, Licht. Azara. N. 79.

Bill and feet brown; olive-green; throat white-brown streaked; belly and vent ferrugineous. Brazils.

T. Crotopegus, Licht. Azara. N. 80.

Bill and feet brown, olive-brown; throat black-brown white-streaked; crop, belly, and vent white. Brazils. *Female*, *T. Jamaicensis*, Lath.?

T. Furmigatus, Licht.

Olive-red; beneath paler; belly and throat whitish; throat brown-streaked; primaries internally brown margined. Brazils.

Junior *T. Variegatus* and *T. Striatus*, Gml.

Orpheus Curvirostris, Swain.

Gray; beneath whitish; throat and breast spotted; vent pale fulvous; bill long curved; length ten inches and a half. Mexico.

Orpheus Cærulescens, Swain.

Bluish; crown and throat paler; ears and sides of the head black; length four inches. Mexico.

T. Flavipes, Spix. t. 67. f. 2. Brazils.

T. Rufiventer, Spix. t. 68. Brazils.

T. Albiventer, Spix. t. 69. f. 1. 2. Brazils.

T. Albicollis, Spix. t. 70. Brazils.

T. Guyanensis, Gml. Pl. Enl. t. 390. f. 1. *T. Jamaicensis*, Vieil.?

Greenish brown; beneath ochraceous, black-streaked; bill and feet brown. South America.

T. Sinensis, Gml. Brisson 2. t. 23. f. 1.

Reddish; head brown-streaked; eyebrows white; tail brown darker streaked; bill and feet yellow. China.

T. Cayanensis, Gml. Pl. Enl. t. 515.

Ash; beneath reddish-gray; vent gray; larger wing coverts and quills black. South America.

The genus *Tanypus* of Oppel has all the characters of the Thrushes; but the *tarsi* are longer. *Turdus* § 4. Temminck.

Tanypus Australis, Oppel. Mem. Acad. Bavière. 1811. t. 8.

Some of these birds are allied to the Butcher Birds both by their manners and the form of their beaks.

Temminck has separated them under the name of *Turdoides* or *Ixos*, chiefly characterised by the beak being shorter than the head.

Guava Flycatcher. Muscicapa Psidii, Gml. *Sonnerat*, Voy. t. 28. *T. Analis*, Hors.

Brown beneath; and eyebrows white; vent yellowish; band under eye black. Phillippine Islands.

Turdus Bimaculatus, Hors.

Brownish-olive; chin and forehead brown; each side of the forehead an orange spot; cheeks, shoulders, and vent yellow; belly white. Java.

T. Cafer, Lath. *Muscicapa Hæmorrhousa*, Lath. Pl. Enl. 563. f. 1. *Merle Curouge*, Vieil. O. A. iii. t. 107. f. 1.

Slightly crested; blackish; rump and belly white, vent red. Africa.

T. Aurigaster, Vieil. *T. Chrysorhoeus*, Tem. *Le Culdor*, Vieil. O. A. t. 107. f. 2. Brown, Ill. Zool. t. 31.

Brown-gray; beneath white; crown, cheek, and throat black; vent golden. Africa.

T. Capensis, Vieil. *T. Nigricans*, Vieil. *T. Le Vaillantii*, Tem. Pl. Enl. 317. *Le Brunoir*, Vieil. O. A. t. 106. f. 1.

Brown; head and throat black; eyelids orange; belly yellowish brown. South Africa.

T. Phœnicopterus, Tem. Pl. Col. t. 71.

Bluish or violet-black; tail and wings dull-black; small wing coverts bright red; length seven inches. Senegal.

T. Disparis, Hors. *T. Concolor*, Pl. Col. t. 137.

Olive-green; head and neck blue; throat crimson; chest and beneath yellow; length six inches. Java.

T. Atriceps, Tem. Pl. Col. t. 147. *Lanius Melanocephalus*, Gml.

Olive-green; head and upper part of neck blue-black; quills and middle of tail-feathers black; belly, vent, edge of secondaries, and tips of tail yellow; length six inches. Java.

T. Azureus, Tem. Pl. Col. t. 274.

Blue; chin, throat, and front of chest brown; head grayish; length eight inches. Java.

Ixos Virescens, Tem. Pl. Col. 382. f. 1.

Greenish-ash; face, orbits, ears, throat, and beneath white varied with greenish ash; length six inches and a half. Java.

Ixos ——? Tem. Pl. Col. 382. f. 2.

Blue-black; throat beneath dull-red; streak over eye; chin, side of throat, edge and secondaries, and tip of two outer tail-feathers white; length seven inches. Java.

Ixos Chalcocephalus, Tem. Pl. Col. t. 453. f. 1.

Dull-lead gray; head and top of neck metallic-black; quills and middle of tail-feathers black; latter white tip. Java.

Turdoides Leucocephala, Ruppel Atlas, t. 4.

Bill black; head white; wings and tail dull pale-brown; the soft feathers of the nape, back, and interscapulars paler; beneath brownish-white; throat white spotted.

Perhaps here should be placed the genus *Chloropsis* of Jardine.

Black-chinned Thrush, Lath. *T. Cochinsinensis*, Gml. *Meliphaga Javanica*, Horsf. Vieil. O. D'Or. ii. t. 77. 78.

Green; lores and crop black; lower jaw with a blue streak; a yellow moon under the throat; bend of the wing shining blue; length five inches and half. India.

Yellow-fronted Thrush, Lath. *T. Malabaricus*, Gml. Jar. Ill. Zool. t. 5.

Green, shining; forehead orange; chin and throat hyacinth; on the crop a golden moon; bend of the wings blue: length six inches and a half. India.

Sonnerat's Thrush, *Chloropsis Sonnerati*, Jardine.

Green; lore, throat, and crop black; a small hyacinth maxillary streak; bend of the wings blue-green; length eight inches. India.

Hook-billed Chloropsis. *Chloropsis Casmarhynchos*, Jard. Ill. Orn. t. 7.

Entirely green; small maxillary streak blue; bend of wings blue-green; beak brownish; apex adunc; length seven inches and a half. India.

It is doubtful whether these birds should be placed with the Thrushes, or the *Meliphagæ.*

Some of the blackbirds which have slender beaks are difficult to be distinguished from the stonechats, such as the

Le Tanfredic, Vail. O. A. t. 111.

Brown; eyebrows and beneath white; throat and rump reddish; cheeks and quills black.

Le Grivetin, Vail. O. A. t. 118.

Brown, beneath pale; eyebrows, edge of wings, secondaries, and tail-feathers white-edged.

Le Culdor, Vail. O. A. t. 119.

Brown; breast and beneath white; eyebrows and throat yellow; mustachios black.

T. Trichas, Gml. Pl. Enl. t. 709. f. 2. Edw. t. 237.

Olive; body beneath yellow; eye-streak black.

Motacilla Subflava, Gml. Pl. Enl, t. 584. f. 2. *Le Citrin*, Vieil. O. A. t. 127.

Red-brown; beneath gray; rump pale; sides of body reddish; tail wedge-shaped. Senegal.

Motacilla Macroura, Gml. Pl. Enl. t. 752. f. 2.

Brown; beneath yellowish-white, black-spotted; eyebrows white; tail long, wedge-shaped. Cape of Good Hope.

Dr. Horsfield places in this genus *Gracula Saularis*, under the name of *Turdus Amœnus*. Cuvier refers it to the *Lanii.* *Turdus Labradorus*, *Palmarum*, *Hudsonius*, and *Noveboracensis*, are Icteri.

T. Speciosus and *Albifrons* are Muscicapæ.

T. Leucocephalus, *T. Ochrocephalus*, *Malabaricus*, *Roseus*, and *Pagodorumi*, are Pastors.

T. Jugularis, *T. Manachra*, *T. Motacilla*, and *T. T. Arundinacea*, are Orioles.

T. Aurocapillus and *T. Calliope*, are Sylviæ. *T. Cyanus* is a female Chatterer. *Le Fluteur*, Vail. is a Malurus; and *T. Orientalis*, a Lanius.

Neither can one distinguish by sensible characters certain blackbirds of Africa, which live in numerous and noisy flocks, like the starlings, and pursue insects or make great havock in gardens, (the Stournes of Daudin or the Pastors of Temminck): one of them is often found in Europe, which is

The Rose-Coloured Thrush, *T. Roseus*, Enl. 251.

Of a shining black; but with the back, croup, scapulars, and breast of a pale rose-colour; the feathers on the head are narrow and elongated into a tuft. It is serviceable in hot climates, by destroying the locusts.

Vail. O. A. t. 96. *Female*, crest shorter; rose colour paler. *T. Sellacis*, Gml. the genus Psaroides of Vieillot.

Cuvier has placed this bird here as a section of Turdus; and its analogous species, with which it has always since been arranged, he has formed into a genus, under the name of *Gracula*; which see.

Others are remarkable for the brilliancy of their plumage, which is usually of a dark brown colour.

They are peculiar to the old continent, and especially Africa. The genus Lamprotornis of Temminck, &c.

T. Auratus, Pl. Enl. t. 540. *Nabirop*, Vail. O. A. t. 84.

Violet; back and wings golden-green; cross-band on inner edge of wings; tail and upper coverts blue. Cape of Good Hope.

T. Nitens, Pl. Enl. t. 561. *Couigniop*, Vail. O. A. t. 90.

Blue; reflecting green, violet, and purple; bill and feet black. Senegal.

T. Morio, Pl. Enl. t. 199. *Le Roupenne*, Vail. O. A. t. 83. 84. *Corvus Rufipennis*, Shaw.

Shining black; primaries red black tipt. Africa.

T. Bicolor, Lin. *Le Spreo*, Vail. O. A. t. 88.

Brown, changing into bright green on the neck and tail; vent, and under wings, white; base of lower mandible yellow; tail wedge-shaped. South Africa.

L'Eclatant, Vail. O. A. t. 85.

General tone of colour, refulgent green, varied with blue, purple, and gold. South Africa.

Corvus Splendidus, Shaw, &c. *Choucador*, Vail. O. A. t. 86.

In colour like the last; tail shorter, with the feathers nearly equal. South Africa.

T. Chrysogaster, Lin. *L'Orambleu*, Buff.

The whole upper part blue; underneath orange; bill, feet, and quills black. South Africa.

Several of the species described as Blackbirds probably belong to this section.

T. Lamprotornis,
Tail graduated, Pl. Col. t. 648. f. 2.

Songster Thrush, Lath. n. *T. Cantor*, Gml. t. 75. Sonnerat, India.

Upper parts greenish-black, with a gloss of blue and violet; quills and tail black.

T. Chalybeus, Hors. Pl. Col. t. 199. t. 1. 2.

Metallic green; feathers of neck long and lanceolate; wings and tail blue; tail rounded; length seven inches. Java. Perhaps *T. Mauritianus*, Gml.

Pigeon Thrush. *T. Columbus*, Lath.

General colour green, very changeable in different reflections of light; rump and vent sometimes white.

T. Leucogaster, Gm. Pl. 6. 48.

Violet; belly white; quills blackish. Africa.

† *Tail graduated; middle feather longest.*

Lamprotornis Metallicus, Tem. Pl. Col. t. 266.

Metallic purple; wing and tail bluish; feathers of head and neck long and lanceolate; bill and feet black; length eight inches and a half. Timor.

Lamprotornis Erythrophiis, Tem. Pl. Col. t. 267.

Slate coloured; band over eyes rigid scarlet; eye spots and ears black; wings and tail green, under and upper tail-coverts and edge of quills yellow; quills crown tipt.

Some have the tail graduated and one-third longer than the body.

T. Æneus, Pl. Enl. t. 220. *Vest-dore*, Vail. O. A. t. 87.

Golden-green, beneath grassy-green; head blackish; shining golden rump; and middle tail feathers purplish; tail wedge-shaped. Senegal.

We must evidently unite to these the *Merle de la Nouvelle Guinée*, with a tail three times the length of the body, with a double crest on the head, which has been treated as a bird of Paradise (*Paradisea gularis*, Lath. and Shaw; *Par. nigra*, Gml., Vail. Ois. de Par. 20 and 21. Vieil. Ois. de Par. Pl. viii.) but solely on account of its singularity, and the incomparable magnificence of its plumage.

This forms the genus *Astrapia* of Vieillot, and the first section of *Lamprotornis* of Temminck. Vieil. Gal. Ois. t. 107.

The Chocards, Pyrrho-corax, Cuv.,

Have the compressed, arched, and sloped beak of the Blackbirds, but their nostrils are covered with feathers like those of the crows, to which they have been annexed We have one,

The *Alpine Chocard.* *Chocard des Alpes, Corvus Pyrrho-corax,* L. Enl. 351.

Black, with the beak yellow, the feet at first brown, then yellow, and in the adult state red, which builds in the clefts of rocks of the higher mountains, whence they descend in winter in large flocks into the vallies. They live on insects, snails, and fruits, and do not disdain carrion.

In India there is another,

The *Sicrin.* *Sicrin,* Vail. Ap. Pl. 82.,

distinguished by three barbless stalks, as long as the body, on each side, among the feathers which cover the ears.

This is the *Corvus Crinelus* of Daudin, *Corvus Sex-setaceus* of Shaw, and the *Pastor Setiger* of Wagner, to which genus it appears to be most allied.

Temminck places in this genus the *Pyrrho-corax Leucopterus,* and also the *Corvus Garrulus,* of Lin., of which Cuvier forms the genus *Regilus,* and places it with the *Hooppoes.*

I find no sufficient character for separating from the blackbirds

The true Orioles, (Oriolus, Lin.)

whose beak resembling that of the blackbirds, is only a little stronger, and whose feet are a little shorter in proportion. Linnæus and his followers have joined them to the *Cassiques,* which they resemble only in colours.

The *Oriole of Europe.* *Oriolus Galbula*, L. Gml. 26.

A little larger than the blackbird. The male is of a beautiful yellow; the tail and a spot between the eye and beak black, the end of the tail yellow; in the female the yellow is substituted by an olive, and the black by brown. This bird suspends its nest skilfully formed on the branches; eats cherries and other fruits and insects in spring.

Oriolus Auratus, Vieil. Vail. O. A. t. 260. *T. Flavus*, Gml. *O. Bicolor*, Temm. Licht.

Yellow; eye-band black; quill black; secondaries, outer edge, yellow; tail black: yellow tipt. South Africa.

O. Galbula, *β*. Lath. *O. Melanocephalus*, Lin. Vail. O. A. t. 263. Pl. Enl. t. 79. Edw. t. 77.

Yellow; head and throat black; a yellow spot at base of primaries; tail all yellow. This appears to be also the

O. Annulatus and *O. Nov. Hispaniæ*, from Seba, t. 55. f. 4. and t. 63. f. 3.

Merla Bicolor, Aldrov. *O. Coudougan*, Vail. O. A. t. 261. 262. *O. Radiatus*, Gml. *O. Larvatus*, Licht. *O. Monachus*, Wagner. Female, *T. Monachra*, Gml.

Yellow; back olive; head and throat black; smaller wing-coverts white-tipt; base of all the tail-feathers black. South Africa.

O. Sinensis, Gml. Pl. Enl. t. 590. *O. Cochin Chi-*

nensis, Bris. t. 33. f. 1. *O. Hippocrepis*, Wag. *O. Galbula*, 8. e. junior. Lath. *O. Maculatus*, Vieil.

Yellow; head-band black: wings and tail black and yellow-tipt. Senegal.

O. Leucogaster, Temm. *O. Xanthonotus*, Horsf. Z.R. t. Pl. Col. t. 214. f. 1. ♂ 3. ♀.

Black; belly white, black streaked; scapulars, rump, vent, and inner tail-feathers yellow; bill red; feet black; length six inches and a half. Java.

O. Arundinarius, Burchel.

Citron-yellow; face black; wings brown; quills and coverts edged with yellow; back and tail greenish-brown; rump yellow. Female? Nest globular, suspended between the stems of reeds. Var. *O. Radiatus*, Gml.? Africa.

Temminck, and more modern authors, have referred here the *Paradise Oriole, Oriolus Aureus*, Lin.

O. Leucopterus is a Tanager.
O. Capensis, and *O. Textor*, are Plocei.
O. Furcatus is an Edolius.
O. Picus, a Dendrocolaptes.

And the other species of Gmelin and Latham are Cassea.

Mr. Swainson (Zool. Jour. i. 478.) has separated from the Oriole, the genus *Sericulus*.

Meliphaga Chrysocephala, Lewin.

N. H. Birds, t. 6. *O. Regens*, Pl. Col. t. 320. Quoy and Gaim. t. 20.

Black; feathers of back of head short, velvety orange;

neck, shoulders, and secondaries yellow, New Holland. Female, brown; back and chest white; lanulated crown; middle of throat and nuchal collar black; belly whitish-brown. See *Paradisea Aurea*, Lath.

The genus *Mimeta* of Capt. King is separated from the Orioles for the same reason.

Green-grackle, Lath. H. 24. *Gracula Viridis*, Lath. *O. Viridis*, Vieil. *O. Variegatus*, Vail.

Olive-green; beneath whitish-black, broad streaked; wing and tail black-brown; edge of wing and tips of tail white.

Mimeta Flavocinctus, King.

Yellow-green; beneath paler; head and back brown-lined; wing and tail black, green, and yellow varied.

Mimeta Meruloides, Vig. and Horsf.

Above brownish, olive-brown streaked; beneath white; crown striately dropped; wing-coverts and secondary quills pale-red edged; tail white tips; length ten inches; both probably varieties of the green grackle.

Buffon with justice has separated from the Blackbirds

The Anteaters, Myiothera, Illig.

which are recognized by their long legs and short tail. They live on insects, and principally ants. They are found in both continents.

Still the species of the old continent are remarkable for the lively colours of their plumage. These are the

BREVES of Buffon (*Corvus Brachyurus,* Enl. 257. 258. Edw. 324. and his *Azurin.*) (*T. Cyanurus,* Lath. Gml. *Corvus Cyanurus,* Shaw,) Enl. 355.

This is the genus *Pitta* of Vieil. and Temminck. The species are all from India, and *Pitta* of Waggler, not Temminck, which is *Ptilinorhynchus* of Kuhl; the species have been recently more divided, as

Corvus Brachyurus, Pl. Enl. t. 258. Edw. 324. *T. Triostechus,* Spa.? *Myiothera Brachyura,* Illiger.

Green; beneath and lines on head fulvous; wing with a white spot; tail black green tipt.

The *Breve des Phillippines,* Pl. Enl. t. 89. is the same as Edwards' with the head of a Thrush. See Vail. Ois. Par. 1. 106.

Corvus Brachyurus, ♂ Lath. Pl. Enl. t. 257. *Pitta Hippocrepis,* Waggler. *Myiothera Velata,* Tem.

Green: beneath yellowish; head blackish brown; nape yellowish; cervical, lunule, and band under the eye black.

Blue-Tailed Thrush, Lath. *T. Cyanurus,* Gml. Pl. Enl. t. 355. *Myiothera Affinis,* Hors. Gall. Ois. t. 153.

Red; brown beneath; yellow belly, blue-banded; back of head and sides of neck with a longitudinal black streak; pectoral band and tail blue; length eight inches. India, not South America.

Pitta Strepitans, Tem. Pl. Col. t. 383.

Corvus Brachyurus, ε Lath. Sonn. Voy. t. 110. *Pitta Superciliaris,* Wag.

Head and nape black; eyebrows greenish blue-edged; throat white; crop and back green; belly reddish; vent red. Malacca.

Pitta Versicolor, Swain. Zool. Jour.

Green, beneath fulvous; rump and wing-coverts cærulean blue; vent red; crown rufous; nape, chin, and abdominal spot, black. New Holland.

Corvus Brachyurus, β. Lath. Pl. Enl. t. 89. *Pitta Melanocephala,* Wag. Edw. 324.

Green; head and neck black; rump and wing-coverts bluish-green; tail beneath rosy; tail black. Said by Cuvier to be another species with the head of a blackbird, but Waggler makes it distinct.

Corvus Brachyurus, η Lath.? *Pitta Brachyura,* Vigors. *Pitta Australis,* n.

Green; beneath fulvous; eyebrow pale fulvous; head, wings, and tail black; throat and wing-spot white; rump blue; middle of belly and vent scarlet. New Holland.

Pitta Erythrogastra, Cuv. Pl. Col. t. 212.

Back and broad pectoral collar green; head and nape reddish chestnut; chin whitish; throat brown; necklace, wing, tail, and upper tail-coverts blue; beneath crimson; 2, 3, and 4, quills with a white spot; secondaries black; length six inches. Manilla.

Pitta Gigas, Tem. Pl. Col. t. 217.

Blue; beneath brownish ash; crown and half collar;

ears and quills black; latter blue tipt; legs long; length nine inches. Sumatra.

Blue-Winged Breve. *Pitta Cyanoptera*, Tem. Pl. Col. t. 218.

Back and scapulars green; wing-coverts and rump blue; head, chin, and neck black; crown and half collar yellowish; throat white; breast yellow; belly and vent red. Quill and tail black; former white-banded; latter blue tipt; length seven inches. Java.

Pitta Angolensis, Vieil.

Head black; dull yellow-green; throat streaked reddish; collar yellow; beak green; small wing-coverts and rump blue. Africa.

Some species have been separated under the name of *Timalia*.

Pitta Pileata, Tem. Pl. Col. t. 76.
Timalia Pileata, Hors. Zool. Jav.

Olive-brown; crown chestnut; chin and throat lined with black; belly dull testaceous. Java.

T. Gularis, Hors. *Motacilla*, Raffles.

Brown; beneath yellowish; head and tail ferrugineous; throat and breast black-streaked. Sumatra.

Timalia Thoracica, Hors. Pl. Col. t. 76.

Olivaceous brown above; underneath testaceous-gray; top of head chestnut; throat and cheeks white; narrow white band from base of bill passes over the eye. Java.

Temminck has separated the genus *Myophonus*, which, like the *Pitta*, belongs to the Old World.

Cyaneous Thrush, Lath. *Pitta Glaucina*, Tem. Pl. Col. t. 194. *T. Cyaneus*, Hors. Java. t.

Deep azure; head, belly, bill, feet, and outer edge of quills and tail feathers black. Java.

Yellow-Billed Thrush, Lath. *T. Flavirostris*, Hors. *Myophonus Metallicus*, Tem. Pl. Col. t. 170.

Black; head, collar, chin, throat, and breast waved with steel; base of tail-feathers white; bill yellow. Java.

The species of the new continent, much more numerous, have browner tints and vary in the force of the beak and proportional length of the tail. They subsist on the immense ant-hills of the woods and deserts of this part of the world. The females are more bulky than the males. These birds fly little, have sonorous, and, in some species, remarkably loud voices; among these, with strong and arched beak, is

The King of the Anteaters (*T. Rex*, Gml. *Corvus Grallarius*, Shaw.) Enl. 702.

The largest and longest legged of all, and shortest tailed. At first sight it looks like a wader. It is about the size of a quail and its grey plumage is agreeably variegated. It lives more isolated than the rest.

The genus *Grallaria* of Vieillot; *Myioturdus* of Boie; peculiar for the base of the thighs being naked.

The genus *Myrmothera* of Vieillot.

T. Tinnicus, Pl. Enl. t. 706. f. 1 *Myiothera Tinnicus*, Illiger.

Brown; beneath white; chest black spotted; tail equal; bill above black, beneath white.

Some species have the bill more straight, not strong: they have an affinity to the *Lanii* with the same bill.

T. Colma. β Lath. *Myiothera Tetema,* Illiger. *Myrmothera Tetema,* Vail. Pl. Enl. t. 821.

Black-brown; crown and nape red. South America.

Myiothera Umbretta, Licht.

Sooty brown; throat whitish; bill and feet slender; length six inches and a half; tarsi nine lines.

T. Formicivorus, Pl. Enl. t. 700. f. 1. (α t. 644. f. 1. 2. ?)

Red-brown; beneath ash; chin, throat, and chest black; surrounded by a varied black and white band. South America.

T. Colma, Lath.? *Myrmothera Colma,* Vieil. Pl. Enl. t. 703. f. 1.

Red-brown; beneath ash; chin and throat black; white spotted.

T. Lineatus, Gml. Pl. Enl. t. 823. f. 1.

Olive-brown; chin, throat, and chest white; chest brown spotted; sides of neck white lined. South America.

Myiothera Campanisona, Licht.

Olive; frontal streak short black; eyebrows and throat white, black-dotted; chest, vent and sides white, black streaked; tail short black tipt; length eight inches. Brazils.

Myiothera Strictothorax, Tem. Pl. Col. t. 179. f. 1. 2.

Green above; light yellow underneath; small dark spots on breast; top of head dark red; sides of head spotted ash colour; wing-coverts edged white; length four inches six lines. Brazil.

Myiothera Mentalis, Tem. Pl. Col. t. 179. f. 3.

Above and beneath like the last, without the spots; head and throat dark ash coloured; length four inches. Brazil.

Myiothera Capistrata, Tem. Pl. Col. t. 185. f. 1.

Dirty-yellow above; brighter underneath; crest nearly black; streak under it yellow; face and throat ash colour; length five inches and a half. Java.

Myiothera Melanothorax, Tem. Pl. Col. t. 185. f. 2.

Crest and back brown; face, breast, and belly light-blue and white; lower belly gray; lesser wing-coverts red with a white spot; and irregular black spots on breast.

Myiothera Superciliaris, Licht.

Sooty; eyebrows white; wing-coverts and tail-feathers black, white tipt; quills entirely black; length five inches. Brazil.

Myiothera Fuliginosa, Illiger.

Slate-black; middle of chest and belly black; wing-coverts black, white tipt. Brazil.

Others have the bill slender and acute, and their tail streaked; they are allied to the Wrens.

T. Bambla. Gml. Pl. Enl. t. 703. f. 2.

Spotted, reddish-brown; beneath ash; wings black with a white cross-band; bill black. South America.

Musician Thrush. T. Arada, Lath. *T. Cantans,* Gml. Pl. Enl. t. 706. f. 2.

Red-brown; black banded; beneath white; cheeks black, white dotted; neck fulvous. Cayenne.

Myiothera Nematura, Licht.

Olive-brown; nape streaked; streak behind the eye; narrow spot and drops on the belly white; tail black; end of each shaft extended wirelike; bill weak; length five inches and a half.

Myiothera Perspicillata, Licht.

Forehead, orbits, and ears black; crown and nape chestnut; back slate; tail olivaceous, beneath slate; middle of throat and belly white; tail short.

Myiothera Loricata, Licht.

Chestnut; eyebrows and tips of black wing-coverts yellow; tail-feathers spotless; length six inches. Like *Pipra Nævia.* Bahia.

Myiothera Squamata, Licht.

Black white spotted; chest scale-like; tail four banded; vent slate colour; length four inches and a half. Bahia.

Myiothera Pileata, Licht.

Gray; crown black; eyebrows white; quill and coverts

black, white edged; tail and middle feathers black, white tipt; rest white black-based. Bahia.

Myrmothera Fuscicapilla, Vieil.

Deep blue; crown brown; cheeks reddish; throat black; belly white.

Chiming Thrush, Lath. *T. Campanella*, Lath. *T. Tintinnabulatus*, Gml. Pl. Enl. t. 700. f. 2.

Brown; rump and belly orange; crown and temples white, black spotted; eyebrows black; chin white; chest flesh coloured, black spotted. South America.

Myrmothera Axillaris, Vieil.

Ashy-blue; chest, quills, and side tail-feathers black; the latter and wing coverts white tipt; axillæ white. Guiana.

Myrmothera Longipes, Vieil.

Reddish-ash; forehead, eyebrows, throat, and belly white; chest, tail, bill, and feet black. Guiana.

Myrmothera Melanoleucos, Vieil.

Feathers black, white edged; wing band white; body beneath white brown spotted; wings rounded, short. Guiana.

Myiothera Malura, Natter. Pl. Col. t. 353. f. 1. 2.

Tail very long, much graduated; bill slender, brownish gray; head black and white varied; wing-coverts black white tipt; cheeks, front of neck, and chest whitish, streaked with black; length five inches and a half. *Female* more brown. Brazil.

Myiothera Rufimarginata, Tem. Pl. Col. t. 132. f. 1.

Back dark-olive; belly yellow wavy; sides of head bluish, with dark waves; large wing-covers red; lesser black and white; tail black and white. Brazil.

Myiothera Ferruginea, Licht Pl. Col. t. 132. f. 2. 3.

Head black, with four white bands; back reddish; loins, wings, and tail black, white spotted; chin white; rump and beneath deep ferrugineous; length five inches. Bahia.

Barred Tail Thrush. *T. Coraya*, Gml. Pl. Enl. t. 701. f. 1. *Spix.* Braz. t. 73. f. 2. ? *Sphænura Coraya*, Licht. *Cichla Coraya*, Wagler. *Campycorlumchus Striotatus*, Spix. *Myiothera Coraya*, Illiger.

Red-brown; crown, cheeks, and neck black; throat and streak under eye white; tail gray, black banded.

White-backed Thrush. *T. Alapi*, Lath. Pl. Enl. t. 701. f. 2.

Brown, beneath ash; neck and chest black; wing-coverts white dotted; back white spotted. Guiana.

Buff-winged Thrush. *T. Fuscipes*, Lath.

Crown black; upper parts dark ash; wing-coverts barred with buff; quills brown; under parts rufous; legs brown. Cayenne.

Spotted Nuthatch. *Sitta Nævia*, Gml. Edw. t. 348.

Head coloured; white spotted; beneath blue ash; white lined; throat white.

Myiothera Ruficeps, Spix. t. 72. f. 1.

Above olive-green; beneath blackish; head above red; quills reddish brown; tail blackish; ocillæ fulvous; hind claw nearly straight. Brazil.

Myiothera Leuconota, Spix. t. 72. f. 2.

Above chestnut brown; beneath reddish; front of the back with two white bands; cheeks and tail black; bill yellowish; feet brown; body six inches and a half; tail two inches and a half long. Brazil.

Myrmothera Cærulescens, Vieil.
Thamnophilus ? Spix.

Bluish; wings and tail black, white spotted; bill brown. Brazil.

Rocky Mountain Ant Catcher, Troglodites Obsoleta, Say. *Myiothera?* Pr. Musig. A. O. t. 1. f. 2.

Dusky brownish undulated with pale; beneath whitish, marked with brown; tail long, rounded, ferrugineous, yellow tipt; bill very slender, slightly curved; tarsi seven-eighths; tail two inches long. North America.

Myrmothera Gutta, Vieil. Gal. Ois. t. 155; also belongs to this genus. Temminck and Illiger have placed in this genus *Pipra Albifrons*, which Cuvier treats as a *Lanius* and *T. Auritus* and *Pipra Nævia*, the latter forming the genus *Conophagus* of Vieillot, which Cuvier refers to the *Muscicapæ*. Buffon placed here *T. Pectoralis Cinnamomeus* and *Rufifrons*, which Cuvier says are *Turdi*.

There may be added the genus *Grallina* ofVieil.

? Lath. Hist. *Grallina Melanoleuca*, Vieil. Gal. t. 151.

Eyebrows, back of neck, chest, and hinder parts black; long band on wing; loins, rump, side tail-feathers white. New Holland.

And the genus *Chamæza* of Vigors, which has the colouring of a Thrush and the form of a *Pitta*.

Chamæza Memloides, Jardine. Illust. Zool. t. 11.

Above brown; beneath reddish; white with long black spots; throat white; rump and tips of tail reddish.

Also should be separated from the Thrushes,

The Cincles, (Cinclus, Bechst.)

whose beak is compressed, straight, with equal mandibles, almost linear, sharpening towards the point, and the upper one scarcely arched.

(*Sturnus Cinclus*, L.) *T. Cinclus*, Lath. Enl. 940.

Legs a little raised; tail rather short, approaching the Anteaters. It is brown, with throat and chest white. It has the singular habit of descending completely into the water without swimming, but walking at the bottom in search of the animalculæ on which it feeds.

The *Cinclus Aquaticus*, Bechst. The *Cinclus Bicolor*, Vieil.; also found in South Asia. *Cinclus Pallasii*, Tem. *C. Unicolor*, Pr. Musig.

Ash brownish; chin ash-brown. North Asia? and America.

Cinclus Mexicanus, Swain.

Cinereous; gray head; and chin brown. Mexico.

Perhaps the genus *Colluricincla* of Vigors and Horsfield should be added in this family.

Colluricincla Cinerea, Lin. Trans. 13.

Ashy above, paler underneath, with the throat and before the eyes white; under the wings brown. Found in New Holland.

And also the genus *Sphecothera* of Vieillot, which has the bill thick, straight, and bald at the base, and curved at the top; the orbits naked; and the first and third quill the longest.

Sphecothera Viridis, Vieil. Gal. t. 148.

Greenish; beneath yellowish; head, bill, and feet black. New Holland.

Africa, and the countries which border on the Eastern Ocean, produce a genus of birds approximating to the Merles, which I shall name

PHILEDON.

Their beak is compressed, slightly arched in its entire length, and sloped at the end. Their nostrils are wide, with a cartilaginous covering, and the tongue terminated by a brush of hairs.

The species for the most part remarkable for some singularity of conformation have been thrown into all kinds of genera by authors.

Some have prominences on the beak, others have fleshy appendages at its base.

Some have portions of skin denuded of feathers on the cheeks.

Even in those which have no part naked, singular arrangements of plumage are at times observed.

The Honey-suckers are confined to the oceanic islands, and Temminck observes the *Diceæ* of Cuvier are not to be separated from them. The notch of the bill is not a very certain character; he also observes, that the *Nectariniæ* are not found in the oceanic islands, and that the *Honey-suckers* are not found in India or Africa.

The genus *Meliphaga* of Lewin is divided into several genera by Vieillot, and placed with the Tenuirostres by all modern authors. The *Philedon* of Desmarest.

Some are peculiar for having a prominence on their bill.

The genus *Tropidorhynchus* of Vigors.

Knob-fronted Honey-eater, Lath. *Merops Corniculatus*, Lath. *Le Corbi Calao*, Vail. O. A. and Ind. t. 24. Lewin. N. H. B.

Brownish-gray; beneath whitish; head, neck, upper part of neck, throat, and narrow collar black, naked; chin, chest, tips of tail white; tail finely brown lined; base of the bill keeled, with a large tubercle. Newfoundland.

Cowled Honey-eater, Lath.? *Merops Monachus*, Lath. Cuv. R. A. t. 4. f. 3. White's Jour. t. at p. 190. Philemon.

Above brownish-gray; nape varied with white; beneath whitish; head black, naked; back of head covered with white feathers; tail-feathers not banded; bill keel subtubercular. Probably young of former.

Others have pendulous peduncles at the base of the bill.

The genus *Creadion* of Vieillot, some of them referred to the genus *Pastor*, by Waggler. The genus *Anthochæra*, Vigors.

Wattle-bee Eater, Lath. *Merops Caruncullatus*, Lath. *Corvus Paradoxus*, Daud. Phillips' Bot. Bay, t. 28. White's Jour.

Back brown-gray, white streaked; head and body beneath whitish-brown streaked; middle of belly yellowish; quills and tail brownish-black, white tipt, side of neck with cylindrical caruncles.

Anthochæra Lewinii, Vigors and Horsf.

Above gray-brown, white streaked; head blackish, white lined, beneath paler; belly yellowish; quill and tail brownish, white tipt; side of neck with short suboval caruncles; length eleven inches; young of former?

Wattled Stare, Lath. 6. *Sturnus Gracula Carunculatus*, Gmel. *Philedon Pharoides*, Desm. Lath. Sys. t. 36.

Referred to *Gracula* by Daud. *Sturnus*, Lath. and Temminck.

Wattled Creeper. *Certhia Caruncullata*, Lath. *Philedon Musicus*, Desm. Vieil. O. Dor. ii. t. 69. ♂. 60. ♀. *Meliphaga*. Temm.

Olive-brown, beneath yellowish-ash; throat fulvous; base of lower jaw with a fleshy-yellow wattle.

This bird is said to sing exceedingly well.

Some have the base of the bill simple, and the skin round the eye more or less naked.

The genus *Philemon*, of Vieil., the *Polochion* of Commerson.

Gracula Calvus, Lin. Pl. Enl. t. 200. O.

Ash ; beneath brown-gray ; head naked ; chest, quills, and tail brown-black.

Placed with *Pastor* by Temminck, and *Æridotheres* by Vieillot.

Merops Molluccensis, Lath. *Philemon Cinereus*, Vieil. *Meliphaga*, Temm.

Gray ; cheeks black ; orbits naked. Mollucca.

Merops Phrygius, Shaw, Zool. viii. t. 20. Lewin. N. H. B. t. 4. *Le Merle Ecaillé*, Vail. O. A. t.

Black ; yellow, varied ; eye-spot and outer tail-feathers yellow. New Holland.

Goruck, Shaw. *Le Goruck*, Vieil. O. Dor. ii. t. 88.

Head, upper and under part of body, and the wing-coverts deep-green, inclining to brown ; most of the feathers edged with white ; space between the bill and the eye, and skin round the eye naked and reddish.

Philemon Chrysopterus, Vieil.

Brown ; wing-spot fulvous ; quill and outer tail-feathers white tipt ; lorum or orbits reddish. Var. ?

Certhia Lunata, Shaw. *Le Fusculbin*, Vieil. O. Dor. t. 61.

Back, wings and tail cinnamon-brown; whole under parts white; upper part of head and back of neck black, marked posteriorly with a white crescent.

Gracula Icterops, Lath, *Philemon*, Vieil.

Black; beneath and wing-band white; orbits yellow-ridged; feet yellow. New Holland.

Philemon Marmoreus, Vieil.

Black, yellow-spotted; orbits naked; beneath gray-white; side tail-feathers yellow-edged. New Holland.

Philemon Viridis, Vieil.

Olive-green, beneath dull-gray; occipital streak white, sides of head bald. New Holland.

The genus *Entomyzon*, Swainson, and part of *Tropidorhynchus*, Vigors and Horsf.

Graculine Honey-eater, Lath. H. *Gracula Cyanotis*, Lath. Sup. *Meliphaga Cyanops*, Lewin, N. H. Birds, t. 4. *Philemon*, Vieil. Ois. Dor. t. 87. *Corvus Graculinus*.

Above olive-green; head and nape black; crop and chest grayish-black; subocular lines from the mouth, the occipital collar, body beneath, and tail-tips white. New Holland.

Merops Cyanops, Lath. *Philemon*, Vieil.

Brown, beneath white; head above and throat black; eye-spot blue; bill black; feet bluish; cheeks naked, Vieil.? feathered, Desm.?

Others have the cheek covered with feathers, and the cheeks, neck, or under the wing is ornamented with long feathers.

Merops Cincinatus, Lath.

Merops Novæ Zealandiæ, Lath. Brown. Illust. t. 9. Shaw. Zool. vii. t. 22. Vail. O. A. t. 92. *Le Cravate Frisée Anthochæræ*, Vigors. *Meliphaga*, Temm. Gal. Ois. t. 183. *Sturnus Crispicollis*, Daud.

Shining black-green; sides of throat and wing-bands white.

Philedon Auriculatus, Desm.

Philemon Erythrotis, Vieil.

Greenish-gray, beneath yellow, varied with ash; crown green-yellow; orbits black; ears yellow. New Holland.

Yellow-tufted Bee-eater, Lath. *Merops Fasciculatus*, Lath. *Merops Niger*, Gml. *Gracula Nobilis*, Merrem Boytr. 1. t. 11. *Gracula Longirostris*, *β*. Gml. *Meliphaga*, Temm.

Shining black; vent and axillary turf-yellow; tail largely cuneate; outer tail-feather white, rest white-tipt. Var. Dixon Voy. t. 19.

The other species of the genus have none of these peculiarities.

Certhia Chrysotis, Lath. *Philemon Chrysotis*, Vieil. *Philedon Xanthotis*, Desm. Vieil. O. Dor. ii. t. 84.

Ash-brown, beneath white; spot behind the ears ovate golden, and another above black. New Holland.

Merops Cucullatus, Lath.
Philemon Cucullatus, Vieil.

Brown-lead colour, beneath white; streaked hood passing between the eyes black; tail rounded. New Holland.

Merops Garrulus, Lath. *Philemon*, Vieil.

Brown, beneath white; vertical band black; spot behind the eyes, and great part of quills yellow; bill and feet yellow. New Holland.

Merops Chrysopterus, Lath.
Philemon Chrysopterus, Vieil.

Brown; wing-spot orange; quill and outer tail-feathers white-tipt; tail wedge-shaped.

Muscicapa Auricomis, Lath.
Swain. Zool. Ill. 1. t. 43. *Philemon*, Vieil.

Olive; crown, body beneath, and eye-spot yellow; eye-streak white. New Holland.

Coracias Sagittata, Lath.
Philemon, Vieil.

Above olive, beneath white-streaked; cheeks ash. New South Wales.

Merops Ornatus, Lath. Suppl.
t. 128. *Philemon*, Vieil.

Blue and green; varied nape; throat and base of quill fulvous; two middle tail-feathers long. New Holland.

Merops Albifrons, Lath. *Philemon Albifrons*, Vieil.

Red, beneath whitish; head above black; forehead

snow-white; quill and tail-feathers spotted. New Holland.

Philemon Nævius, Vieil.

Deep-gray, beneath pale-ash; feathers black-edged; crown and cheeks black; an immature bird? New Holland.

Merops Auritus, Lath.
Philemon, Vieil.

Red; beneath whitish; streak behind the eye, quill, and tail black. New Holland.

Merops Olivaceus, Shaw. Vieil. O. Dor. 1. t. 5. is a *Philemon* of Vieillot, and a *Nectarinia* of Cuvier.

Olive; yellow spot on side of head; beneath olive-yellow; quills and tail brown.

Gracula Plicatus, Lath. *Philemon*, Vieil.

Black; chest-band black; beneath and double wing-band white. New Holland.

Gracula Melanocephalus, Lath.
Philemon, Vieil.

Bluish-gray; beneath white; head black.

Certhia Ignobilis, Sparm. Mus. Carls. t. 56.

Sooty-black, beneath ash, with white elliptical lines; length eight inches.

T. Melanops, Lath.

Ferrugineous; crown and beneath brown.

Certhia Atricapilla, Lath. Pl. col. t. 336. f. 1.

Olive-green; head and cheeks, and spot on sides of

chest black ; occipital band and body beneath white ; five inches. New Holland.

Meliphaga Mystacalis, Temm. Pl. col. t. 336. f. 2.

Deep-gray ; head, nape, and top of the back, streaked, black and white, beneath white ; sides splashed with deep gray ; streak on side of neck black ; six inches. Manilla.

Meliphaga Maculata, Temm. Pl. Col. t. 29. f. 1.

Back and lesser wing-coverts dark-olive ; eye in an ashy spot with a yellow spot behind, and white streak under ; rest of the bird olive waved.

Meliphaga Reticulata, Temm. Pl. col. t. 29. f. 2. *T. Maxillaris,* Lath.

Brown, beneath bluish-white ; crown and maxillary band white ; feet yellow.

T. Leucotis, Lath.

Green, beneath yellow ; crown ash ; throat and chest black ; ear-spot white ; bill and feet black.

T. Lunulatus, Lath.

Brown, beneath white ; both lunulated with black.

T. Melinus, Lath.

And perhaps most of his Polynesian thrushes belong to this genus, a list of which is placed at the end of the thrushes.

Cuvier has placed the *T. Cochin-Chinensis,* the *P. Nigricollis* of Vieil. in this place ; it forms the genus *Chloropsis* of Jardine, with the thrushes.

Temminck places in this genus *Certhia Sanguinea* and

Cardinalis, both of which Cuvier calls *Nectarinia*, and the former Vieillot refers to *Petrodroma*.

The following species referred to by Cuvier have been removed by other authors.

Certhia Novæ Hollandiæ, Lath. Vieil. O. D. t. 57 and 71. *Melitreptus*, Vieil. White's Voy. t. 16. 65.

Black, beneath white streaked; eyebrows and ears white; tail and quill yellow-edged. New Holland.

Certhia Melanops, Lath. *Certhia Mellivora*, Vieil. *Melitreptus*, Vieil.

Brown, beneath white; band across eye, and descending to each side of chest, black. New South Wales.

Merops Spiza, Merrem. *Certhia Spiza*. Lin. Pl. Enl. t. 578.

Green, beneath blue; head and throat black.

C. Cærulescens, Lath. *C. Cærulea*, Cuv. Vieil. O. Dor. t. 3. *Melitreptus*, Vieil.

Brown, beneath flesh-coloured; throat and crop gray-blue; quill and tail blue-black. New Holland.

C. Cuculata, Shaw, Vieil. O. Dor. t. 60. *C. Seniculus*, Shaw, Vieil. t. 50? *Melitreptus*, Vieil.

Head black; throat yellow; back and wing-coverts bluish-ash; quill and tail black. New Holland.

C. Xanthotis, Shaw, Vieil. O. Dor. t. 84.

Gray-brown above; white underneath; yellow spot behind the ear, black speck between it and the eyes;

quills and tail-feathers edged bright yellow; tongue strongly pencilled at the top.

C. Australasiana, Shaw, Vieil. O. D. t. 55.

Above deep brown; beneath white; lower part of abdomen dusky; throat and chest with slight longitudinal streaks; white streak over each eye; tail-feathers edged yellow, white tipped; length six inches.

Tufted-eared Creeper. *C. Auriculata*, Shaw, Vieil. O. Dor. t. 84.

Blackish olive above; throat olive, bright yellow, and gray; top of head greenish yellow; length seven or eight inches.

C. Graculina, Vieil. O. Dor. t. 87.

Rufous brown above; crown of head black; a naked yellow skin from the mouth round the eyes, with a white bar across the top of the head; underneath white; length twelve or thirteen inches.

C. Coccinea, Shaw, *C. Mexicana*, Gml. Vieil. O. D. t. 77, 78. Pl. Enl. 643.

Red, crown paler; throat and crop green; quills bluish tipt.

The Martins, (Gracula, Cuv.),

are another genus bordering on Turdus, inhabiting Africa and the countries adjoining the Indian sea. Their beak is compressed, very little arched, and slightly sloped. Its commissure forms an angle as in the Sturni. The feathers of the head are almost always narrow, and there is a naked space around the eye. They have also the habits of Sturnus, and fly like it in large flocks in pursuit of insects.

The genus *Pastor* of Temminck and the genera *Acridotheres* and *Delophus* of Vieillot. Cuvier has placed some species referred to this genus as a section of *Turdus*. It may be divided into two sections; first, those with thick gracula-like bill: the genus *Acridotheres* of Vieillot.

One of their species (*Paradisæa Tristis,* Gml. *Gracula tristis,* Lath. et Shaw. *Gracula Gryllivora,* Daud.) Enl. 219.

Is become celebrated by the services it has rendered in the Isle of France, by destroying grasshoppers. It is, moreover, omnivorous, nestles in palm-trees, and is easily tamed and trained. It is the size of a thrush, brown, with blackish head, a spot towards the edge of the wing, the abdomen and end of the lateral caudal quills white.

Pastor Fuscus, Tem. MSS.

Back dull sooty brown; belly paler; speculum of wing, white; vent, and tail black, white tipt; head and quills black; bill yellow. India.

Pastor Temporalis, Tem. Catal.

Cheeks naked, red; head and streak over ears pure white; collar black; another near the back white; scapulars and wings black brown; chest and belly white; wing-coverts white edged; tail-ends white; length eleven inches. Bengal.

Pastor Corythaix, Waggler.

Crest erect compressed shining black; a square on each side the eye and another outer; the jaws white: quills reddish brown; occipital crown feathers truncated; size of *Parad. Tristis.* Java.

Gracula Calva, Gml. Pl. Enl. t. 200.

has been referred to this genus; but Cuvier places it as a Meliphaga.

Corvus Crinitus, Daudin, *Le Sicrin,* Levaill. O. A. t. 82., a *Pastor* of Waggler, and a *Pyrrhocorax* of Vieillot and Cuvier.

Pagodo Thrush. T. Pagodarum, Gml. *T. Melanocephalus,* Wahl. Men. Copenh. iii. t. 8. Vail. O. A. t. 95. f. 1. Tem. *T. Malabaricus,* Gml.

Crested gray head; body beneath, quills, and tail black; belly white streaked; vent white.

T. Gingianus, Lath. Vail. O. A. t. 95. f. 2. *Gracula Grisea,* Daud.

Orbital spot naked, behind acute; above iron-gray; crown and cheeks black; beneath reddish; quills purplish-black; primaries white based; four wing-coverts on each side reddish tipt; length six inches and three quarters. Coromandel and South Africa.

Coracisa Docilis, Gml. Reis. t. 42.

Bill slightly inclined, yellow; claws rose coloured; orbital spot naked; whitish head, and upper part of neck white; belly, vent, and quills black; primaries white based; tail black, white tipt; allied to former. South Asia.

Gracula Melanopterus, Daud. *Pastor Candidus,* Tem. MSS. *P. Tricolor,* Hors.

Shining white; spurious wing quills, and tail metallic

black; tail white tipt; bill and feet yellow; length eight inches and a quarter. Java.

Upupa Capensis, Gml. Pl. Enl. t. 697. (badly coloured). *Upupa Madagascariensis*, Shaw. Vieil. O. Dor. t. 3. Vail. Prom. t. 18. *Coracina Cristata*, Vieil.

Crest erect compressed; head, neck, and beneath white; tibia, back, rump, wings, and tail pale fuscous, powdery; nape grayish; length ten inches. Madagascar (not Africa.)

Gracula Cristatella, Lath. *Pastor Griseus*, Hors. not Waggler, Edw. t. 19. Pl. Enl. t. 507.

Crested black; base and tip of primaries white; bill yellow.

Oriolus Sinensis, Gml. Pl. Enl. t. 617. *Pastor Turdiformis*, Waggler.

Wing-coverts, rump, top of side tail-feathers, and beneath white; head, neck, chest and back ashy; quills, and middle, and base of outer tail feathers greenish black. China.

Pastor Jalla, Hors. Zool. Java.

Forehead, crown, napes, sides, and front of neck black; ears, lore streaks beneath, rump, and oblique band on scapulars white; back, wings, and tail brownish black; orbits yellow; length nine inches. Java.

Sturnus Contra, Gml. ♀ *Sturnus Capensis*, Lath. Pl. Enl. t. 280.

Head and neck violet-black; occular spot large, and in an occipital band, beneath white; back, wings and tail blackish brown; outer tail-feathers white edged and larger wing-coverts white tipt.

Pastor Ruficollis, Waggler's Syst.

Head, and back of neck, and beneath white; side of neck ferrugineous; back dull violet; humerus, secondaries, and tail dull grassy-green; spot and band on wing white beneath; feet and bill black.

Sturnus Dauricus, Pallas, Art. Stockh. 1778. t. 7. f. 1. *Gracula Sturnina*, Gml.; *female*. *T. Leucocephalus*. Gml. *Sturnus Cericeus*, Gml. Brown Illust. t. 21.; *young*. *T. Dominicanus*, Pl. Enl. t. 627. f. 2.

Violet-black above; beneath ashy-white; head and neck bluish gray; crown with a violet-black streak; bill and feet blackish.

Turdus Sinensis, Gml. is probably of this division.

Sturnus Zeylanicus, Gml. *T. Ochrocephalus*, Gml. Brown Ill. t. 22.

Crown and cheeks yellow; body beneath ash; quills and tail dull green; chin-streak white. Ceylon.

The genus *Delophus* of Vieillot. Wattled.

Cockscomb Stare, Lath. 7. *Gracula Caruncula*, Gml. *Sturnus Gallinaceus*, Lath. *Gracula Larvata*, Shaw. *Le Porte Lambeaux*, Vail. O. A. t. 93, 94. Naturfosch ii. t. 21.

Ashy; orbits naked, a double wattle, and an erect, bifid, membranaceous crest; when young, wattle smaller; length six inches. Cape of Good Hope.

The Lyres, (Mænura, Sh.)

which their size has occasioned some to refer to the Gallinacea, belong evidently to the passerine order from their feet, with separated toes (except the first articulation of the external and middle); from their beak, triangular at base, elongated, a little compressed and sloped; toward the point the membranous nostrils are large and partly covered with feathers as in the jays. They are distinguished by the large tail, in the male, very remarkable for the three kinds of feathers which compose it; viz. the twelve ordinary ones, very long, with fine and very separated barbs; two more, in the middle, furnished on one side only with serrated barbs, and two external ones curved like an S, or like the arms of a lyre, the internal barbs of which, large and serrated, are like a broad ribbon, and the external, very short, grow broad only towards the end. The female has but twelve quills of the usual structure.

This singular species (*Mænura*, Sh. Vieillot, Ois. du Par. Pl. xiv, xv.) inhabits the pebbly districts of New Holland; its size is something less than that of a pheasant.

The *Parkinson* of Shaw's Leverian Museum. The *Parkinsonius Mirabilis*, Bechstein, Trans. Lath. Syn. *Megapodius Mænura*, Waggler. Waggler has placed this bird as a section of the genus Megapodius; and Temminck refers one genus to the Passerine, and the other to the Gallinaceous order.

The Manakins, (Pipra, Lin.)

are a small genus of America, with compressed beak, more high than broad sloped, large nasal fosses, and short tail. They are, in some respects allied to the ant-eaters, if their feet were not short, and if they were not otherwise distinguished from all other dentirostres, by having their two external toes united at nearly half their length. In other respects, the short beak and general proportions have a long time caused them to be considered like our titmice. We should put at their head, and in a separate group,

The Rock Manakins, (Rupicola,)

which are large, and bear on their heads a double vertical crest of feathers arranged like a fan. The adult males of the two known species are of the finest orange, and the young of an obscure brown. These birds live on fruits, scratch the earth like hens, and make their nests with dry wood in the deep caverns of the rocks. The female lays two eggs.

They are confined to South America.

Rock Manakin, or *Hoopoe*, Hen. *Pipra Rupicola*, Lin. Pl. Enl. t. 39. f. 47.

Crest erect, purple edged; body saffron; red wing-coverts truncated; length eleven inches and a half. Surinam.

Peruvian Manakin. Pipra Peruviana, Pl. Enl. t. 745.

Saffron-red; larger wing-coverts ash; quills and tail black. Peru.

Others are found in the Indian Islands; they form the genus *Calyptomena* of Raffles.

Calyptomena Viridis, Hors. Pl. Col. t. 216.

Beautiful green. Java.

The True Manakins, (Pipra. Cuv.)

are small, and all remarkable for lively colours. They inhabit, in small flocks, the humid forests.

Divided into the *Pipra* of Vieillot, which are confined to South America, which have the bill short, rather broad at the base, and compressed at the end, the third and fourth quill the longest.

Of these some have the tail even.

Pipra Pareola, Lath. Pl. Enl. t. 687. f. 2. 303. f. 2. *P. Superba*, Pallas, Spic. 1. t. 3. f. 1.

Black; crown erectile, yellow-red; back pale-blue; primaries brownish. Brazil.

Female, olive-green; beneath glaucous. Brazil.

Pipra Erythrocephala, Lin. Pl. Enl. t. 34. f. 1. *P. Aurocapilla*, Licht.

Black; crown and thighs fulvous. Brazil.

Differs from *Manacus Rubrocapilla* in the weaker bill and short tail.

Pipra Aureola, Lin. Pl. Enl. t. 34. f. 3. and t. 302. f. 2.

Black; head and chest scarlet; quill with a white spot; face fulvous; belly reddish; *female* olive.

Pipra Serena, Lin. Pl. Enl. t. 324. f. 2.

Black; forehead white; rump blue; belly fulvous. Brazil.

Pipra Gutturalis, Lin. Pl. Enl. t. 324. f. 1.

Black; throat white.

Pipra Leucocapilla, Lin. Pl. Enl. t. 34. f. 2.

Black; crown white. Brazil.

P. Manacus, Lin. Pl. Enl. t. 302. f. 1. 303. f. 1.

Black; beneath white; aurical and wing spot white.

P. Strigilata, Pr. Max. Pl. Col. t. 54. f. 1, 2.

Above green-waved; quills ash, beneath buff, with deeper waves; male with a crimson crest.

Pipra Erythrocephala, β Lin. *Manacus Rubrocapilla*, Briss. *P. Erythrocephala*, Licht. Pl. Col. t. 54. f. 3.

Black; crown and thighs red; *female*, olive.

Pipra Pileata, Natt. Pl. Col. t. 172. f. 1.

Back dark red; under parts yellow; cap black; wing-coverts green; tail yellow and black. Brazil.

Pipra Chloris, Natt. Pl. Col. t. 217. f. 2.

Back and head olive; beneath yellow; wings black and olive.

Pipra Galeata, Licht.

Black; frontal crest erect; crown, nape, and middle of

back scarlet; feathers beneath yellowish; *female,* olive; wings and tail brownish; frontal feathers erect; length six inches and a quarter; tarsi nine lines. S. Paulo.

Pipra Galeata, Spix. Braz. t. 7. f. 2.

Fine black, horn-like, occipital turfs; throat, cheeks, and thighs scarlet; beak strong; length three inches and a half. Brazil.

Pipra Nigra, Vieil.

Black head; crested red; bill and feet black. Peru.

Pardalotus Cristatus, Vieil.

Occipital crest red; body beneath yellow; above olive-green; bill, base and tip black, middle horny; feet black; length three inches. South America.

Pipra Gutturosa, Desm. Tann. t. 10.

Above black; beneath white; bill black; feet yellow; feathers of the crop long, slender; *female,* reddish; beneath paler. Guiana.

Pipra? Plumbea, Vieil. *Pico de punzo obscura aplomado.* D'Azara. n. 111.

Lead coloured; quills and tail black, bluish edged; bill black; feet brown.

Pipra Pectoralis, Lath.

Blue-black; belly ferrugineous; pectoral lunule golden, bill and feet pale.

Pipra Cyanocephala, Vieil.

Olive-green; beneath yellow; crown blue; quill and tail black, green edged; bill and feet black. Trinity Island.

Pipra Coronata, Spix. Braz. ii. t. 7. f. 1.

Small; coal-black; crown and nape blue; bill short, pressed, slender; body three inches and a half; tail one inch and one third long. Brazil.

Pipra Filicauda, Spix. Braz. ii. t. 8. f. 1. ♂ f. 2. ♀.

Moderate; beneath yellow; above black; head and nape purple; *female*, green; tail-feathers lengthened filiform; body five, tail four inches and a half long. Brazil.

Pipra Herbacea, Spix. Braz. ii. t. 8. f. 1.

Moderate; bill very slender; shining herbaceous green; belly yellowish white; body three, tail one inch long. Brazil.

Pipra Elata, Spix. Braz. ii. t. 80. f. 2.

Small; olive-green beneath; pale yellowish green on the sides; orange in the middle; wing-coverts yellow tipt; body three inches and a half, tail an inch and a quarter long. Brazil.

Is it *Sylvia Elata* or *Cristata*, Lath.? and the *Pardanalotus* of Vieil. with a very short, strong bill, dilated on the sides, and rather blunt, and the first and second quill the longest? They are confined to the Indian and Oceanic islands of the Old World.

Striped-headed Manakin, Lath. *Pipra Striata*, Lath.

Back, grayish-brown; rump fulvous; head black, white-streaked; wings and tail black, white-streaked; eye-streak yellow-white; throat yellow; chest and belly white, varied with yellow. New Holland.

P. Punctata, Lath. Pl. Col. t. 78. Gal. Ois. t. 73. Nat. Misc. t. 111.

Olive-gray varied with fuscous; head and wings black, white-spotted; eye-streak white; rump scarlet, beneath white; throat yellow; *female*, head yellow-spotted; New Holland.

Sylvia Hirundinacea, Lath. *P. Gularis*, Lath. n. 5. Lewin's Birds, N. H. t. 7. Shaw, Nat. Misc. t. 114.

Black-blue, beneath scarlet; belly white; bill pale. Pacific Ocean.

S. Superciliosa

Body above chestnut, beneath yellowish-white; spot above eye white; quill brown; tail black, side one white-tipt; bill and feet brown.

P. Desmarestii, Leach's Zool. i. t. 94.

Above shining-black blue; throat and chest red; belly white. New Holland.

Pardolotus Cristatus of Vieillot, appears to be a *Pipra*. Tail, two middle feathers longest.

P. Caudata, Lath. Spix's Braz. t. 6. f. 1. ♂. 2. ♀.

Blue; head, wings, and tail black; crown scarlet; long tail; feathers pointed; *female*, dull-blue; crown red; body five inches and a half, tail two inches and a half long. Brazil.

P. Longicauda, Vieil. *Pico de pungo cola de pala*, Azara, n.

Throat, wings, and tail black; crown red; two middle

tail-feathers long, blue ; feet reddish ; length six inches and a half.

P. Militaris, Shaw's Misc. t. 849.
P. Rubrifrons, Vieil.

Black, beneath white ; forehead and rump red ; tail, two middle feathers longest, pointed ; bill blackish ; feet yellow.

P. Melanocephala, Vieil.

Head, primary quills, and tail black ; rump, wing-coverts, and tail above red ; cheeks and throat ashy ; body beneath white ; two middle tail-feathers longer, acute ; bill brown ; feet gray. South America.

And the genus Phibalura of Vieillot, which has the bill very short, strong, and conical, convex, the tail slender, very long, and forked ; only one species is known, found in South America, appearing to unite the *Pipra* with the *Tanagers*.

Phibalura Flavirostris, Vieil. An. et Gal. Ois. t. 74. *P. Crisopogon*, Ill. MSS. Pl. Col. t. 118.

Varied, black and reddish above ; crown, quills and tail black ; occiput and throat red-brown ; back of neck, chest black and white ; belly spotted black and white. Brazil.

P. Cristatus, Swain. Zool. Ill. i. t. 31.

The genus *Pachycephala* of Swainson is peculiar for the puffed-out feathers of the head; the bill is broad based, and with a few weak bristles at the base ; the wings are rounded, and the tail moderate and nearly equal.

Black-crowned Thrush, Lewin's Birds, N. H. t. 10.
T. Gutturalis, Lath. Suppl.

Olive-yellow; head and pectoral spot black; crop white; nuchal collar, chest, belly, and vent yellow; called Thunder Bird. New Holland.

Orange-breasted Thrush, Lewin's Birds, N. H. t. 6.
Musc. Pectoralis, Lath.

Gray; throat, eye-streak, and pectoral spot black; crop white; belly ferrugineous; wings and tail blackish-brown; externally gray-edged. New Holland.

Pachycephala Striata, Vig. and Horsf.

Above olive-gray, slightly brown-streaked, beneath whitish, with broader brown-streaks; wings and tail brown; *female,* above gray, beneath yellowish-white, brown-streaked; throat whitish; length six inches and a quarter. New Holland.

The WARBLERS, (MOTACILLA, Lin.)

form a family exceedingly numerous, remarkable by the straight slender beak, like an awl. When it is a little depressed at the base, it approaches to that of the Flycatchers. When compressed, and with a point slightly curved, it approximates to the Straight-beaked Shrikes.

Naturalists have attempted to divide them as follows:

The *Stonechats,* (*Saxicola,* Bechst.)

have the beak a little depressed, and a little broad at

the base, which approximates them to the last small tribe of Flycatchers. They are lively birds, tolerably high on the legs. The species of this country nestle on the ground, or under, and eat nothing but insects.

We possess three:

The *Stonechat*, (*Motacilla Rubicola*, Lin.) Enl. t. 678, 1.

A small brown bird, with red breast, black throat, with white on the side of the neck, wing, and croup. It flies continually over the bushes and briars, with a small cry like the clack of a mill, whence its name.

Also *Motacilla Ischecantschia*, Gmel., and perhaps Vail. O. A. t. 180. f. 1. 2.

Whenchat, Lath. The *Tarier*, (*Mot. Rubetra*,) Enl. ib. 2.

resembles much the Stonechat, but its black, instead of being under the throat, is on the cheek. It is a little larger, and more attached to the ground.

The *Wheat-ear*, (*Mot Œnanthe*,) Enl. 554.

The croup, and half the lateral plumes of the tail, white. In the male, the upper part is ash-colour, the under reddish-white, the wing and band over the eye black. In the female, all the upper part is brownish, and the under reddish. This bird remains in the fields when ploughing, to take the worms which the share exposes.

Also called *Fallaw Sniech*, and *White-tail*.

The *Rousset Wheat-ear*, Lath. *Mot. Stapazina*, Gml. from Edw. t. 11.

Ferrugineous; orbits, wings, and tail brown; tail outermost white-sided. Southern Europe.

Stapazina, Ray. *Vitiflora Rufescens*, Briss. t. 25. f. 4. Ed. t. 31. (hinder fig.) *Saxicola Aurita*, Temm. *Sylvia Stapazina, β*, Lath. *Sylvia Albicollis*, Vieil.

Reddish, beneath whitish; eye-streak black; tail-feathers, two middle black, outer white, black fringed at tip. South of Europe.

Leucomela, and *Black and White Warbler*, *Mot. Leucomela*, Pallas. N. C. Petrop, xiv. t. 22. f. 3. Falk. Voy. t. iii. t. 30. *Mus. Melanoleuca*, Pallas. l. c. xiv. t. 15.

Black; crown, nape, rump, belly, and tail greater part white. South Russia.

Temminck also places in this genus, observing that it has all the habits of the Stonechats, *T. Leucurus*, which Cuvier calls a blackbird. See Savig. *Descrip. d'Egypte*, t. 5. f. 1.

Allied to the Stonechat there are, the *Luzonia Warbler*. *Mot. Caprata*, Lin. Pl. Enl. t. 235. *Saxicola Fruticola*, Horsfield.

Black; rump, vent, and wing-cover spot white; length four inches and a half. Java.

Sooty Warbler, *Mot. Fulicata*, Lath. Pl. Enl. t. 185. f. 1.

Violet-black; vent chestnut; wing-cover spot white; length six inches. Phillippine Islands.

Phillippine Warbler, *Mot. Phillippensis*, Pl. Enl. t. 185. f. 2. *Le Patre*, Vail. O. A. t. 180.

Violet-black, beneath and head red-white; chest black; outer tail-feathers reddish, white-edged; length six inches and a half. Phillippine Islands.

Temminck observes that Vaillant's bird does not differ from the European Stonechat.

Sibyl Warbler, *Sylvia Sperata*, Lath.
Traquet Familier, Vail. O. A. t. 183.

Brownish-green; beneath and rump red-gray; two middle tail-feathers blackish; side-feathers obliquely halved, fuscous yellow. Cape of Good Hope.

Black-hooded Wheat-ear. *Sylvia Pileata*. *Traquet Imitateur*, Vail. O. A. t. 181. *S. Imitatrix*, Vieil.

Red-brown; head and chest black; forehead, throat, eyebrows, rump, and side tail-feather, from the middle to the lower, white. Cape of Good Hope.

Mot. Leucorhoa, Pl. Enl. t. 583. f. 2.

Red-brown, beneath yellowish-white; chest reddish; rump and base of tail white. Senegal.

Le Traquet Montagnard, Vail. O. A. t. 184.

Adult entirely black, except belly, shoulders, and the edges of the tail quill feathers, which are white. When young, nearly all the feathers, which when adult are black, are blue.

Sylvia Nigra, Vieil. Vail. O. A. t. 189.

Crown white; body, bill, and feet black.

Sylvia Formicivora, Vieil. *Le Fourmillier*, Vail. O. A. t. 186.

Brown; throat, crop, and chest reddish; small wing-coverts white-spotted. Cape of Good Hope.

Sax. Superciliaris, Licht *Jan. Fredric,* Vail. O. A. t. 111.

Olive-gray brown above; throat yellow; crest mottled; belly and vent white; a white patch over the eye.

Sax. Thoracica, Licht.

Crown slate-colour; back olive; oblique band on side of head ending in a broad pectoral band, black; throat and middle of belly white; hypochondria ferrugineous; quills black; secondaries and wing-coverts ferrugineous-edged; tail black; length five inches and a half. Cape of Good Hope.

Sax. Moesta, Licht.

Throat, neck, and middle of back black; forehead, eye-brows, chest, belly, vent, and rump white; nape and coverts, ashy; quills brown, white-edged; tail, base under the coverts red, rest black; length six inches and a half, tarsi one inch. Egypt.

Sax. Lugens, Licht.

Throat, neck, middle of back, and wing-coverts black; crown, nape, chest, belly, and rump white; vent isabella; quills black; base of inner web white; secondaries white-tipt; tail-feathers white, with a black subapical band; two middle feathers, from the middle to end, black; length six inches and a half, tarsi ten lines; young like the Wheat-ear, but throat blackish, and tarsi shorter.

Mot. Solitaria, Lewin's Birds, N. H.

Above fuscous-brown; forehead, chest, and belly ferrugineous-red; throat whitish; length five inches. New Holland.

Sax. Jardinii, Vigors and Horsf.

Blackish-gray; belly white; wings and tail black; wings white-banded; tail-feathers, middle excepted, white-banded; tips slender, white-tipt; length six inches and three quarters. New Holland.

The *Mot. Cyanea* of Gmel. Lath. Syn. ii. t. 53. has the bill of the Stonechats, but differs in its long legs.

Sylvia Saxicola Obscura, King, Zool. Jour.

Black-brown; wings short and rounded; tail short, feet long, strong, and pale in colour.

The Rubiettes (Sylvia, Wolf et Meyer. Ficedula, Bechst.)

have the beak only a little more narrow at the base than the preceding. They are solitary birds, which nestle generally in holes, and live on insects, worms, and berries.

We have here four species:

The *Redbreast*, (*Mot. Rubecula*, Lin.) Enl. 361. 1.

Gray-brown above; throat and chest red; belly white; nestles near the ground in woods; is inquisitive and familiar; some remain in winter, take refuge in habitations, and are easily tamed.

The *Blue-throated Warbler*, (*Mot. Suecica*, Lin.) Enl. 361, 2. and 610. f. 1. 2. 3.

Brown above; throat blue; chest red; belly white; more rare than the preceding; nestles on the borders of woods and marshes.

This species is named *Sylvia Cyanecula* by Meyer.

The *Redstart,* (*Mot. Phœnicurus,* Lin.) Enl. 351. 1. 2.

Brown above; throat black; chest, croup, and lateral quills of the tail red; nestles in old walls, and has a sweet song, which has some of the modulations of the Nightingale.

The *Red-tail Warbler,* (*Mot. Erithacus,* Lin. *M. Titys,* Retz. *M. Gibraltariensis, Atrata,* Gm.) Edw. 29.

differs from the preceding in having the chest black, as well as the throat: it is much more rare.

M. Atrata, and *Gibraltariensis,* are the old male, *M. Tithys* is the female.

Blue Warbler, Mot. Scialis, Lin., Edw. t. 24. Catesby, t. 47.

Blue, beneath reddish; belly white; primaries black-tipt; length six inches. Carolina.

This species is the type of the genus *Scialis,* Swainson, of *Œnanthe* of Vieil., and *Saxicola* of Prince Musignano.

Ruby-throat Thrush. T. Calliope, Lath. Suppl. t. front. *T. Camtschatkensis,* Gml. *Mot. Calliope,* Pallas.

Ferrugineous, beneath yellowish-white; throat crimson-white and white-edged; lores black; eyebrows white. An Accentor, according to Temminck.

The Warblers, (Curruca, Bechst.)

have the beak straight, slender throughout, a little compressed in front. The upper crest curved a little towards the point.

The most celebrated bird of this sub-genus is,

The *Nightingale*, (*Mot. Luscinia*, Lin.) Enl. 615. 2.

Reddish-brown above; whitish-gray underneath, the tail a little more red. Every body knows this songster of the night, and the melodious and varied sounds with which it charms the forests. It nestles in trees, and only sings until the young are hatched. The care of their subsistence then occupies the male as well as the female.

The eastern part of Europe produces a race a little larger, the breast slightly varied with grayish tints. (*Mot. Philomela,* Bechst.)

The *Silky Warbler*, *Sylvia Sericea*, of Natterer and Temminck; it is rather smaller than the Nightingale, more silky, and the tail is slightly rounded. Spain and Gibraltar.

These three species, or races, differ in the comparative length of the primary quills.

The other species bear, in common, the name of Warblers. They almost all have an agreeable song, gaiety of habits, flit continually in pursuit of insects, nestle in the bushes, and, for the most part, near the edge of waters, in reeds, &c.

I place at their head a species almost large enough to have been still put in the genus of the Thrush.

Reed Thrush, (*River Nightingale*, &c. *T. Arundinaceus*, Lin.) Enl. 513.

Reddish-brown above, yellow under; throat white; a pale mark over the eye; somewhat smaller than the

Mavis; beak almost as much arched; nestles in reeds, and eats little but aquatic insects.

Reed Wren, (*Mot. Arundinacea*, Lin.)

Like the preceding in habits and colours, but one-third smaller.

To these may be added,

S. Galactotes. Temm. Pl. Col. t. 251. f. 1.

Bright-red above; outer tail-feather black; spot, &c., white-tipt; eyebrows white; beneath yellowish-white; length six inches and a half. South of Spain.

S. Fluviatilis, Meyer.

Above olive-brown, spotless; beneath white; olive-streaked; belly white; lower tail-coverts white-tipt; hind claw long, arched; length five inches. Austria.

Perhaps *S. Luscinoides*, Savi Bul. Sec. viii. 105.

S. Certhiola, Temm. *T. Certhiola*, Pallas.

Olive-brown, brown-spotted; throat white, brown-spotted; belly reddish; tail long; hind claw very long, arched; bill strong; length five inches. Russia.

These last species are allied to the *Anthi*, by their long hind claws, and strong bills.

Sedge Warbler. *S. Phragmitis*, Bechst. Naum. Voy. f. 107. Sepp. Voy. t. 53. Jun.

Gray-olive, brown-streaked; cheek with a black and white band; beneath reddish-white; length four inches and a half.

Bog Warbler. *S. Palustris*, Bechst. t. 26. Naum. Voy. f. 105.

Olive-brown; wings ash-edged; cheek with a yellow

streak; bill base broader than high; length five inches. Germany.

The *Warbler of the Reeds*, (*Mot. Salicaria*, Gml. Enl. 581. 2.) Still smaller than the River Nightingale, with shorter beak in proportion; olive-gray above, very pale-yellow under; a yellowish cast between the eye and beak.

The *Spotted Warbler. M. Nævia*, Albin iii. 266. No. ii. Pl. 53.

Inhabits also reeds; is the smallest of the aquatic kind; fawn-colour, spotted with blackish above; whitish, spotted with fawn underneath; spotted with gray on the chest.

The *Grasshopper Warbler. S. Locustella*, Lath. Pl. Enl. t. 581. f. 3. *S. Albini*, Albin, t. 266. Penn. Brit. Zool. t. 9. f. 5.,

is a foreign and distinct species, with red bill and feet. Noseman's figure is a young Sedge-warbler.

A variety, not spotted on the breast, has been named *Mot. Schœnobanus*.

The *Aquatic Warbler* of Lath. *Mot. Aquatica*, Gml. *S. Schœnobanus*, Scopol. *S. Salicaria* of Bechstein, Nauman Voy. t. 106.

A very distinct species, common in Germany.

Mot. Schœnobanus, Lin., is a variety of the *Hedge Sparrow*.

Cetti's Warbler. S. Cetti, Marmora.

Deep brown; wings and tail blackish; beneatn white;

sides reddish; tail very broad, rounded; length five inches. Sardinia. A Malurus?

S. Ruppeli, Tem. Pl. Col. 245. f. 1.

Slate-coloured; crown and throat black; streak under the eye and beneath white; wings and tail blackish-brown; outer tail-feathers white, with a black spot. Candia.

S. Melanopogon, Temm. Pl. Col. 245. f. 2.

Bill very slender, edges inflexed, deep-brown; beneath paler, crown and streaks on back black; streak over eye and throat white; tail much graduated, blackish. Rome.

Among the species most attached to dry soils are first distinguished,

The *Black Cap*, (*Mot. Atricapilla*, Lin.) Enl. 580. 1. 2.

Brown above; whitish underneath; a black hood in the male, red in the female, when it is *M. Mosquita*, Gml.

Orpheus Warbler. (*Sylvia Orphea*, Tem.) Enl. 579. f. 1.

One of the largest; ashy-brown above; whitish under; white at the edge of the wing; the external quill of the tail two-thirds white; the remainder marked with a spot at the end; the others with an edging.

The *Gray Warbler*. (*Mot. Silvia*, Lin.) White-Throat of the English Brit. Zool. Pl. 5. f. 4.

Smaller and more gray than the foregoing; the beak more slender, but the white spots similarly disposed.

This is the *Silvia Cinerea* of Latham, Pl. Enl. t. 579. f. 3. and t. 581. f. 1.

The *Babbling Warbler.* (*Mot. Curruca,* L.) Enl. 380. f. 3. Noseman II. Pl. 97.

Above reddish-gray-brown; white under; the white of the tail like the two preceding; the quills and wing-coverts edged with red.

This species has been described under the names of *Curruca Garrula,* by Brisson; *Motacilla Dumetorum,* by Gmel.; *Mot. Garrula,* by Retz; and White-breasted Warbler, by Latham. Trisch. Voy. t. 2. f. A. Naum. t. 34. f. 70. and also the *Lesser White Throat, Sylvia Sylvicella* of Latham.

The *Passerine Warbler,* (*Mot. Passerina,* Gmel.) Lath. Syn. Sup. Pl. cxiii. Noseman II. p. 72.

Uniformly ashy-gray-brown; white under.

The *Hawk-like Warbler,* (*Mot. Nisoria,* Bechst.)

A little larger than the Passerine, of the same colour, only some grayish waves on the sides, and some spots under the base of the tail *.

There are also found in Europe,

Blackhead Warbler, Silvia Melunocephala, Lath. Pl. Col. t. 245. f. 3.

Greenish-ash; beneath, gray; orbits naked; crown

* N. B. The descriptions of the Warblers are so vague, and the figures so bad, that it is almost impossible to determine the species. Each author arranges them differently. The reader may, therefore, depend on our descriptions, but not absolutely on our synonymy.

It is, perhaps, needless to remind the reader that the above note of the Baron applies only to the species he has mentioned, and not to those we have ventured to insert in an inner margin. These, though it is hoped they are correctly quoted in general, must be necessarily subject to no small degree of uncertainty.—Ed.

black; *female*, crown blackish-ash; bill strong; length five inches. Sardinia.

Sarda Warbler, Sylvia Sarda, Temm. Pl. Col. t. 24. f. 2.

Blackish-ash; orbits naked; crown and throat blackish-ash; in *female*, pale-ash; bill short, feeble; length five inches. Naples.

Pettichaps, Motacilla Hortensis, Gml.? *Sylvia Hortensis*, Bechst. Pl. Enl. t. 579. f. 2. Naum. Vogl. t. 33. f. 68.

Gray-brown; orbits white; throat whitish; chest and sides reddish; belly white; length five inches and a half. South Europe.

Spectacled Warbler, Sylvia Conspicillata, Marmora. Pl. Col. t. 6. f. 1.

Vinous red; head ash; orbits white, black-edged; wings blackish; throat white, beneath reddish; tail white tipt; length four inches and a half. Sardinia.

Dartford Warbler, Sylvia Dartfordiensis, Lath. *Motac. Provincialis*, Gml. Pl. Enl. 655. f. 1.

Dusky-brown; cheeks ash; throat, neck, and breast ferrugineous. South of England and France.

Subalpine Warbler, Sylvia Subalpina, Temm. Pl. Col. t. 251. f. 1. 2 *Sylv. Leucopogon*, Meyer.

Ash; sides of neck and chest vinous; belly white; wings black-ash; outer tail-feathers white-tipt; length four inches and a half. Turin.

Sylvia Cisticola is probably a *Malurus*.

In the Old World,

Chestnut-bellied Warbler, Motacilla Erythrogastra, Pallas, Nov. Com. xix. t. 16. 17.

Black; beneath chestnut; crown ash; wing-spot white; thighs black; length seven inches. Caucasus.

Motacilla Caffra, Lin.

Olive; throat and tail ferrugineous; eyebrows white. Cape of Good Hope.

Black-jawed Warbler, Motacilla Nigrirostris, Gml.

Olive-brown; beneath white; chest red, black streaked; lore and throat red-yellow; maxillary streak blackish; length seven inches.

Buff-faced Warbler, Motacilla Lutescens, Gml.

Ferrugineous-brown; beneath reddish-white; forehead and throat yellowish; ears dull-red; length six inches.

Blue-tailed Warbler, Motacilla Cyanura, Pallas, Iter.

Yellowish-ash; beneath, and eye-brows yellow-white; wing and tail brown; rump and edge of tail-feathers blue. Siberia.

Daurian Warbler, Mot. Arcola, Pallas.

Black; crown ash; forehead and wing-spot white; beneath and side tail-feather foxy; two middle ones black. Siberia.

Murine Warbler, Mot. Murina, Gml.

Mouse-colour; beneath and eye-streak white; head, neck, and centre of belly black.

White-crowned Warbler, Mot. Albicapilla, Gml.

Green; beneath whitish; throat vertical and subocular spot white; length seven inches. China.

Sylvia Flaviventris, Burchel.

Pale mouse-colour; throat and breast whitish; belly yellow; quills and tail brown, white edged. Africa.

Pink Warbler, Brown, Illust. t. 33. *Mot. Caryophyllacea,* Gml.

Pale pink; wing and tail dull; bill and feet red. Ceylon.

Olive Warbler, Brown, Illust. t. 14. *Mot. Olivacea,* Gml.

Olive; beneath white; face yellowish. Ceylon.

Green Indian Warbler, Lath. Edw. t. 15. and 79. Brown, Illust. t. 36. *Mot. Typhia,* Lin. *Mot. Zeylonica,* Gml.

Green; beneath yellowish; crown, and nape, and wings black; wing band, two cut, white.

Referred to the genus *Elgithina,* by Vieillot.

Scapular Wagtail, Lath. *Jöra Scapularis,* Hors. Zool. Java.

Greenish-yellow; quills blackish, externally yellow; internally white edged; belly and chest yellow. Java.

Perhaps the same as former.

Cingalese Warbler, Mot. Cingalensis, Gml. Brown, Illust. t. 32. *Syl. Cingalensis,* Lath.

Green, variegated; beneath yellow; neck fulvous; length four inches and a half. Ceylon.

Ærithina Atricapilla, Vieil. Dict. Vail. O. A. t. 140. f. 1. 2.

Head black; upper parts olivaceous; throat, breast, belly, and vent yellow; tail tipped with white. Ceylon.

China Warbler, Mot. Sinensis, Gml.

Green; beneath flesh coloured; ears pale; tail feathers mucronate; length six inches. China.

Bourbon Warbler, Mot. Mauritiana, Gml. *Sylvia Borbonica*, Lath. Pl. Enl. t. 705. f. 1. 2.

Gray-brown; beneath, yellowish gray; quills and tail-feathers gray edged. Madagascar.

Madagascar Warbler, Mot. Livida, Gml. Pl. Enl. t. 705. f. 3.

Bluish-gray; vent whitish; quill and tail black; two outermost tail-feathers white. Madagascar.

Citron-bellied Warbler, Mot. Flavescens, Gml. Pl. Enl. t. 582. f. 3.

Brown, beneath yellow; cheek whitish; quills and tail brown; length four inches and a half. Senegal.

S. Rufigastra, Lath. Pl. Enl. t. 582. f. 1.

Olive-brown, beneath yellowish-red; quill and tail brown; length three inches and three-quarters. Senegal.

Undated Warbler, Mot. Undata, Gml. Pl. Enl. t. 582. f. 2.

Black; edge of feathers and rump red, beneath white; quills cuneate; tail brown; length four inches. Senegal.

Dusky Warbler, Mot. Fuscata, Gml. Pl. Enl. t. 584. f. 2.

Brown, beneath gray; sides reddish; quills and tail darker; tail elongated. Senegal. These are perhaps *Maluri.*

Mot. Cyane, Pallas, Risc. iii.

Deep-blue, beneath snow-white; eyebrows black; side tail-feathers white. Russia.

White-chinned Warbler, Mot. Bonariensis, Gmelin.

Black, beneath furrugineous; throat, lores, middle of belly, and tips of tail white; length five inches and a half. Borneo.

Taylor Warbler, Mot. Sutoria, Gml. Ind. Zool. t. 8.

Entire yellow; length three inches. India.

Mot. Ischecantchia, Gml.

Blackish-brown, beneath ferrugineous; head black; nape whitish; crown and oblong wing-spot white; back black. Siberia.

Mot. Littorea, Gml. Iter. iii. t. 19. f. 1.

Dull-green, beneath yellowish; quills and tail blackish. Caspian Sea.

Mot. Longirostra, Gml. Iter. t. 19. f. 2.

Ash, beneath black; bill long. Caspian Sea.

Mot. Ochirura, Gml.

Head gray; nape and front of back black; throat and chest shining black; belly yellow. Persia.

Mot. Sunamisica, Gml.

Reddish-ash; chin and throat black; chest and belly

reddish; quills white-tipt; vent white; middle tail-feathers brown, side ones fulvous. Persia. These three may be *Saxicola*.

Equinoctial Warbler. *S. Æquinoctialis*, Lath.

Testaceous brown, beneath white; rump pale; tail obsoletely-banded. Island of Nativity.

Black-necked Warbler. *S. Nigricollis*, Lath.

Pale-gray, beneath flesh-coloured; crown, nape, wings, and tail black; head somewhat crested; bill and feet yellow. India.

Plumbeous Warbler. *S. Plumbea*, Lath.

Lead-coloured beneath ash; quill and tail dull.

S. Cambaiensis, Lath.

Blackish-brown, beneath shining-black; belly and vent reddish ferrugineous; wing-coverts white; length six inches. India.

S. Guzurata, Lath.

Greenish, beneath white; crown chestnut; quill and tail brown. India.

S. Asiatica, Lath.

Brown, beneath yellowish; head and neck black; lores and throat white; tail long. India.

S. Palpebrosa, Temminck, Pl. Col. t. 293. f. 3.

A circle of downy feathers round the eye, dark yellowish green above; throat light yellow; belly white. Bengal.

Javan Warbler, Lath. *S. Javanica*, Horsfield.

Olive-green; head gray-lead colour; forehead and

chin pale fulvous; eyebrows white; belly olivaceous-yellow; length four inches and a half. Java. Near *S. Hippolais.*

Chret Warbler, Lath. *S. Montana*, Horsf.

Brownish-olive; wings and tail pale-brown, beneath brownish-testaceous, bill somewhat depressed, blunt; feet and tail long.

Citrine Warbler, Lath. *Mot. Citrina*, Gml.

Yellow, black-streaked, beneath and rump yellow; cheeks, neck, and chest white; tail short, blunt, yellow-tipt; length three inches and a half. New Zealand.

Long-legged Warbler, Lath. *Mot. Longipes*, Gml.

Pale-green, beneath ashy; forehead, cheeks, and sides of neck ash; eyebrows white; tail very short; length four inches and a half. New Zealand. Var.? *Sylv. Minima*, Lath.

S. Macloviana, Gamot. Ann. Sci. Nat. 1826. 39.

Head and rump brown; body above ash, beneath gray-white; tail-feathers and quill brown, white-edged; throat ferrugineous.

Rusty-side Warbler, *S. Lateralis*, Lath.

Greater part of head and wings, lower part of back, and all, except two middle tail-feathers, green; rest blue-gray. New Holland.

Latham describes twelve other New Holland species, in his Supplement. He has referred some of them to

Meliphaga, and the rest as probably also belonging to this genus.

Some of the Warblers of the Oceanic Islands have the nostrils linear, and the first four quills nearly equal, and the longest; the bill has no bristles, and is slender and arched; the eye is surrounded by a white edge. They form the genus *Zosterops* of Vigors and Horsfield.

White-eyed Warbler. *M. Maderaspatana*, Lin. *M. Madagascariensis*, Gml. Brisson, iii. t. 28. f. 2.

Greenish, beneath whitish; throat and vent fulvous; eyelids white; length three inches and a half. Madagascar.

Zosterops Dorsalis, Vig. *S. Annulosa*, β. Swain. Zool. Ill. t. 16.

Yellowish-green; back ash; streak above and before the eye black; beneath yellowish-white; throat pale-yellow; side of belly ferrugineous; length four inches and a half. New Holland.

Some species have the bill short, nostrils large, closed behind with a membrane; wings very short, rounded; legs long; middle toes very long; hind claws long. They form the genus *Brachypterix*, Horsf.

Mountaineer Warbler, Lath. *Brachypterix Montana*, Horsf. Z. R. t.

Bluish-gray, beneath paler; belly whitish; wings very short; quills and tail brown, gray-edged; length six inches. Java.

Batavian Warbler, Lath. *Brach. Sepiaria*, Horsf.

Fulvous-olive, beneath paler; chin, middle of belly whitish; quill and tail bay; length five inches. Java.

In America is found, the *Golden-crowned Thrush. M. Aurocapilla*, Lin. Pl. Enl. t. 192. f. 2. Wils. A. O. ii. t. 14. f. 2.

Olivaceous; crown brownish-orange, margined each side by a black line; beneath white; breast spotted with blackish. Woods, North America; migratory. *T. Coronatus*, Vieil. The genus *Seiürus*, of Swainson.

S. Tenuirostris, Swain.

Above olive-brown, beneath pale-yellow, with triangular blackish spots; stripe above the eye pale. Mexico.

Water-thrush. Syl. Noveboracensis, Lath. *T. Aquaticus*, Wils. A. O. ii. t. 23. f. 5.

Olive-brown; beneath and line over the eye yellowish-white; breast spotted with blackish. North America; migratory. *T. Motacilla*, Vieil.

Yellow-rump Warbler. M. Coronata, Lin. Pl. Enl. t. 709. f. 1. t. 731. f. 2. Wils. A. O. ii. t. 17. f. 4. v. t. 45. f. 3.

Slate, streaked with black, beneath white; breast spotted with black; crown, sides of the breast, and rump yellow; wing with two white bands; tail black; three outer tail-feathers spotted with white. *M. Canadensis*, Lin. and *M. Cincta*, Gml. In winter, brownish-olive, beneath dirty-white. *M. Umbria*, and *M. Pinguis*, Gml. North America.

Palm-warbler, M. Palmarum, Gml. Pr. Musig. A. O. ii. t. 10. f. 2.

Brown-olive; crown rufous; line over the eye and all

beneath rich yellow; breast streaked; two outer tail-feathers white on the inner tip; in winter duller and paler. West Indies

M. Maculosa, Gml. Edw. t. 255. *S. Magnolia*, Wils A. O. iii. t. 23. f. 2.

Crown ash; rump and beneath rich yellow; breast spotted with black; wings with two white bands; tail black; outer tail-feathers white in the middle of their inner web. North America.

Cape May Warbler. S. Maritima, Wils. A. O. vi. t. 51. f. 8. Pr. Musig. A. O. i. t. 3. f. 3. ♀.

Yellow-olive, streaked with black; crown and line through the eye black; cheeks and beneath yellow; breast spotted with black; wings with a broad white band; three outer tail-feathers with a spot of white; *female*, dull-olive, beneath whitish, spotted with dusky. North America.

Canada Flycatcher. M. Canadensis, Lin. Pl. Enl. t. 635. f. 2. Wils. A. O. iii. t. 26. f. 2. *S. Pardalina*, Pr. Musig.

Cinereous-brown; crown dappled with black; beneath and line over the eye yellow; breast spotted with black; tail spotless. North America. *Muscicapa*, Wils. in autumn. *M. Cœrulescens*, Gml.

Hooded Flycatcher, M. Mitrata, Gml. Pl. Enl. 666. f. 2. *S. Cucullata*, Lath. Wils. A. O. iii. t. 26. f. 3.

Yellow-olive; head and neck black; forehead, cheeks, and body beneath yellow; three outer tail-feathers white on one half of their inner web. North America.

Black-throated Warbler. M. Virens, Gml. Wils. A. O. ii. t. 17. f. 3.

Yellowish-green; front, cheeks, sides of the neck, and line over the eye yellow; beneath white; throat black; wings with two white bands; tail dusky; three outer feathers marked with white. North America.

Chestnut-sided Warbler. M. Icterocephala, and *M. Pensylvanica,* Lin. Wils. A. O. i. t. 14. f. 5.

Crown yellow; beneath white; sides from the bill chestnut; wings with two yellow bands; three outer tail-feathers marked with white. North America. Pl. Enl. 731. f. 2. is young. *M. Coronata,* Lin.

Bay-breasted Warbler. S. Castanea, Wils. A. O. ii. t. 14. f. 4.

Forehead and cheeks black; crown, throat, and sides under the wings chestnut; wings with two white bands; three lateral tail-feathers marked with white. North America. Is *S. Ruficapilla,* Lath., a var. of plumage? See Gal. t. 164.

Black-poll Warbler. M. Striata, Gml. Wils. A. O. iv. t. 30. f. 3. ♂. vi. t. 54. f. 4. ♀.

Crown black; cheeks and beneath white; wing with two white bands; tail blackish; three outer tail-feathers marked inside with white; female and young dull yellow-olive, streaked with black and slate, beneath white; cheeks and sides of chest yellowish. North America.

Hemlock Warbler. S. Parus, Wils. A. O. v. t. 44. f. 3.

Black, with a few yellow-olive streaks; head above

yellow, dotted with black; line over the eye, sides of neck, and breast yellow; belly paler, streaked with dusky; wing with two white bands; tail black; three outer feathers white internally. North America.

Spotted Yellow Warbler. M. Tigrina, Gml. Edw. t. 257. lower fig. *S. Montana*, Wils. A. O. v. t. 41. f. 2.

Yellow-olive; front, cheeks, chin, sides of neck yellow; breast and belly pale yellow, streaked with dusky; wings with two white bands; tail rounded black; two outer feathers white internally. North America.

Blue-green Warbler. S. Rara, Wils. A. O. iii. t. 27. f. 2.

Green; lores, line over eyes, and all beneath pale cream colour; wings with two white bands; tail notched, brownish-black; three outer tail-feathers white externally. North America.

Prairie Warbler. S. Discolor, Vieil. O. A. Sept. t. 98. *S. Minuta*, Wils. A. O. iii. t. 25. f. 4.

Olive; beneath yellow, spotted with black; wings with two yellow bands; tail brownish-black; three outer feathers broadly white-spotted; a black crescent under the eye. North America.

Black-throated Blue Warbler. M. Canadensis, Lin. Pl. Enl. t. 65. f. 2. Wils. A. O. t. 15. f. 7.

Slate-coloured; beneath white; cheeks and throat black; a white spot on the wing; three lateral tail-feathers with white spot on the inner web. North America. In autumn, *M. Cœrulescens*, Gml.

Connecticut Warbler. *S. Agilis,* Wils. A. O. v. t. 39. f. 4.

Yellow-olive; beneath yellow; throat pale-ash; *fem.* throat dullish. North America.

Kentucky Warbler. *S. Formosa,* Wils. A. O. ii. t. 25. f. 3.

Olive-green; beneath and line over eye yellow; crown deep-black, spotted behind with light ash; lores, and a spot curving down the neck black. North America.

Autumnal Warbler. *S. Autumnalis,* Wils. A. O. iii. t. 23. f. 4.

Olive-green; back streaked with dusky beneath, and cheeks dull yellowish; belly white; wing bifasciate with white; tail blackish, white-edged; three lateral tail-feathers white tipt. North America.

Pine Swamp Warbler, Sylvia Pusilla, Wils. A. O. v. t. 43. f. 4. *S. Sphagnosa,* Pr. Musig. *S. Palustris,* Shaw.

Deep green-olive; beneath pale ochreous; wings with a triangular spot of yellowish white; three lateral tail-feathers with a whitish spot on the inner web. North America.

Cærulean Warbler, Sylvia Cærulea, Wils. A.O. ii. t. 17. f. 5. Pr. Musig. ii. t. 11. f. 2. *S. Agurea,* Steph. *S. Bifasciata,* Say.

Greenish-blue; beneath and line over eye white; wings bifasciate with white; tail black; tail feathers with a white spot. North America.

Blue-gray Flycatcher, Mot. Cœrulea, Lin. *Muscicapa Cœrulea,* Wils. A. O. ii. t. 18. f. 5.

Bluish-gray, beneath bluish-white; tail longer than the body, rounded black; outer tail-feathers nearly all white; two next white tipt. North America. Young *Mot. Cana,* Gml.

Small-headed Flycatcher, Muscicapa Minuta, Wils. A. O. vi. t. 50. f. 5.

Dull yellow-olive, beneath pale yellow; wings and tail dusky brown; wing-coverts white tipt; two lateral tail-feathers with white spot on the inner web. North America.

Maryland Yellow-throat, Sylv. Trichas, Lath. *S. Marylandica,* Wils. A. O. t. 8. f. 1. ♂ ii. t. 10. f. 2. ♀.

Green-olive, beneath yellow; front and wide patch through the eye black; bounded above by a bluish-white; *female,* dull olive: beneath dull yellow. The genus *Trichas* of Swainson.

Mourning Warbler, Sylvia Philadelphia, Wils. A. O. ii. t. 14. f. 6.

Deep greenish-olive, head slate; breast with a crescent of alternate white and black lines; belly yellow. North America. "Var. of *S. Trichas*"? Pr. Musig.

Blue-yellow-backed Warbler, Parus Americanus, Lin. Pl. Enl. t. 731. f. 1. *Sylvia Pusilla,* Wils. A. O. iv. t. 28. f. 1.

Bluish; interscapulars yellow-olive; throat yellow; belly white; wings with two white bands; side tail-feathers, inner side marked with white; *male,* forehead yellow with a black crescent; breast tinged with orange;

young, brownish-gray; beneath dirty white. North America. *M. Ludoviciana*, Gml. and *S. Torquata*, Vieil. ?

Sylvicola Inornata, Gm. Phil. Jour.

Above olive-green, beneath white; sides of head, ears, and throat ashy; wing with two yellow bands. Vera Cruz.

Yellow Red-poll, Mot. Petechia, Lin. Edw. t. 250. lower, Wilson, A. O. iv. t. 20. f. 4.

Yellow-olive, streaked with dusky; beneath and line over the eye yellow; breast streaked with dull red; crown reddish; wings and tail blackish edged with olive; no red cap except in summer. North America. Is it distinct from *M. Æstiva* ?

Blackburnian Warbler, M. Blackburniæ, Gmel. Wils. A. O. iii. t. 28. f. 3.

Head striped with black and orange; throat and breast orange, bounded by black spots; wings with a large white spot; three side tail-feathers white on the inner web. North America. Mexico.

Blue-eyed Yellow Warbler, M. Æstiva, Gml. Pl. Enl. t. 5. f. 2. (not 1.) *S. Citrinella*, Wils. A. O. iii. t. 15. f. 5.

Greenish yellow; forehead and beneath yellow; breast and sides streaked with dark red; side tail-feathers interiorly yellow. North America. Young greenish yellow; throat white; *M. Albicollis*, Gml.

Yellow-throated Warbler, M. Pensilis, Gml. *S. Flavicollis*, Lath. Pl. Enl. 686. f. 1. Wils. A.O. ii. t. 12. f. 6.

Light slate; frontlet, ear-feathers, lores, and above the

eyes black; throat and breast yellow; belly and line over the eye white; wings and tail black, varied with white. North America.

Patagonian Warbler, Lath. Dixon. Voy. t. p. 359. *M. Patagonica*, Gml.

Ash, beneath white-spotted; eyebrows white; wing-spot and bands brown; outermost tail-feather white; length nine inches. Terra del Fuego.

Rufous-tailed Warbler. M. Ruficauda, Gml.

Rufous-brown, beneath white; throat red-edged, brown spotted; wing-coverts and tail brown; length five inches and a half. Cayenne.

Yellow-bellied Warbler. M. Fuscicollis, Gml.

Greenish-brown; chest and belly yellow-foxy; wing brown, reddish-edged. Cayenne.

Rusty-headed Warbler. M. Borealis, Gml.

Olive-green, beneath yellow; forehead, cheeks, and throat ferrugineous; side tail-feathers white-tipt: length five inches. Kamtschatka.

Magellanic Warbler. M. Magellanica, Gml.

Yellowish-brown, black-waved; beneath yellow-ash, black, cross-streaked; length four inches and a half. Terra del Fuego.

Grisly Warbler. M. Grisea, Gmel. Pl. Enl. t. 64. f. 1. 2.

Gray-ash, beneath and eye-band white; crown and chest black; length four inches and a half.

St. Domingo Warbler. M. Albicollis, Gml. Brisson. ii. t. 26. f. 5.

Olive-green, beneath yellow-white, brown-streaked; inner-half of side-feathers yellow; length five inches. St. Domingo.

Green and White Warbler. M. Chloroleuca, Gml. Brisson. iii. t. 26. f. 2.

Olive-green, beneath yellow; head and neck above ash; inner part of side tail-feathers half yellow; length four inches and a half. St. Domingo.

Worm-eater Warbler, Ray. Sloan. Jam. t. 265. f. 2. *Brown-throated Warbler, M. Fuscescens,* Gml.

Brownish, beneath red and gray varied; eye-band and crop deep-brown; length five inches. Jamaica.

Jamaica Warbler. M. Dominica, Lin. Brisson. ii. t. 27. f. 3.

Ash, beneath white; spot before the eyes yellow, behind white; beneath black; length four inches and a half. Jamaica.

Orange-headed Warbler. M. Chrysocephala, Gml. Pl. Enl. t. 58. f. 3.

Red-brown, beneath white; face and throat fulvous; wing-coverts varied black and white; tail black; bill black. Guiana.

Rufous and Black Warbler. M. Multicolor, Gml. Pl. Enl. t. 391. f. 2.

Black, beneath white; neck and chest, side of the tail

from the base to the middle, and band on the wings, red; length five inches. Cayenne.

Hang-nest Warbler. M. Calidris, Lin. Edw. t. 121. f. 2.

Greenish-brown; beneath fulvous; streak above and beneath the eye black; quills yellow tipt; nest pensile. Jamaica.

Banana Warbler. M. Bananivora, Gml.

Backish-gray, beneath and rump yellowish; eye-band black; eyebrows, wing-spot, and tips of tail white; length three inches and a half. St. Domingo.

Simple Warbler. M. Campestris, Lin. Ed. t. 122. f. 1.

Gray; head greenish-ash; belly whitish. Jamaica.

Black-throated Tanager. M. Guira, Lin. *T. Nigricollis*, Gml. Edw. t. 351. f. Pl. Enl. t. 720. f. 1.

Green, beneath and rump yellow; cheeks and throat black, girt by a yellow line. Brazil.

Motacilla Gularis, Miller. t. 30. 1.

Ferrugineous, beneath white; throat, wing, and tail black. South America.

? *Long-billed Warbler. M. Kamtschatkensis*, Gml.

Olive-brown; belly, middle, white; forehead, cheeks, and throat, pale-ferrugineous; bill long. Kamtschatka.

Awatcha Warbler. *M. Awatcha*, Gml.

Brown; beneath white; chest black-spotted; side tail-feathers yellow-based; lore yellow. Kamtschatka.

S. Dorsalis, King. Zool. Jour. iii. 428.

Black; back and scapulars red; quill and tail brown; bill and legs black; length four inches and a half. S. America.

Temminck has separated some of the American Sylvia as a genus, under the name of *Hylophilus;* the character is not yet given.

Sylvia Plumbea, Swain. Zool. Ill. iii. t. 139. *S. Venusta*, Temm. Pl. Col. t. 2. 293. f. 1.

Head and back blue; tail and wings varied with black; throat, belly, and across the back yellow; vent and bars on wings white. Brazil.

Hylophilus Thoracicus, Temm. Pl. Col. t. 173. f. 1.

Patch of greenish-yellow on the thorax and flanks; neck and throat ash-colour; top of head, back, wings, and tail green. Brazil.

Hylophilus Poicilotis, Temm. Pl. Col. t. 173. f. 2.

Top of head and occiput red; forehead paler; cheeks striped black and white; throat ash; upper part and tail green; belly yellow. Brazil.

Some of the American species have the habits of Creepers (*Certhiæ*), the genus *Niniotello* of Vieil.

Pine-creeping Warbler. *S. Pinus*, Lath. Wils. A. O. iii. t. 19. f. 4.

Olive-green, beneath yellow; vent white; wing with

two white bands; tail brown; three outer feathers with a broad white spot near tip; lores not black.

The *Mot. Varia*, Linnæus, is the type of the genus, but Cuvier refers it to *Certhia*, and Pr. Musig. to *Sylvia*. The latter ornithologist places here the genus *Dacnis*, which Cuvier considers as a section of *Cassicus*. Swainson has again established the genus under the name of *Vermivora*.

Bechstein separates from the other Warblers, his

ACCENTOR,

which is the *Alpine Warbler*, Buff. (*Mot. Alpina*,) Enl. 668, because its slender beak, but more exactly conical than that of the other Warblers, has its edges a little re-entering.

It is an ash-coloured bird, white throat, picked out with black, with ranges of white spots on the wing, and a lively red on the sides. It stays in the pasture-grounds of the High Alps, where it chases insects, and from which it descends into the villages in winter to find grains.

This is also the *Collared Stare* of Latham, *Sturnus Mauritanus* and *St. Collaris* of Gmelin.

I believe I have observed a similar beak on our Hedge-Sparrows, (*M. Modularis*, Lin.) Enl. 615. 1.—the only species which remains with us in winter, and which enlivens this season a little by its agreeable song. It is fawn spotted with black above, slate-ash below. In the summer it goes northward, and into the mountain woods. In winter, it contents itself with grain for want of insects.

To this the *Syl. Schænobænus*, or *Red Warbler*, of Lath. Penn. Brit. Zool. t. 51. f. 3. 4. Fisch. t. 21. f. 2. 13. might perhaps be added.

In Europe is also found,

Mountain Accentor. Accentor Montanellus, Temm.

Reddish-ash; cowl and streak under eye black; eye-streak yellowish; wing two-banded, beneath yellowish-brown spotted; length five inches and a half. South of Europe.

Temminck refers to this genus *M. Calliope* of Pallas, a *Sylvia* of Cuvier.

We may also distinguish some foreign slender-beaks, with a very slender beak, compressed almost as much as in the Blackbirds, and a long and wedged tail. These, preceding naturalists had left among the Titmice. Some of their species construct nests of cotton and other filaments, arranged with considerable art.

The genus is particular for two or three bristles on the side of the mouth; it is MALURUS of Vieillot: all the species are confined to the Old World, especially Africa and Oceania; only one is found in Europe, which has been called a Warbler.

Sylvia Cisticola, Temm. Pl. Col. t. 6. f. 3.

Crown, nape, back, and wing-coverts pale-brown, with blackish brown stripes; loins and back pale brown, uniform; beneath reddish white, spotless; tail short, graduated, blackish brown; side-feather black spot at end; length four inches. Portugal. India, Gen. Hardwicke.

Sylvia Macroura, Gml. *Le Capocier*, Vail. O. A. t. 129. 130. Pl. Enl. t. 752. f. 2.

Brown, beneath yellowish white, black spotted; eyebrows white; tail, wedge-shaped, long; length six inches. Cape of Good Hope.

Long-tailed Warbler, Sylvia Longicauda, Gml.

Greenish olive; crown reddish; quills brown; tail wedge-shaped. China.

Malurus Galactoides, Temm. Pl. Col. t. 65.

Reddish brown, fuscous brown streaked; beneath whitish; shaft of tail-feathers brown. New Holland.

A *Megalurus* of Vigors and Horsfield appears to unite the two groups.

Malurus Clamans, Ruppel. Atlas, t. 2. f. 1.

Forehead and crown varied black and white; body above helvola, beneath yellowish; wing-coverts black, white limb; length 4″, tarsus 9‴.

Malurus Gracilis, Ruppel. Atlas. t. 2. f. 6. *Sylvia Gracilis*, Licht. Lat. Savigny, Egypt, t. 5. f. 4.

Above olive-gray, beneath whitish; crown and nape black, with obscure, oblong sooty spots; length 5″. Egypt. (The feathers are dark sooty, with broad pale margins.)

Malurus Palustris, Vieil.

Brown, beneath ferrugineous; throat blue; tail long, wedge-shaped; tail-feathers pennate, disjointed; bill blackish brown; feet brown. New Holland. Allied to *Musc. Malachura.*

T. Brachypterus, Lath. Suppl.

Pale-brown, inclining to ash, beneath breast obscurely waved; wings very short. New Holland.

Muscicapa Malachura, Lin. Trans. iv. t. 21. Vail. O. A. t. 130. f. 2.

Ferrugineous brown, beneath paler; streak before the eye and eyebrow pale blue; throat gray; beard of tail-feathers loose. New Holland.

Superb Warbler, Lath. Phil. Bot. Bay. tap. 159 ♂ 159. ♀. *Motacilla Cyanea*, Lath. ii. 142. White Jour. t. at 256. Lath. H. t. 106.

Head, subocular streak, and muchal lunate stripe silky blue; eye-streak, nape, throat, chest, and back silky black; belly white; quills and tail fuscous; tail rounded; *female*, above fuscous brown, beneath whitish.

Lambert's Warbler, Malurus Lambertii, Vigors and Horsfield, White Jour. tap. p. 256. fig. infer.

Head-streak extending to the nape and middle of back, silky blue; throat, chest, nape, back, and rump silky black; scapulars reddish brown; belly white; quills and tail brownish; tail graduated; *female*, brownish; beneath white; length five inches and a half. New Holland.

White-wing Tailor Bird, Malurus Leucopterus, Quoy. and Gaim. Frey. Voy. t. 23. f. 1.

Deep blue; crested; scapulars and wing-coverts brown; wing shining blue. New Holland.

Orange-rump Warbler, Lath. *Muscicapa Melanocephala,* Lath. Supp.

Head, front of neck, and chest brownish black; back scarlet; quill and tail brown; belly whitish. New Holland.

Brown's Tailor Bird, Malurus Brownii, Vigors and Hors.

Head, slight crest, front of neck, wing-coverts, and tail-feathers black; back scarlet; quills fuscous brown; body thirty-three inches and three-quarters long.

Exile Warbler, Lath. H. *Malurus Exilis,* Lath. MSS.

Above rufous brown, with broad brown streaks; beneath paler; quill and tail-feathers brown; tail white-tipt; length four inches.

Flaxen Warbler, Motacilla Subflava, Gml. Pl. Enl. t. 584. f. 1. *Le Citrin,* Vail. O. A. t. 127.?

Reddish brown; beneath gray; rump pale; sides of body reddish; tail wedge-shaped; length four inches and a half. Senegal.

Malurus Superciliosus, Le Double Sourcil, Vail. O. A. t. 128.

Brown above, brownish white underneath; black streak over, and another under the eye.

Sylvia Lateralis, β Lath. *Malurus Hirundinaceus,* Vieil. Shaw. Nat. Misc. t. 114.

Body above black; crop and chest scarlet; belly white, with a broad, long, black streak; vent fulvous; bill blackish; feet pale.

Meriones Maculatus, Vieil.

Above brown; beneath whitish, black-spotted; tail ash; tips with black and reddish white; bill and feet brown. New Holland. Mus. Paris.

The *Sylvia Magnifica* of Temminck belongs also to this division.

The genus *Dasyornis*, of Vigors and Horsfield, has all the habits of these birds; but the front of the forehead above the bill has some peculiar projecting bristles, and the texture of the feathers is very soft and loose.

Southern Bristle Bird, *Dasyornis Australis*, Vigors and Horsfield.

Above fuscous brown; beneath paler; crop and middle of belly whitish: quills and tail rufous brown; length seven inches and a half. New Holland.

African Warbler, *M. Africana*, Gm. *Sphænura Tibicen*, Licht. *Le Merle Fluteur*, Vail. O. A. t. 112. f. 2.

Crown red, black-streaked; feathers of back and nape ashy; of loins and wings red-edged; tail-feathers long linear; scape black; web red, beneath ashy; sides black-lined; length eight inches. Cape of Good Hope.

Cuvier speaks of this species at the end of his Notes on the Thrushes, and again at this section of *Sylvia*, and Temminck placed it both in *Sylvia* and *Malurus*.

The genus *Acanthiza* of Vigors and Horsfield has the same kind of bristle at the bill, wings, and legs, but their tail is short and rounded, or nearly even, and the bill is rather short and more depressed. They appear to be confined to Oceania.

Dwarf Warbler, var. A.? Lath. Gen. Hist. vii. p. 134. No. 101. *Acanthiza Nana*, Vigors and Hors.

Olive-green; beneath yellow; forehead and check whitish yellow; quills and tail olive-brown; tail black-banded near the tip; bill and feet yellowish; length three inches and a half.

Golden Crest-like Warbler, Acanthiza Reguloides, Vigors and Hors.

Olive-green above; beneath yellowish; white forehead; front of occiput ferrugineous; rump, and base, and tip of tail fulvous yellow; middle black; length three inches and three-quarters. New Holland.

Acanthiza Frontalis, Vigors and Hors.

Fuscous brown; beneath paler; forehead, throat, and chest white; rump reddish; length of body four inches and a half; tail two inches.

Acanthiza Pyrrhopygia, Vigors and Hors.

Fuscous brown; beneath whitish; rump red; tail, subapical; band black; tip white; tail long, subgraduated; length five inches; allied to *Malurus Exilis.*

Acanthiza Buchanani, Vigors and Hors.

Olive-green; front of head white-lined; beneath whitish; throat and chest brown-lined; tail black; rump scarlet.

Dwarf Warbler, Lath. H. *Motacilla Pusilla,* Lath. White Jour. tap. p. 257. *Acanthiza Pusilla,* Vig. and Hors.

Fuscous brown; forehead variegated fulvous; beneath whitish; chest and throat brown-streaked; rump reddish; middle of tail brown band; tips pale. N. Hol.

The WRENS, or FIG-EATERS, (REGULUS, CUV.) have the bill completely in a very sharp cone, and, even when viewed from the top, its sides appear a little concave. They are small birds, which sojourn on trees, and pursue gnats through the branches. We have three in Europe.

The *Gold-crested Wren,* (*M. Regulus,* L. Enl.651. 3.) the smallest of our European birds; olive above, yellowish-white beneath; head black, marked with a beautiful golden-yellow spot, the feathers of which are capable of erection; it makes on the trees a nest like a ball, the aperture of which is on the side, suspends itself to the branches in all directions, like the titmice, and approaches habitations in winter.

The Yellow *Wren Warbler,* (*M. Trochilus,* Lin.) Enl. ib. 1.
a little larger than the last, of the same colour, but without crest, of similar manners, but a prettier song. It removes in winter.

Lesser Pettichaps, (*M. Hypolaïs,*) Bechst. III. xxiv. A little larger, with a more silvery belly.

The foreign fig-eaters are very numerous, and often clothed with agreeable colours.

Pensile Warbler, Lath. *M. Pensilis,* Gml. *Le Cou Jaune,* Enl. 686. 5.

Above deep-gray; head grayish-black; throat, neck in front, and breast yellow; sides of neck spotted with black; bill dusky; length five inches. St. Domingo.

Yellow-poll Warbler, Lath. *M. Æstiva,* Gml. *Le Figuier Tacheté,* Enl. 58. 2.

Olive-yellow above, fine yellow beneath; neck and breast spotted reddish; bill black; length four inches.

Orange-bellied Warbler, Lath. *M. Fulva,* Gml. et *Ludoviciana. Figuier à gorge jaune,* Enl. 731. 2.

Olive-brown above, beneath to breast yellow, inclining to brown on the last; rest rufous. Louisiana.

Maurice Warbler, Lath. *M. Mauritiana*, Gml. Enl. 705. 1. *Le Figuier Bleu*, Buff.

Above blue-gray, beneath white; bill blackish; quills and tail black, white-edged. Isle of France.

Le Plastron Noir, Vail. 123. 1 et 2.

A black collar of crescented form at the bottom of the neck, space between this and throat white; above olive-gray, whitish-yellow beneath; *female* without collar. South Africa.

The Troglodites, Cuv.

do not differ from the fig-eaters but in having the beak still more slender and slightly arched.

Divided into two sections, the first, or True Wrens, have the bill slender at the base, the hind toe equal to the inner, the spurious feather moderate.

We have but one in Europe.

The *Common Wren*, (*M. Troglodytes*, L.) Enl. 651. 2. named in many places *Roitelet*.

Brown, radiated crosswise with blackish; with some whitish on the throat and edge of the wing; the tail rather short and elevated. It nestles on the ground, and sings agreeably even in the depth of winter.

This is the *Troglodytes Europeus*, Stephans. *Trog. Hyemalis*, Vieil. The *Winter Wren* of Wilson. A. O. t. 8. f. 6.

The *Brown Warbler?* Brown Illust. t. 18. *House Wren, Sylvia Domestica*, Wilson, A. O. t. 8. f. 3. *Sylvia Furva*, Lath.? , *Troglodytes Œdon*. Vieil.

Brown banded with black; beneath dull grayish, obsoletely banded; tail long, rounded. North America.

Boie refers *Motacilla Modularis*, (*Accentor* of Cuv.) to this genus.

Buenos Ayres Wren. *Sylvia Platensis*, Lath. Pl. Enl. t. 730. f. 2.
Red varied with black, beneath white; quills and tail banded. South America.

The second section form the genus *Thryothorus* of Vieillot: the base of their bill is broad, the hind toe long and slender, and the spurious feather long. Lives in watery places in America.

Great Carolina Wren, *Certhia Caroliniana*, Wilson, A. O. iii. t. 12. f. 5. *Sylvia Ludoviciana*, Lath. Pl. Enl. t. 730. f. 1. *Troglodytes Arundinaceus* and *Thryothorus Lateralis*, Vieil.
Reddish brown; wings and tail black-barred, beneath light rusty; eyebrows yellowish. Pennsylvania.

Marsh Wren, *Certhia Palustris*, Wilson, A. O. ii. t. 12. f. 4. *Thryothorus Arundinaceus*, Vieil.
Dark brown; crown black; neck and back black, white streaked; eyebrows white, beneath silvery-white; vent brownish. United States.

The Wagtails, (Motacilla, Bechst.) unite, to a still more slender beak than that of the Warblers, a long tail, which they raise and lower incessantly, elevated legs, and, particularly, scapulary feathers, long enough to cover the end of the folded wing, which gives them an analogy with most of the waders.

The Wagtails Proper, (Motacilla, Cuv.) have the claw of the thumb curved like the other Warblers.

White Wagtail, (*Mot. Alba et cinerea, L.*) Enl. 652.

Ashy above, white under; neck and chest black, with a coif on the occiput.

When young *M. Cinerea*, Gml. Pl. Enl. t. 674. f. 1. and t. 652. f. 2. the complete winter plumage,—the Albine variety *M. Albida*, Gml. Jacq. Heyt. 8.

Mournful Wagtail, M. Lugubris, Pallas.

Above, throat, and chest black; eyes, ears, belly, and two outer tail-feathers white.

Sometimes breeds with the former. Middle of Europe and Russia.

Green Wagtail, Brown Illust. *M. Viridis*, Gml.

Pale-green, beneath white; head ash; tail and wing ash, white-edged. Ceylon.

M. Aguimp, Temm. Vail. O. A. t. 178. *Le Bergeronnette à Guimpe. M. Capensis*, Licht.

Shining-black; eyebrows, throat, and belly white; chest-band black; two outer tail-feathers and wing-band white.

Pied Wagtail. M. Maderaspatana, Gml. Bay. Syn. t. 1. f. 1. and f. 6. Vail. O. A. t. 184.

Black; belly white; wing-band white; tail white, two middle feathers black.

All the species of this genus are peculiar to the Old Continent, *M. Hudsonica* of Lath. not being a Wagtail.

M. Variegata, Vieil. Vaillant Ois. Afr. t. 179.

Head and back olive-brown, beneath the same varied with yellow, and a black stripe across the breast; quills black varied with yellow and white. Cape of Good Hope.

Also consult *M. Atricapilla* and *Cœrulescens*, Lath.

Suppl.; *M. Melanops*, Pallas; *M. Indica, M. Afra, M. Tschutschensis*, Lath.

M. Capensis, of Gmelin, according to Lichtenstein, is the young of *M. Aguimp*, Temminck.

The Budytes, (Budytes, Cuv.)

have, with the characters of the last, the thumb-claws elongated, and a little arched, which approximates them to the *Pipits*, or *Field-larks*. They remain in pasturages, and hunt insects among the flocks.

The *Yellow Wagtail. Bergeronnette de Printemps,* (*M. Flava,*) Enl. 674. 2. Edw. t. 258.

is ash-coloured above, olive on the back, yellow underneath, an eye-brow, and two-thirds of the lateral quills of the tail white.

The *M. Chrysogastra*, Bechst., and *Yellow Wagtail*, Edw. t. 258.

It does not change its colour with the season, like the other species.

M. Melanocephala, Lich.

Like *M. Flava*, but forehead, ears, and nape black.

M. Boarula, Lin. *M. Melanope*, Pallas, Pl. Enl. t. 28. f. 1. young hen. Edw. t. 259. ♂. *M. Sulphurea*, Bechst.

Above ash; rump yellow-olive; eye and neck-band white; throat black; beneath pale-yellow; wing and middle tail-feathers black, greenish-white edged; outer tail-feathers internally white; length seven inches and a half.

Yellow-headed Wagtail, Lath. *M. Citreola*, Pallas, Talk. Voy. iii. t. 29. *M. Scheltobriusk*, Lepech. Voy. i. t. 8. f. 1.

Crown, cheeks, and beneath lemon-yellow; occipital

band broad, black; above and sides ashy; larger wing-coverts white-edged and tipt; tail and quills blackish; two outer tail-feathers white; length seven inches.

Cape Wagtail. *M. Capensis*, Lin. Pl. Enl. t. 28. f. 2. Brown, beneath white; chest-band brown; eyebrows white; tail black; side tail-feather obliquely white. Cape of Good Hope.

Pipits, or Field-larks, (Anthus, Bechst.) have been for a long while united to the Larks, on account of the long kind of claw; but their slender and sloped beak approximates them to the other slender-beaks. At the same time their quills and secondary coverts, as short as usual, do not allow us to confound them with the Wagtails. Some, whose claw is sufficiently marked, perch willingly.

The *Field-lark*, (*Alauda Trivialis et Minor*, Gml.) *Anthus Arborius*, Bechst. Enl. 660. 1.

Brown-olive, above grayish, underneath spotted with blackish on the chest; two pale transverse spots on the wing.

The *Anthus Breviunguis*, Spix. Braz. t. 76. f. 1. belongs also to this division.

Others have on the thumb the complete claw of a lark. They most usually remain attached to the ground.

The *Meadow-lark*. *Alauda Pratensis*, Gm. *Anthus Pratensis*, Bechst. Enl. 661. 2. Geoff. Ois. Egypt. t. 5. f. 6.

Olive-brown above, whitish underneath; some brown spots on the breast and sides; a whitish eye-brow; the edges of the external quills of the tail white.

It sojourns in humid or inundated meadows, nestles in the rushes or tufts of grass. It grows singularly fat in autumn by eating grapes, and is then in great request in many of our provinces, under the names of *Becquefigue* and *Vinette*. Also found in Nubia.

The genus *Enicurus* of Horsfield and Temminck, has the bill-base broad, suddenly compressed, tapering, abruptly curved; hind-claw strong-curved; tarsi slender, elevated; tail forked; and the habit of the Wagtails. They are peculiar to India.

M. Speciosa, Horsf. 3. R. t. *Enicurus Coronatus*, Temm. Pl. Col. 113.

Black; crown crested; belly, rump, band on wing, outermost tail-feather entirely, the rest at the tips, white; tail very long, forked; length ten inches and a half.

Enicurus Velatus, Temm. Pl. Col. 160.

Neck, throat, upper part of the back, wings, and tail ashy black; tail-feathers tipped ash white; underneath, and lower part of back, and eye-spot, white; top of head brown. Java.

The *Alauda Mosellana*, Gml. is best distinguished from them by the shortness of the hind toe.

Willow Lark, Penn. Br. Zool. t. 2. f. 4. *A. Rufescens*, Temm. *Anthus Campestris*, Meyer. Pl. Enl. t. 661. good. Trisch. t. 15. f. 2. A.

Above Isabella gray; feathers brown-streaked; throat yellowish, beneath whitish; length six inches and a half.

Dusky Lark, Lewin. Br. B. iii. t. 94. *Al. Obscura*, Montagne. *Al. Obscura*, Gml. *An. Montanus*, Koch. *A. Aquaticus*, Bech. *An. Rupestris*, Nelson. *Al. Campestris Spinoletta*, Gml. Pl. Enl. t.

661. f. 2. *Meadow Lark*, Lath. *Al. Spinoletta*, Lin. *An. Spinoletta*, Pr. Mus. *Al. Rufa*, Wils. A. O. v. t. 42. f. 1.

Above gray-brown; feathers darker in the centre; smaller wing-coverts white-tipt; beneath white; chest and flanks ash-streaked. Europe and America.

An. Richardi, Vieil. Ency. Méth. Zool. Jour. t. *Al. Lusitania*?

Bill strong; tarsi very long; hind-claw much longer than the toe, slightly arched; brown; feathers pale-edged, beneath white. English.

The genus *Corydalla*, Vigors!

Al. Capensis, Lin. Pl. Enl. t. 504. f. 2. *L'Alouette Sentinelle*, Vail. O. A. t. 195.

Three side tail-feathers white-tipt; throat yellow, black-edged; eyebrows yellow. Cape of Good Hope.

Al. Rufa, Gml. *A. Fulva*, Lath. Pl. Enl. t. 738. f. 2.

Blackish-brown; nape, back, and scapulars orange; wing and tail dark. South America.

Rufous Lark. *Al. Rufa*, Gml. Lath. Pl. Enl. t. 738. f. 1.

Blackish-red, divided; body beneath and throat white; two outer tail-feathers white-edged. S. America.

African Lark. *Al. Africana*. Pl. Enl. t. 712.

Red-brown, varied with white; beneath white, brown-spotted; wings and tail brown. A lark of Cuvier.

Red Lark, *Al. Ludoviciana*, and *An. Ludovicianus*, *Al. Rubra*, Gml. Edw. t. 297.

Dull-brown, beneath reddish-fulvous, brown-spotted; cheeks blackish; eyebrows pale-red. North America.

A. Australis, Vigors and Horsfield.

Olive, reddish-brown, variegated with fuscous-brown;

beneath yellowish-white, brown-streaked; eyebrow-spot fulvous; throat white; quills and tail-feathers fuscous-brown, the two outer white-edged; length six inches. New Holland.

An. Pallescens, Vigors and Horsfield. Above varied pale-red and brown, beneath whitish; chest brown, scarcely spotted; quills and tail-feather fuscous-brown, two outer white-edged; length five inches. New Holland.

An. Minimus, Vigors and Horsfield. Above olive-green, varied with brown; head brown, white-streaked; beneath greenish-white, brown-streaked; tail-feathers, middle excepted, brownish-black, white-tipt; length four inches. New Holland.

An. Fuliginosus, Vigors and Horsfield. Above olive-green, beneath paler, black-streaked; quills and tail pale-brown; tail black-banded, white-tipt; length four inches. New Holland.

An. ? *Rufescens*, Vigors and Horsfield. Pale-brown, clouded with fuscous-brown, beneath paler; throat white; rump reddish; quills and tail brownish; length six inches. New Holland.

An. Chi, Spix. Braz. t. 76. f. 2. *Chii* Azara, n. 146. Like *A. Pratensis*, but much smaller, and tarsi longer; hind claw long, nearly straight. Brazil.

The genus *Megalurus* of Horsfield differs from the *Anthi*, in the leg and bill being stronger. The Doctor has indicated one species.

Javan Pipit. Megalurus Palustris, Horsf. Lin. Trans. Brown; back and head varied with gray; underneath white, with gray tinge on breast. Java.

Generic Characters of Birds.

ORDER PASSERES Pl. I.

Fam. 1. Dentirastres.

1 *Lanius*
2 *Psaris*
3 *Tanagra*
4 *Tachyphonus*
5 *Pyranga*
6 *Tan Ramphocelinus*
7 *Tyrannus*
8 *Ampelis*
9 *Gymnoderes*
10 *Edolius*

London Pub. June, 1828, by Whittaker & Co. Ave Maria Lane.

Generic Characters of Birds

ORDER PASSERES Pl. II.

Fam. 1. Dentirostres

1 Turdus
2 Philedon
3 Gracula Cuv.
4 Mænura
5 Rupicola
6 Pipra
7 Sylvia
8 Malurus Vieill.
9 Regulus Cuv.
10 Troglodytes Cuv.
11 Anthus Bech

London. Published by Whittaker & Co Ave Maria Lane, June 1829.

SUPPLEMENT

ON

THE DENTIROSTRAL FAMILY

OF THE

ORDER PASSERES.

But few observations can be offered, in a general way, on this immense order of the feathered race. It comprehends more species than all the others put together, and though they vary considerably in size and strength, yet they exhibit so great an analogy in other particulars, that they must be classed together. In the muscular stomach, the two small cæcums, the capacity of singing, the complication of the lower larynx, the conformation of the sternum, they all, with few exceptions, generally resemble each other. Their aliment consists of fruits, grains, and insects. Some few give chase to the smaller birds, and one group subsists on fish. They exhibit, of all other birds, the greatest variety and ingenuity in the construction of their nests. All, with the exception of a single group, are monogamous. The male, in a great majority of the species, administers food to the female, while she hatches the eggs, and partakes the cares of incubation. Both feed the little ones in the nest; the latter do not quit it until they can fly with perfect ease, and even after their departure they are for some time nourished by their parents, until they acquire the complete capacity of providing for themselves.

As the Passeres are so very numerous, and are divided into five families, or principal sections, differing materially in some respects from each other, notwithstanding their general relative similitude, it has been thought most advantageous to insert our supplementary observations at the end of each of these families in the text. The reader will thereby be relieved from too tedious a detail of generic characters and specific descriptions, and will be the better enabled to confine his attention to a single portion of the order at one time.

The first genus of the Passeres which claims our attention is that of the SHRIKES. Its characters and divisions we have seen in the text. As it was originally constituted by Linnæus, it comprehended species extremely different from each other, which have since been referred to more suitable genera, or have served as the types of several new genera.

They are naturally divided into three sections, as Levaillant originally divided them. The first have the longest wings and strongest beak. They fly well, and are much given to the chase. The second have the wings shorter and rounded: their bill is weaker, and their disposition more mild. They quit the bushes less frequently, where they remain concealed the greater part of the day. Those of the third section have the body more compact and heavy, the tail very short, and the beak feeble.

Notwithstanding the dismemberments which have been made from this genus, it still contains a great number of species, some of which lead to the thrushes, and others to the warblers, in an insensible manner, chiefly through the species of the third section. Their habits, too, and insectivorous diet point out their natural relation with other groups of the Passeres.

The shrikes are spread over the entire globe, and everywhere exhibit similar dispositions, habits, and modes of existence. Of small size, but armed with a strong and crooked beak, of a fierce and courageous disposition, and of a sanguinary appetite, they bear much affinity to the birds of prey. Naturally intrepid, they defend themselves vigorously, and do not hesitate to attack birds much stronger and larger than themselves. The European shrikes can combat with advantage, pies, crows, and even kestrills. They attack and pursue these birds with great ferocity, if they dare to approach their nests. It is even sufficient if any of them should pass within reach. The male and female shrikes unite, fly forth, attack them with loud cries, and pursue them with such fury, that they often take to flight without daring to return. Even kites, buzzards, and ravens will not willingly attack the shrike. They are habitually insectivorous, and also pursue small birds. They will cast themselves on thrushes, blackbirds, &c., when these last are taken in a snare. When they have seized a bird they open the cranium, devour the brain, deplume the body, and tear it piecemeal.

The prudence to foresee and provide for the wants of the future, is another of their qualities. That they may not fail of those insects which form their subsistence, and which only make their appearance at a determinate epoch, some shrikes form kinds of magazines, not in the hollows of trees, nor in the earth, but in the open air. They stick their superabundant prey on thorns, where they may find it again in the hour of need.

Falconers have taken advantage of the character of these birds, and occasionally trained them to the chase. Francis the First of France, according to the account of Turner, was accustomed to hunt with a tame shrike, which used to speak, and return upon the hand. The Swedish hunters, availing themselves of the habit of the Gray Shrike of uttering a peculiar sort of cry at the approach of a hawk, make use of it to discover the birds of prey which this kind of cry announces.

Though we have said that the shrike genus is extended over the entire globe, we believe South America must be excepted. The South American birds which have been called shrikes belong to other divisions, and it would appear that this genus does not pass beyond the Floridas, Louisiana, and the North of Mexico.

As a complete enumeration of species is made in the text and additions, we shall only notice here those which have any peculiar points of interest.

The *Cinereous Shrike (Lanius Excubitor)* is spread over all Europe, very common in France, though not so frequently found in England. It remains in woods and wilds during the summer, but on the appearance of winter will approach the habitations of man. It constructs its nest in the embranchments and furcations of lofty trees in solitary forests, and sometimes in thick and thorny hedges. This is composed of hay without, of small fibrous roots and moss interlaced together; and the small branches of neighbouring trees are introduced, and twisted to form its seat and basis. The interior is profusely furnished with feathers, down, and wool. The female lays from four to six eggs of a grayish white, spotted with pale green olive, and ash-colour. The young are born naked, and are never covered with down.

The parents evince the greatest tenderness for their offspring, tending them carefully during the entire period of infancy, and never quitting them until spring. These birds are seen to fly during the autumn and winter in small flocks, each composed of a single family. These companies never unite together. This sort of family division renders the shrikes easily cognizable at a distance. They are also distinguished by a piercing cry which may be thus expressed *troüi, troüi*, which may be heard very far off, and which they repeat incessantly, perched on the summit of trees or flying. Their mode of flying is peculiar. It is neither oblique, nor horizontal at the same elevation, but continually up and down, by successive springs and undulations. They are always seen perched on the extremity of the most lofty and isolated branches of trees and thickets, a position which their peculiar mode of chase seems to require; for, as they fly with difficulty, and always drop perpendicularly on their prey, they thus secure an elevated situation for that purpose which they could not obtain by attempting to rise from the ground. Dropping thus upon their victim, they force it to the earth, where it is instantly seized and torn in pieces. In this manner the Cinereous Shrike catches small birds, field-mice, and other little quadrupeds. The destruction of these last is an advantage to the farmer, and accordingly we find in many countries this bird is spared and regarded, from this circumstance, and also because it

Bullocks Mus.

Griffith sc.

GEOFFROY'S SHRIKE.

L. PLUMATUS.

London Published by G.B. Whittaker March 1828.

destroys a number of pernicious insects, and never in the slightest degree injures the harvests. According to Gmelin and Latham, the Cinereous Shrike is found in North America. The last mentioned writer declares it to be frequent at Hudson's Bay, where it breeds, making the nest half way up a pine or juniper tree, in April. It is called then *Wapaw*, *Whisky John*, or *White Whiskey John;* and Latham says, it is also found as far south as Georgia, and known by the name of big-headed mocking-bird. But M. Vieillot declares that this is a different species from the *Lanius Excubitor,* and has denominated it *L. Borealis.* His reason for this distinction is, that the bird in question has the first remex rather shorter than the fifth; the second and third equal and the longest of all; while in the Cinereous Shrike the first and fifth are equal, and the second longer than the third. Those found in the United States, retire in spring into the dense forests, and build their nests in the fork of a small tree, composing it of dried herbs and white moss, with plenty of feathers within. The female lays five or six eggs, of a dirty white, or pale ash-colour, marked towards the large end with gray and red stripes.

M. Vieillot states, that the shrike which most approaches to the Lanius Excubitor in the New World is the *Lanius Ludovicianus* described by Brisson under the name *pie-grieche de la Louisiane.* The first mentioned naturalist, however, considers this also as a distinct species, though exhibiting many relations to Excubitor. It differs, however, in the deeper colour of the upper part of the body, and in the beak, which is more robust, and armed with a more decided tooth. The male also has a black forehead. This species is numerous in the southern parts of the United States, and travels in families during the autumn. The Americans call it the butcher bird. It lays five or six eggs spotted with brown.

We have given a figure of *Geoffray's Shrike.* It is of the size of a thrush, with a bill somewhat stout, straight, flat, and hooked at the point, with a slight notch; head crested, with the feather pointing backwards, and white. This species inhabits Africa. M. Le Vaillant seems to think it more analogous to the stares. Our figure is from a specimen which was seen some years ago in Riddle's Museum, Leadenhall-street. How numerous are the instances which our own collection of drawings alone would establish of species which have been of late described and named by foreign naturalists, which have existed and been disregarded in our own country years back!

The shrike, also on the opposite side, is from the Museum in Paris. It belongs to the crested division called Vanga. Its general colour is black, but there is a large indented white patch on the neck, and two white lunated spots, one above and the other behind the eye, and the large wing-coverts are dark-brown.

Our figure, which Major Hamilton Smith refers to the *Lanius Emeria* of Shaw, and the Great Bulbul of India, is of one of these species which, in the present state of arrangement of this countless order, it is very difficult properly to allocate. Mr. Swainson, in his excellent observation on the family of the Laniadæ, or Shrikes, proposes a new genus which would include this bird. "In some species," he says, "of this family, the bill is smaller, the nuchal bristles less conspicuous, and those of the rictus much shorter. We are thus prepared for the transition which here takes place into the genus *Brachypus*, a name by which I propose to distinguish the short-legged thrushes of Linnæus and of modern writers. These birds are exclusively confined to Africa and India, and are so strikingly distinguished from the true thrushes, that it is somewhat singular their peculiarities should not have been noticed long ago. Their tarsi are remarkably short, their bills are weak, and the nuchal bristles scarcely perceptible. In short, it is in this genus that all the habits of the *Edolianæ* gradually disappear, and bring us to a small group of genuine thrushes found in Africa, having lengthened tarsi, a graduated tail, and other characters assimilating to the *Meruladæ*."

C. Hamilton Smith Esq. del.

GREATER BULBUL.

LANIUS EMERIA _ Shaw.

London Published by G.B. Whittaker March 1828.

C. Hamilton Smith Esq. del.

Mus. Paris.

CRESTED SHRIKE.

VANGA CRISTATA_Veil.

London Published by G.B. Whittaker, March 1828

The present bird clearly belongs to the small intermediate group thus pointed out.

The head, neck, and throat are violet-black, with a crest, not long, but inclining forward, and behind the eye is a large subquadrangular red patch; on the sides of the neck and upper part of the breast there are various lunated black and white patches; the wings, back, and upper side of tail-feathers are of a delicate ash colour, with the edge of each feather lighter than the rest; beneath the bird is white, with a slight tinge of ash; the vent is red.

We must now dismiss the shrikes proper; for notwithstanding the number of species, there is nothing more in their conformation, manners, or habits, to entitle them to any further notice here.

The species which compose the genus *Langrayen*, or Swallow Shrikes, are found in Africa, India, and Australasia. Little is known of them beyond their exterior. With long and pointed wings, sometimes exceeding the tail in length, they have the mode of flying peculiar to the swallows, perpetually and rapidly chasing the insect tribes which appear to constitute their principal source of subsistence. According to Sonnerat, they add to this attribute, all the courage of the shrikes, and do not hesitate even to attack the raven. From this the Baron has denominated them Swallow-Shrikes, and *Ocypterus*, from the conformation of their wings.

Of the CASSICANS nothing is known with any certainty except their forms and colours; they are all natives of Australasia and Polynesia.

The BECARDS have many relations with the shrikes and tyrants, and were originally classed by naturalists with the former. But they do not possess the generic characters of the shrikes, as a simple comparison is sufficient to prove. (See Text.) The name *Becarde* was given them by Buffon, from the thickness and length of their bill. Their forms are not so elegant as those of the shrikes, and their body is thicker and longer; they are natives of South America.

To this subdivision seems referable the Spotted Psaris, so named by our respected friend Major Hamilton Smith. The whole upper part of the head is black; the upper wing-coverts and back are cinereous, and the quills and tail are black. The whole under part of the bird is white, which on the chin, throat, breast, and anterior part of the abdomen is spotted, or striped, with dark drop-like patches.

Of the habits of the Choucaris absolutely nothing is known, and the same observation is applicable to the Bethyles, which are termed Pillurion by M. Vieillot.

We pass on to the Tanagers. This genus appears to have been a sort of depot for all the birds with conical and notched bill, which could not conveniently be classed elsewhere until the appearance of M. Desmarest's history of the Tanagers, from which he has justly excluded a number of pretended species, and exhibited no small degree of merit as a classifier. According to M. Vieillot, the only birds which should be ranged under this genus, are the *Tanagers* proper, and the *Euphonian Tanagers.* All the rest should be referred to groups already known, or purposely created. The preservation of the name of tanager to these, says this naturalist, is only calculated to create confusion, and even if the term, which is supposed to signify rich in colours, be applicable to some of them, it is equally applicable to an immense number of others of the feathered race.

The tanagers live on berries, insects, and small grains; they seek their food in thickets, among brushwood, on plants and trees, many of them hopping about on all the branches, in search of insects, like the warblers. Most of the tanagers are remarkable for the richness and brilliancy of their colours; accordingly M. D'Azara gives them a Spanish name expressive of this attribute, *Lindo*, which both in Italian and Spanish means *spruce*, *neat*, *elegant*, &c. But, as we find it to be fre-

C. Hamilton Smith Esq.[r] del.[t]

Mus. Paris.

THE SPOTTED PSARIS.

PSARIS VIRGATA. Hamilton Smith.

London, Published by G.B. Whittaker. March, 1828.

quently the case in the feathered kingdom, this external beauty is not accompanied by any corresponding melody or power of voice: very few indeed of the tanagers possess agreeable notes. Their movements are rapid and abrupt, their flight lively, and their natural disposition active and inconsiderate. They rarely descend to the earth, and when they do, they proceed by jumps, not walking. Some frequent the interior of large forests, where they are attracted by certain berries of which they are extremely fond. Others usually sojourn on the borders of woods, and others in the dry grounds, where they conceal themselves in bushes and briars; others again prefer the summits of trees, and many visit rural habitations, where they frequent the gardens and the meadows. Such species love the society of their fellows, and unite in flocks more or less numerous: others live in families, some in pairs, and some even solitarily. The tanagers which are stationary in the torrid zone, hatch at different seasons, but they lay a smaller number of eggs than the natives of temperate climates.

America is the country of the tanagers; and the greatest number of species are found in the Equinoxial part of that vast continent. Certain authors have imposed this name on birds of Africa, the East Indies, and even the Caucasian mountains; but it is at the least extremely doubtful that they appertain to this genus.

The *Tanagra Canora* possesses an agreeable voice, and is accordingly occasionally kept in a state of captivity. The *Tanagra Striata* frequents rural habitations, and does much mischief in the gardens of Paraguay by destroying leguminous plants, oranges, grapes, and other kinds of fruit. Buffon has given it the name of *Onglet*, from a small concentric groove, exhibited on the lateral facet of each claw.

The *Tanagra Musica* is called, in the districts of St. Domingo, the organist, or musician, because in its song it runs through all the tones ascending from the bass to the treble. It is extremely mistrustful, and escapes the fowler by turning round the branches with extraordinary dexterity.

The *Tanagra Cayana*, called *Dauphinois* by the Creoles of Cayenne, is very common in that country. It inhabits open places, approaches the habitations, and lives on fruits. It destroys the bananas, and gayavas in great quantities; it also carries devastation into the rice-fields in the period of their maturity. It is only, in fact, in the rice-grounds that these birds unite in any numbers; for ordinarily they are seen only in couples. They have no song or modulation of voice, and generally utter but a short cry.

The *Tanagra Tatao*, *Paradise Tanager* of our text, is called *Septicolor* by Vieillot. The figures of Buffon to which we have referred, are, even according to his own confession, defective. The first was taken from a bird dried at the fire, and to which the tail of some other bird was added. The other is from a skin, but badly preserved.

This tanager is about the size of a canary-bird; the bill and feet are black, and the tail a little forked; the wings when folded extend about half its length. Some individuals are handsomer than others, and the colours of the female are, in general, less brilliant than those of the male. The lower part of the back in the male is of a very brilliant red, which the young does not assume until maturity.

These birds, which fly in numerous flocks, appear in September, in the neighbourhood of Cayenne, and in the inhabited portion of Guiana, remain there six weeks, and return in April and May. They are attracted, it is said, at these two epochas by the fruit of a very large tree which they never quit. It is stated that they are never seen on any other trees; an asserted fact, which, to say the least of it, appears doubtful.

The *Turdus Palmarum* is a species, rare in Guiana, but very common in St. Domingo. In this island it abounds in lofty and dense thickets. It also frequents wood, and, notwithstanding its name, does not appear to give one tree the preference to another. Perhaps it received this name in Guiana, from the accidental circumstance of being occasionally seen on the palm tree. It lives on berries and insects.

C. Hamilton Smith Esq.re del. Plymouth.

AZURE TANAGER.

TANAGRA. ?

London Published by G. B. Whittaker, July, 1828.

We insert a figure of a Blue-headed Tanager, with black stripes, from a specimen in Drew's Collection at Plymouth. The bill and legs are black, and the former has a line of deeper black at its base. The head, neck, and breast are azure-blue, with the sides of the neck marked with several lunated black stripes; the belly and vent are white, covered, like the side of the neck, with black patches; the smaller wing-coverts, lower part of the back, and insertion of the tail are like the head; the rest is black. We cannot identify this with any of the described species, though there are several to which it seems to approximate.

The FLYCATCHERS, in general, are of a wild and solitary character. Their physiognomy is sombre and distrustful, and not without a certain expression of ferocity. As they are obliged to seize upon their prey in mid-air, they are almost always perched upon the summit of trees, and rarely descend to the ground. As they are chasers of flies, their true country must be in the southern regions of the globe. Accordingly, for three or four species which are known in Europe, we reckon in Africa, a great number, also in the warm climates of Asia, and Australasia, and still more in America. In this last continent we find the larger species which have been denominated *Tyrants*. As nature has increased the growth, and multiplied the number of insects in the New World, so has she opposed to them enemies more numerous and more powerful. It is a trite observation, but one which the study of nature illustrates at every step, that all in this world is balanced : when evil exists there will always be found some equiponderating good, and it rarely happens that any one species, or genus, is suffered to multiply and extend, to the serious prejudice of another. We see, it is true, every where a great destruction of life, but we also see an equivalent reparation; we must not take a circumscribed or conventional view of the grand operations of nature. What are myriads of lives to that power, which, by a single volition, can call myriads of myriads into existence? To that principle which is itself the perpetual well-spring of all life, and in which, universal creation lives, and moves, and has its being?

We cannot do better here than avail ourselves of the picture drawn by the eloquent naturalist of France, of the advantages derived to man, from all the insectivorous races of the feathered kingdom.

"Without them, without their assistance, vain would be the efforts of man to destroy or banish the clouds of flying insects by which he would be assailed. Innumerable in quantity and rapid in generation, they would invade our dominions, fill the air, and devastate the earth, did not the birds restore the equilibrium of living nature, by the destruction of her superfluous products. The greatest inconvenience of warm climates is the continual torment caused there by the insect tribes. Man and the quadrupeds cannot defend themselves against them. They attack with their stings; they oppose the progress of cultivation, and devour the useful productions of the earth. They infest with their excrements or their eggs, all the provisions which are necessary to be preserved. Thus we find that the beneficent birds are not even sufficiently numerous in such climates, where, nevertheless, their species are by far the most multiplied. How happens it, that in our temperate climates we are more tormented with the flies in the commencement of autumn, than in the middle of summer? Why in the fine days of October do we see the air filled with myriads of gnats? Because all the insectivorous birds, such as swallows, nightingales, warblers, &c., have deserted us. This short lapse of time, during which they have too prematurely abandoned our climate, is sufficient to cause us to be more incommoded with the multitude of insects, than at any other season. What then must be the consequence, if, from the moment of their arrival; if, during the entire summer; if, in short, for the whole time of their sojournment among us, we continue to make their destruction a source of amusement?"

Without pursuing the order of the text, we shall here notice what is most remarkable in the different groups and species.

Among the Flycatchers proper, *The Spotted Flycatcher* (*Grisola*), arrives in France in spring; inhabits forests, orchards, &c., and prefers sheltered and shady haunts. It subsists on winged insects which it seizes in its flight. Its life is solitary; it has a sombre melancholy air, expressive of a sort of stupid inquietude; but it flies lightly, and its general movements are brisk. It seizes its prey by a quick and sudden turn, and rarely misses the insect which it marks as its victim. Its favourite prey are the diptera and tetraptera; but it seldom attacks the coleopterous insects. According to Latham, this flycatcher is also frugivorous, and destroys an immense quantity of cherries. In Kent they call them *cherry-suckers* from this circumstance.

This species nestles indiscriminately in trees and bushes, and most frequently in the hollows of trees and the holes of old walls. It both constructs and conceals its nest equally ill; the materials which it employs are moss, fibres, hair, and wool. The number of its eggs is four or five, white, and marked with reddish spots. The male and female partake equally the cares of incubation. As any degree of cold which banishes the winged insects, deprives these birds of the means of subsistence, they depart for the south before the first setting in of cold weather, and they are never seen in France after the end of September. Aldrovandus, indeed, says that they do not emigrate; but this can only be understood as referable to Italy, and other warmer climates. They are numerous in the southern parts of Europe, but rare in the north. According to Latham they are common enough in southern Russia.

The *Muscicapa Azurou* is found in the country of the Great Namaquois. The cry of the male may be expressed by the syllables, *piet*, *piet*, *pieret*, *pieret:* these birds build their nest on the mimosas, construct it in a furcation, and attach it solidly to the branches which surround it. They compose it of stalks of the liana, turned with much art, and give it a very considerable depth. The female lays five or six eggs of an olive green colour with red points. These points are greatly multiplied towards the large end, where they form a kind of zone.

The *Muscicapa Pristinaria* is another African bird, which the colonists at the Cape call *Molinar* (the Miller) from a fancied resemblance between the song of the male, and the sound of a handmill used in this colony for grinding corn. Its cry may be thus expressed: *gre r r r r r r r r rar, gre r r r r r r r r rar, gre r r r r r r r r rar.* This sound it utters without interruption wherever it is found, and thus reveals the place where it is concealed. Without this noise it would be difficult to discover it, as it remains constantly in the thickest bushes. This species is very numerous in the neighbourhood of the river of *Uywenhoc.*

The *Muscicapa Aëdon* is remarkable for the sweetness of its song. It is true, that this is not the only flycatcher which has been remarked for this attribute, and received the epithet of musical. It is, however, at the least questionable, if all the birds so called, are in reality belonging to this genus. The bird in question inhabits the rocks and vallies of Oriental Tartary; and Pallas, to whom we are indebted for its discovery, informs us that it sings during the night in a strain not inferior to the nightingale. This last mentioned bird is not found in the same country.

We shall give the substance here of M. Vieillot's observations on the *Black Flycatchers* of Europe, as we think them of importance towards the discrimination of species.

There are few birds which have occasioned, and do still occasion, more mistakes than those which, in the same year, exhibit different liveries, or whose colours vary in each season. In many systems of ornithology we find the same species repeated two or three times, as distinct ones, in consequence of the male assuming many dissimilar plumages, both before he is clothed in the covering of maturity, and after he has quitted it. This occurs even among the aquatic birds, both waders and web-footed, and still more in the order now under our survey. Among the sylvan birds it has, indeed, been doubted, whether a double moulting ever takes place; but the males of a great number of European species, have, in spring and summer, different colours from those which they bear in autumn and winter. With some they pass from an obscure shade, to tints of deeper brilliancy; while in others a perfect contrast takes place. This last is the case with our black fly-catchers of Europe: the gray tint of their wintry plumage changes in spring, first to a pale black, and finally to a lustrous black on the upper parts. The white of the under parts grows more pure, and finally assumes a snowy brilliance; all this takes place without any fresh moulting. This metamorphosis, and the very different livery of the females and the young, have given rise to the creation of spurious species.

We take the present opportunity of extracting the opinions of Dr. Fleming concerning these changes in the colour of the clothing of animals. The observations in question, originally appeared in the Edinburgh Encyclopædia, but we borrow them from the Doctor's excellent work 'The Philosophy of Zoology,' a work which, for cautious induction, close thinking, and sound and comprehensive views, is assuredly unparalleled in our language by any production on the same subject. We are aware that no praise of ours can enhance its merits; but it would be ungrateful to withhold our acknowledgments of the pleasure and profit which we have derived from its perusal.

"It has been supposed by some, that those quadrupeds which, like the alpine hare and ermine, become white in winter, cast their hair twice in the course of the year: at harvest when they part with their summer dress, and in spring when they throw off their winter fur. This opinion does not appear to be supported by any direct observations, nor is it countenanced by any analogical reasonings. If we attend to the mode in which the hair on the human head becomes gray as we advance in years, it will not be difficult to perceive that the change is not produced by the growth of new hair of a white colour, but by a change in the colour of the old hair. Hence there will be found some hairs pale towards the middle and white towards the extremity, while the base is of a dark colour. Now, in ordinary cases, the hair of the human head, unlike that of several of the inferior animals, is always dark at the base, and still continues so during the change to gray: hence we are disposed to conclude from analogy, that the change of colour in those animals which become white in winter, is effected, not by a renewal of the hair, but by a change in the colour of the secretions of the rete mucosum, by which the hair is nourished, or, perhaps, by that secretion of the colouring matter being diminished or totally suspended.

"But as analogy is a dangerous instrument of investigation in those departments of knowledge, which ultimately rest on experiment or observation, so we are not disposed to lay much stress on the preceding argument which it has furnished. The appearances exhibited by a specimen of the ermine now before us, are more satisfactory and convincing. It was shot on the 9th of May, 1814, in a garb intermediate between its winter and summer dress. In the belly and all the under parts, the white colour had nearly disappeared, in exchange for the primrose-yellow, the ordinary tinge of those parts in summer. The upper parts had not fully acquired their ordinary summer colour, which is a deep yellowish-brown. There were still several white spots, and not a few with a tinge of yellow. Upon examining those white and yellow spots, not a trace of interspersed new short brown hairs could be discerned. This would certainly not have been the case, if the change of colour is effected by a change of fur. Besides, while some parts of the fur on the back had acquired their proper colour, even in those parts numerous hairs could be observed of a wax-yellow, and in all the intermediate stages, from yellowish-brown, through yellow, to white.

" These observations leave little room to doubt that the change of colour takes place in the old hair, and that the change from white to brown passes through yellow. If this conclusion is not admitted, then we must suppose that this animal casts its hair at least seven times in the year. In spring it must produce primrose-yellow hair, then hair of a wax-yellow, and, lastly, of a yellowish-brown. The same process must be gone through in autumn, only reversed, and with the addition of a tint of white. The absurdity of this supposition is too apparent to be further exposed.

" With respect to the opinion which we have advanced, it appears to be attended with few difficulties. We urge not in support of it the accounts which have been published, of the human hair changing its colour during the course of a single night; but we think that the particular observations on the ermine warrant us in believing, that the change of colour in the alpine hare is effected by a similar process. But how is the change accomplished in birds?

" The young ptarmigans are mottled in their first plumage, similar to their parents: they become white in winter, and again mottled in spring. These young birds, provided the change of colour is effected by moulting, must produce three different coverings of feathers, in the course of ten months. This is a waste of vital energy, which we do not suppose any bird in its wild state capable of sustaining, as moulting is the most debilitating process which they undergo. In other birds of full age, two moultings must be necessary. In these changes the range of colour is from blackish-gray, through gray to white, an arrangement so nearly resembling that which prevails in the ermine, that we are disposed to consider the change of colour to take place in the old feathers, and not by the growth of new plumage, this change of colour being independent of the ordinary annual moultings of the birds.

"Independent of the support from analogy which the ermine furnishes, we may observe that the colours of other parts of a bird vary according to the season. This is frequently observable in the feet, legs, and bill. Now, since a change takes place in the colouring secretions of these organs, what prevents us from supposing that similar changes take place in the feathers? But, even in the case of birds, we have before us an example as convincing as the ermine already mentioned,—it is a specimen of the little auk (*alca. alle*,) which was shot in Zetland, in the end of February, 1810. The chin is still in its winter dress of white, but the feathers on the lower part of the throat have assumed a dusky hue. Both the shafts and webs have become of a blackish-gray colour at the base and in the centre, while the extremities of both still continue white. The change from black to white is here effected by passing through gray. If we suppose that, in this bird, the changes of the colour of the plumage are accomplished by moulting, or a change of feathers, we must admit the existence of three such moultings in the course of the year: one by which the white winter dress is produced, another for the dusky spring dress, and a third for the black garb of summer. It is surely unnecessary to point out any other examples in support of our opinion on this subject. We have followed nature, and our conclusions appear to be justified by the appearances which we have described."

This has been the reason why we find some confusion and diversity among naturalists, in classifying the flycatchers of which we speak. In Brisson and Buffon we find them marked under these names, *Gobe-mouche Noir, ou de Lorraine*, *Traquet d'Angleterre*, and *Bec-figue*, as three distinct species. The *Black-collared Flycatcher* in Latham, Gmelin, and Meyer is a variety of that without collar, and the *Becafico* a particular race. Other naturalists make but a single species of these three birds, considering the collared flycatcher as a male in very advanced age. M. Vieillot considers that there are two black flycatchers, which must be separated specifically, or, at least, regarded as two permanently distinct races; the exterior difference between which is, that the male of the one has a white collar on the upper part of its neck, while the male of the other never exhibits any such mark at any age. Both have a covering which varies in colour in the course of the year. At one season they are black and white, at another grayish-brown and grayish-white, at a third their plumage presents a mixture of all these different colours. The white collar which distinguishes one of these races, is apparent only during the season of reproduction, and is merely indicated afterwards by a faint trait of this colour, frequently interrupted by gray; but the feathers which compose it are always white from their base to beyond the middle, while in the males which have no collar these feathers are gray only at their origin, black in the remaining part during the summer, and entirely gray after the moulting. This observation, made by M. Vieillot on a dozen males, has determined him not to unite these two flycatchers, either as individuals or as varieties of the same species. Moreover, the same naturalist has remarked that, in the collared race, male, female, and young, the first remex is longer than the fourth, while in the others it is either a little shorter, or of equal length.

He adds, that differences may also be remarked in their mode of life. These two species or races, are seldom found at the same times, in the same places. In Lorraine, where they have been most minutely observed by the Count de Riocourt, the collared flycatcher alone is seen during the season of reproduction, while the other is at that time only on its passage thither. Moreover, this last is but seldom found in that country, while the former is very common. M. Vieillot says, that the collared flycatcher, on the contrary, is not found in the neighbourhood of Paris, but that the other is frequent enough, and sometimes even propagates there. He has made the same observation in Normandy, in the forest of Lyons, where the last-mentioned bird resorts on his passage in spring, and where some couples remain during the summer. This led M. Vieillot into the opinion, that these two birds did not traverse the same districts in their northward or southward passage. Bechstein also informs us that their disposition and habits are dissimilar. M. de Riocourt has remarked that the collared flycatcher remains constantly during the summer on the top of the highest trees, and watches the insects to seize them on the wing, while the other pursues its prey in thickets, and on the edges of roads: but in rainy weather, and especially during the back season, the first are obliged to seek their food under the bushes, because the winged insects are then rare on the tops of trees.

M. Vieillot confesses that the young and the females of these two races so closely resemble each other, that it is almost impossible to avoid confounding them, without having regard to the proportions of the first and fourth quills of the wing. He particularly instances females, as he has verified this fact on individuals of that sex taken on the nest.

The males of these two races, with the exception of the young before the first moulting, do not differ from the female in the after season, except by a tint of gray, something more brownish, and not all shaded with red on the upper parts; also by their wings and tail being of a more blackish-brown. The collared males are then distinguished by the feathers which compose this collar being white almost to the point, as has already been observed. From these details M. Vieillot considers it to result, that France possesses three distinct flycatchers, viz., the flycatcher properly so called, the collared flycatcher, and the black flycatcher without collar. According to Bechstein and Meyer, there is a fourth species in Germany, where it is rare. Sparman declares that there is also a fifth in Sweden, but it has been proved to be a bird of a different genus.

We shall enter into a few more particulars of the two species of which we have been treating.

The *Black Flycatcher*, which we call the *English Flycatcher*, (*M. Atricapilla*, Lath.) nestles in the hollow of a tree or on the thickest branches. There is so great a general resemblance between this bird and the collared flycatcher, which is the pied flycatcher of our text, that it is not surprising that they have been united by most ornithologists. But the collar is certainly a very distinctive attribute, and by no means peculiar to the aged males, as some naturalists pretend. In fact, it is not only seen on them, but also on the young during the winter, which would, otherwise, resemble the bird immediately under our consideration. But this last exhibits at all times another material difference, the first quill of the wing being shorter than the fourth, or equal, while it is always longer than the fourth in the collared or pied flycatcher. These two birds also differ from each other in their disposition and cry. One is distrustful and suffers itself to be approached with difficulty, while the other is so little so, that one may come so near it as to kill it with a stone.

Though Bechstein seems to have been right in separating these two birds, yet it is probable he was in error in considering the becafico as a different species. It is more likely that it is a male of the black flycatcher, as the other is in its winter clothing, or a female, or a young one. M. Vieillot had two females exactly like the becafico. The male of the one had a collar, that of the other none. An additional proof that the becafico, is nothing more than one of these flycatchers, may be found in Aldrovandus, who describes it a second time at the moment of its metamorphosis when he says it was neither the becafico, nor atricapilla, and he, therefore, called it the varied becafico.

The Collared or Pied Flycatcher, is, as we have mentioned, distinguished chiefly by the collar. The winter plumage of the male is the same as that of the female at all seasons, and the female is destitute of the collar. A symptom of collar is often seen on the young males, but very narrow.

When the two flycatchers (this and the last) are in their autumnal plumage, they are known in Lorraine (males, females, and young,) under the names, *Mûrier* and *petit pinson des bois*, and in the southern countries under that of *bec-figue* or *becafico*. They arrive there towards the end of spring in numerous flocks, and disperse in all directions. But during summer they live in pairs only. The pied flycatcher makes its nest in the hollow of a tree, composes it of moss and the hairs of animals. The eggs are three or four in number, of a bluish green, spotted with brown. The male utters a plaintive cry, like that of a pullet. Its song is agreeable and melodious, having some resemblance to that of the red-breast, but is not so well sustained. It may be considered but as a single couplet of that bird's performance. This flycatcher is not destitute of courage, and will dispute precedence not unfrequently with the blue titmouse and other small birds. It attacks with so much impetuosity, that it always remains master of the contested object, which seems wonderful on the part of a bird, whose bill is but weak against those which have this organ more thick and robust. This, however, is a fact which has been verified by M. de Riocourt in the forests of Lorraine.

Buffon, in noticing the various liveries of the pied flycatchers, says that the autumnal or winter plumage of the male does not differ from that of the female, and that it then resembles the *mûrier*, vulgarly called *petit pinson des bois*. He adds, that in the second state, when these birds arrive in Provence, the male is altogether like the *bec-figue*. This statement would lead one to imagine that these two liveries were different, seeing that the author makes a distinct species of the *bec-figue*. But the fact is, that this second state is exactly the same with the first, the *bec-figue* being nothing else than the *mûrier*, or *petit pinson des bois*, as Buffon himself actually assures us in the same article.

We do not find, amid the multiplied species of the flycatcher, anything more worthy of the attention of the readers of this

C. Hamilton Smith Esq.^e del.

Drew's Collection.

LUMACHELLI QUERULA.

QUERULA LUMACHELLI.

London. Published by G.B. Whittaker. March. 1828.

part of our work than what we have now presented to them. We shall, therefore, pass on to the COTINGAS, under which head we shall notice the common *Gymnocephalus* of the Baron.

This genus of birds, which, under the Latin denomination of *Ampelis*, is composed of eleven species in the thirteenth edition of the *Systema Naturæ*, by Gmelin, and of fourteen in the *Index Ornithologicus*, of Latham, now forms a more extended family, divided into six sections, viz., the *Piauhaus*, the *Common Cotingas*, the *Echenilleurs*, the *Jaseurs*, the *Procnias*, and the *Gymnoderes;* all of which have the bill depressed like that of the flycatchers, but a little shorter in proportion, tolerably broad, and slightly arched. The *Piauhaus*, thus named on account of their cry, and well designated under the Latin word *querula*, are those which have the bill most pointed; insects constitute their principal aliment, and they hunt their prey principally in the woods.

Of these birds we insert the figure of one under the name of *Lumachelli Querula*, from a specimen in Drew's Collection, Plymouth. The head and upper part of the back are black and green with a metallic lustre; the lower part and tail are black; the wing-coverts are partly blue and partly brown; and the epaulette is composed of distinct red, blue, yellow, and green spots.

The common cotingas, properly named *ampelis*, have the bill rather weaker, and besides insects, they search out in humid places, berries, and tender fruits. M. Le Vaillant even pretends that they are wholly frugivorous. The *procnias*, under which name Illiger forms a distinct genus, and which was first given by Hoffmansegg, have the bill weak, depressed, and slit even to below the eyes. They are also distinguished by caruncles on the forehead, or a naked skin under the throat, and their regimen is more particularly insectivorous. The gymnoderes, of which but a single species is known, have rather a stronger bill than the last: the neck exhibits naked parts, and the head is covered with feathers. The species belonging to these four sections are found in South America.

The Echenilleurs, (*Ceblephyris*, Cuv.) and the Jaseurs (*Bombycilla*, Br. or *Bombycivora*, Temm.) are known by other characters very remarkable, but taken from parts different from those on which the distinction of genera is usually established. The first have the stalks of the uropygial feathers a little prolonged, stiff, and piquant; and with the second the end of the stalk of the secondaries of the wing enlarges into an oval and smooth disk. The former live in Africa and India, and are insectivorous; the latter feed on berries. The species which is most extended is erratic, and traverses in flocks the different countries of Europe.

We shall treat at present of the four first-mentioned sections, in which we shall include the gymnocephalus. The characters most generally applicable to the birds comprised in them are, bill more or less depressed, from the upper to the under part, widened at base, and presenting a form almost triangular; upper mandible narrow and curved at point; lower one a little flattened underneath, with sharp point; nostrils very wide, almost orbicular, situate at the base of the bill, half closed by a membrane, and covered with silky hairs or feathers; tongue short, cartilaginous, narrow and bifid; wings moderate; tail composed of a dozen feathers; tarsi reticulated, three toes in front, the external joined as far as the second phalanx; thumb as long as the middle toe and more strong.

There are among the cotingas some species, whose plumage exhibits nothing very remarkable, and others in which it is even very dull except at the season of reproduction. But at this period many among them display a profusion and variety of the most brilliant and dazzling colours. Such species constitute a principal ornament of most collections. America is the only part of the world in which they are found; nor do they extend beyond Brazil to the South, nor beyond Mexico to the North. The cotingas, however, are not sedentary; but the only object of their little voyages is to arrive in certain places at the epoch in which the fruits they subsist on are mature. In Guiana, the spots in which they most delight, in those seasons when they are seen near habitations, are humid places. It is an error to suppose that they are destructive to the rice-grounds. From the peculiar conformation and absence of solidity in their bill, it is impossible that they can be granivorous birds. According to Sonnini the inhabitants do not eat their flesh, and if the stuffed specimens often arrive in Europe in a bad state, this is not the reason. It is rather, because the feathers not being very adherent, the tender skin requires a degree of care in its preparation, which is not always bestowed upon it in America. The size of the cotingas varies from that of the raven to that of the song thrush. The colours of the females are, in general, much less rich than those of the males; their plumage is, indeed, frequently dull and dusky. The habits of these birds and the facts concerning their reproduction are very imperfectly known; many species, however, are known to make their nests on the loftiest trees, and lay four or five eggs. Mauduyt, in the 'Encyclopédie Méthodique,' testifies his surprise that no attempt has been yet made to bring those beautiful birds alive to Europe. He thinks this might be done by substituting for the berries, which constitute their ordinary food, crumbs of bread moistened, sap of the sugar-cane, and even half melted and softened sugar. But the probability is that this plan would not succeed, as the great majority of these birds are both insectivorous and frugivorous, and it is very likely that such experiments have been made in their native country without effect, as they are never seen there in a state of captivity.

Among the cotingas, the most remarkable is one belonging to the division of procnias; the *Caranculated Chatterer*, Lath. *Ampelis Carunculata*, Gml. This singular species, says M. Le Vaillant, is known at first sight by a sort of feathered caruncle which it has on the forehead (not on the beak, as Buffon avers.) This caruncle, the nature of which is muscular, rounded, and altogether wrinkled, hangs negligently and indifferently from one or the other side of the beak at its base. Buffon assures us, that this bird has not only the faculty of elevating this caruncle, but also, that when the bird is animated by any passion, the caruncle swells, is elongated, and rises perpendicularly by means of air introduced through an aperture wrought in the palate, and corresponding with the tube of the caruncle, where the bird can retain the air. This error of Buffon is the less surprising, as all the carunculated cotingas which he saw were prepared in such a way, as to lead directly to this supposition; in fact, the preparers of birds in Cayenne, from which all the specimens of this species in the French cabinets came, are accustomed to run a small stick, or an iron wire, forcibly through the palate and cranium of these birds, into the caruncle, for the purpose of keeping it upright. Buffon supposed this part to be hollow naturally, whereas it is only made so by art. M. Le Vaillant verified this point on an individual brought entire from Surinam in spirits of wine. In cutting it in two he found the caruncle of this ampelis was precisely of the same nature as that of the turkey, with this only difference, that it is covered with small, convex, rounded, and stiff feathers. This gives to this part, when elongated and erect, the appearance of those fine branches of madripore, which are covered with small white shells, and may be seen in many collections. We are even ignorant, says M. L., if this bird possesses the faculty of erecting this part at will, or if, like the caruncle of the turkey-cock, it is only capable of elongation. It is possible, that the muscles of which it is composed may produce either effect; but it is very certain that there is no communication between the palate and the caruncle, which is situated precisely at the origin of the forehead. There is even in this place a slight sinking, and the upper part of the frontal bone is furrowed, throughout its entire length, by a cavity which appears to divide it into two equal portions. This is

C Hamilton Smith Esq. del
Plymouth Mus.

Griffith sc.

BLACK-HEADED CHATTERER.
AMPELIS MELANOCEPHALUS.

quite perceptible by passing the finger along the bird's head. This cavity may be destined to receive the caruncle in its horizontal elongation; if so, it would appear that, instead of erecting itself perpendicularly, it only extends and lies along the head.

In the individual examined by M. Le Vaillant, the caruncle was of a conical form, almost ten lines in length, and four of circumference in the base, and terminating in a point. It could be drawn out nearly two inches. In its natural state the feathers touched; but, drawn out thus, they left a space between them. The plumage of this bird, in its perfect state, is of a dazzling white over all parts of the body.

The Black-headed Chatterer does not appear to have been figured. We presume the opposite bird to be the same species as that described by Prince Maximilian, under the name of *Procnias Melanocephalus*, which M. Temminck refers to his genus *Casmarhinchos*.

The specimen here figured is in the Museum of the Athenæum* at Plymouth. The head, neck, and throat are entirely black; the anterior part of the back is lightish green-yellow; across the middle of the black, that colour assumes a darker shade, but it again becomes light as it approaches the tail; the wing-coverts are nearly black with a yellowish edge to each feather; the tail-feathers are dusky green; as is also the whole lower part of the bird, though a shade lighter than the tail.

Prince Maximilian's *Procnias* is said to be yellowish green underneath, with darker transverse stripes, which do not appear in the present specimen. In all other respects they appear to be the same.

* We cannot pass by the present opportunity of bestowing a word, however humble, in commendation of provincial societies, similar to the Athenæum at Plymouth. The sciences, especially those which are grounded essentially on observation, are materially assisted by local exertions; while the members of such societies have an honourable object worthy the attention of liberal minds, while disengaged from the necessary avocations of life.

The *Gymnocephalus* is about the size of a crow, and is remarkable for nothing so much as its naked head, and neck not much furnished with feathers. Our author, finding an analogy between its bill and that of the tyrants, has placed it at the end of the flycatchers; but M. Le Vaillant considers it as belonging essentially to the cotingas, by its bill, feet, and identity of habits. The amplitude of its wings has been considered a distinctive character from the cotingas. This M. L. says, is only apparent, arising from the shortness of the tail. Illiger has also placed it in the genus ampelis; but, says M. Dumont, if the head is feathered in early life, (which M. Le Vaillant himself affirms;) if the nostrils are then covered like those of the great cotinga, to which this naturalist approximates it; and if the nudity of the head in age be owing to some peculiar habit and circumstances, resembling those which produce a similar despoliation in crows, it may be necessary to remove it from the cotingas, to which, moreover, its mode of subsistence is not very conformable.

The Jaseurs are classed by Latham and Gmelin with the cotingas of Brisson. M. Vieillot makes a distinct genus of them, and they form a sub-genus of the cotingas in the 'Règne Animal.'

Of the two species, with which we are acquainted, one inhabits Europe, the other America; they are erratic birds, and travel in numerous flocks, but remain in pairs only during hatching-time. They are so extremely fond of the society of their own species, that from the moment the young can provide for themselves, all those in the same district unite and form very considerable flights. They are baccivorous birds; all kinds of berries suit them, but they prefer soft fruits full of juice. When such food is rare, they live on insects. They will take flies on the wing with as much address as the flycatchers. The American species nestles on trees; the hatch consists of four or five eggs; they lay usually twice a year. The mode of propagation in the European species is unknown.

The jaseurs of Europe are erratic, and authors are not agreed as to the native country of the species. It has been supposed that it inhabited Bohemia, and it has received a name from thence ; but it only takes that country in its passage, as it does many others. It is ranged among our birds, though but rarely seen here.

It is occasionally observed in France, but only in the depth of the severest winters. These birds, according to Latham, appear in great numbers in the neighbourhood of Edinburgh in winter, and disappear in spring. They frequent Italy, but rarely at present, though formerly they used to arrive there in considerable flocks. They pass in great numbers through the various countries of Germany, but do not remain there during the summer. It is not exactly known in what country they nestle. Some say in the neighbourhood of St. Petersburgh. Linnæus assumes that they breed in countries to the north of Sweden ; but we have no details whatever on this subject. The jaseurs (so we must call them, as the word chatterer is applied to all the cotingas) do not always follow the same route in their migrations, nor do they visit the same countries every year. They are generally seen in the same places but once every three or four years, and sometimes there are intervals of even six and nine years between their visits. This species is spread even through Siberia and other northern climates of Asia, and is very numerous in those regions. Berries, grapes, and other fruits constitute their food. This bird, however, is not nice, and is very much prone to gormandize. It will eat all kinds of insects; but will never touch grain, unless it is pounded. It soon grows accustomed to the cage, and does not appear to regret its liberty for the first few months; but when the fine weather approaches, it grows uneasy, and, if it cannot escape, soon dies of ennui and disgust.

Except during hatching time, the jaseurs of Europe love society, and unite in great flocks during the winter and part of the spring. Those seen alone at these periods, are birds which have lost their way. Being of a stupid nature, they allow themselves to be approached easily, and give into all kinds of snares. There is scarcely any bird more silent, which renders both its French and English names somewhat ludicrous misnomers. But this is nothing new in ornithology. It only utters from time to time a futile cry, as thus: *zi, zi, zi.* Prince Anersperg says, that it has a very agreeable song during the love season; but it is quite certain that the American jaseur, or chatterer, has no such thing at any time.

Both the American and European species are considered good for eating.

The DRONGOS of Africa, observed by M. Le Vaillant, live in society, and assemble towards the decline of day. They are very turbulent, and utter piercing cries. They live on insects, and principally bees, whence they are denominated by the colonists of the Cape, *bey-vreter* (*bee-eaters*), and by those who are witnesses of their nocturnal meetings, without knowing the cause, they are called *deywels vogel,* (*devilish birds.*) They nestle on trees, and lay from four to five eggs. Drongos are also found in various parts of India, which, having the same external characters with the African drongos, have probably the same habits. It is useless to dwell on them any further, as we can add nothing interesting to the details of the text.

We now come to the great genus of the THRUSH. There are two natural divisions in this genus, designated by the Latin names *Merula* and *Turdus,* and in French, *Merles* and *Grives.* To the former of these we shall give the English appellation of *Blackbird,* to the latter that of *Thrush,* properly so called. A third division has been formed of the *Mocking-birds,* in French *Moqueurs* *.

* In thus giving the familiar denomination of a species to a group, or subdivision of animals, we only follow the system pretty generally adopted by naturalists, and particularly by French naturalists at present. It is not our business, as humble cômpilers, to attempt innovation or reform;

Though the plumage, and even many of the habits, of these birds, present remarkable differences, there are no essential

but we cannot avoid observing that this system is by no means unobjectionable. It is very well calculated to create confusion in the mind of the student of natural history. As long as the Linnæan system of division was adhered to, there could be no confusion in this way. The name of a species might safely be given to a genus, the species itself being properly distinguished by a peculiar epithet; but when naturalists saw the necessity of creating sections and subdivisions in the Linnæan orders and genera, it would have been as well if they had also seen the necessity of characterising such groups, not by trivial, but by scientific names—not by names formed from their vernacular and fluctuating idioms, but by names taken from those languages, which long prescription, intrinsic excellence, stability, and universality among scholars have consecrated to the use of science. The contrary practice has arisen from an overweening national vanity, which it would be flattery to excuse as patriotism—from that aspiration after universal empire, which should receive from the nations of Europe as effectual a check in the scientific, as it has experienced in the political world. The observations of Mr. Vigors on this subject are so admirable to the purpose, that our readers will thank us for transcribing them:

"This attempt at superseding the use of scientific names, by the introduction of French names, is beginning to be carried to an extent, which leaves no doubt of the ultimate object in view. In almost all professed works of science, it is the French word that is quoted, and not the scientific. In the very 'Dictionnaire[1] before us, the same language furnishes the title of every article to which we are to refer, whether belonging to a genus or a species: it is *Perroquet* we must consult, not *Psittacus*. The French word is everywhere the protagonist of the piece, and if the scientific name is at all introduced, it is in the character of an humble companion in the suite of synonymes. If this practice is not met by us with decided opposition in the outset, it will gain a head, against which we shall in vain endeavour to contend. I do not oppose this mode of nomenclature on the narrow ground of every language having an equal right with the French to become the language of science, but upon the broad principle, that there should be but one common language in science —that every nation should unite in one universal mode of nomenclature which could be generally understood—and that naturalists should endeavour to imitate the harmony observable throughout the objects they cultivate, by the only means in their power, however humble these may be—a corresponding harmony in their language. In choosing this common language, it is unnecessary to contend for the superior claims of that which is founded on classical authority. Time and science have equally

[1] 'Dict. des Sciences Naturelles.

ones in those parts of the body from which generic characters should be derived. From the time of Linnæus, the blackbirds, thrushes, and mockers have been comprised under the common denomination of *Turdus*. Their usual aliment consists of berries, insects, and worms. The bill, in general, is of equal breadth and elevation at the base, and afterwards laterally compressed; the upper mandible is convex, and sloped inwards towards the point, which is curved, without forming a crotchet, or being notched so decidedly as the laniadæ. The lower mandible is straight; the nostrils are ovoid, partly covered with a naked membrane, and situated near the origin of the beak; the angles of the mouth are furnished with hairs at intervals, the alignement of which is compared by Meyer to that of the teeth of a rake; the tongue is cartilaginous, and cleft at its extremity; the tarsus is longer than

sanctioned the use of it. No modern terms, however important to the nation which furnishes them, could be otherwise than trivial, and even ludicrous, in the eyes of others, in comparison with words derived from a Greek or Roman source. The contentions that so frequently break out among the chief introducers of these familiar terms, sufficiently proves the instability of the foundation on which they wish to erect their nomenclature; and it certainly is from no blind partiality that I would bestow a preference on such words as *Plyctalaphus*, *Macrocercus*, *Pezaporus*, or even *Palæornis*, over such names, although sanctioned by the pen of a Buffon, as *Crick*, and *Papeguis*, *Perruches*, and *Perriches*."—'Zool. Jour.' No. ix. Jan. 1827.

Had it suited the purpose of Mr. Vigors, he might have remarked further, that French writers carry this rage for *Gallicising* into almost all subjects, as well as natural history. Nor is it entirely the growth of the present day, though it has latterly assumed an alarming luxuriance. It is a long time since the French have travestied all the proper names of classical antiquity. In anatomy and comparative anatomy they have translated literally into French the Latin terms, which sometimes produces an effect sufficiently ludicrous, as, for instance, *crura cerebelli*, "*Jambes de la cervelle*," &c. It is at all times a serious impediment to the foreign student, desirous of availing himself of their works. Even when they are forced to use the scientific term in the singular number, they take care to Frenchify it as far as possible, by adding an *s* to form the plural. All this absurdity would not be worth remarking, but for the serious impediment which it opposes to the extension of science.—E. P.

the intermediate of the three front toes, and has the exterior toe cemented to its base; the internal toe is free; the first remex very short, and the others variable in their respective length.

In consequence of the disposition of colours in the plumage of these birds, Montbeillard made a separation of the thrushes from the blackbirds, as in the former the breast is dappled, or speckled (in French *grivelé*, whence the name *grive*); and in the latter, the colours are either uniform, or distributed in large masses. Among the first, the sexes offer but few differences; among the second they are much more marked. The moulting, which appears to be generally simple, also occasions some changes in the spots and bands; but this effect takes place in both families. Relatively to manners and habits, the thrushes proper are, in general, erratic birds, and, when they emigrate, form numerous assemblages, especially the red-wing and field-fare. The blackbirds, on the contrary, live generally isolated, or in families, and are so sedentary that they will not quit their peculiar districts, where, if they are not disturbed, they will nestle every year, and not unfrequently on the same bush or tree, and even repairing the old nest, when it is not too far gone. If they do remove a little, according to the season, it is only to descend from the mountains into the plain, or to pass from a place become too dry, and destitute of fruits, into some neighbouring spot where fruits and water are more abundant. Some naturalists set down as a mark peculiar to the blackbirds only, the vertical motion of the tail up and down, which is very frequent with them, and almost always accompanied with a trembling of the wings, and a short interrupted cry. This, however, has also been observed with the field-fares, particularly those of Canada, whose cry then resembles that of the common blackbird.

The order in which Montbeillard has described the birds of this genus, is, first treating of the thrushes proper, and mockers, and then of the blackbirds. M. Vieillot has divided the great genus *Turdus* into three sections, the first of which is devoted to the thrushes proper, the second to the blackbirds, and the third to the mockers. M. Temminck, in the first edition of his 'Manual of Ornithology,' divided the birds of the same genus into three sections, according to their manners and habits, under the denomination of *Sylvains*, *Saxicoles*, and *Riverains*, (woodland, rock, and river-haunting birds.) Those of the first section, nestle and live in woods, bushes, parks, gardens, emigrating in troops, and subsisting almost entirely on berries, except at the epoch when they are bringing up their young, in which their principal aliment consists of insects. Those of the second section inhabit precipitous cliffs, and the rocky portions of the highest mountains, in the clefts of which they live in solitude, and have thus some relations with *Saxicola*, but differ from that sub-genus in the colour of the caudal quills, the majority of which are red, and the two intermediate ones black, while the tail of the true saxicola, most generally exhibits large masses of white. Those of the third section do not quit humid places, and live among reeds, and their nourishment principally consists of flies and aquatic insects. This last section comprehended the *Turdus Arundinaceus* of Linnæus; but MM. Meyer and Cuvier, considering that these river-birds exhibited more relations with the numerous species of sylvia which inhabit the water side, have united the last-mentioned species to sylvia; and M. Temminck, in imitation of them, has suppressed his third section.

Turdus and *Sylvia* present in their general attributes so much analogy that it is scarcely possible to trace between them a line of distinction. We accordingly find that many naturalists range with turdus species which others class with the sylvia and motacilla of Linnæus. The *turdus coronatus* of Latham is, for instance, a motacilla with Gmelin, and the *turdus triochos* of Gmelin is a sylvia with Latham. The passage of one genus to another is so nearly imperceptible, that it is next to impossible to draw the line. "A spotted warbler," says M. Vieillot, "is to my eye nothing but a thrush in miniature." From all this, it really appears that one of these two genera must be purely artificial, since we can pass from one to the other without being enabled to seize any tangible point of difference between them. The same is the case with the *loxia* and *fringilla* of Linnæus, and with a great many other genera, as the Baron has most clearly proved in the 'Règne Animal.'

If we consult again the habits, manners, and instinct of the birds which compose the meruline group, we shall find many which do not differ in this respect from sturnus. Among others, we may particularly remark this affinity in the African species, described by M. Le Vaillant.

Thrushes proper. In all systems of ornithology the thrushes and blackbirds have been united in the same genus, according to the generic characters common to both. Montbeillard, as we have above observed, has divided the genus into two families. His remark on this occasion is worth notice. "The generality of mankind," says he, "appear to me to have acted more wisely than naturalists in giving distinct names to things that are really distinct." The French name *grive* has, then, been properly used to distinguish the birds of this genus which have the plumage marked with spots pretty regularly disposed.

Four species of the thrush live in our climates: the thrush properly so called, the missel, the redwing, and the fieldfare. The two former pass the entire year in France, and also in the southern parts of this country. They have a very agreeable song, especially the thrush proper, which is also called the song-thrush. Dr. Latham seems to think that this bird shifts its quarters in winter, in the North of England and Scotland. It probably leaves the country, or retires to the thick and solitary woods. Both these species are distinguished by never uniting in flocks for the purposes of migration. Their plumage has many traits of conformity in colour and distribution.

The redwings and fieldfares seldom appear among us until autumn, remain during the winter, and live in large flocks. They scarcely ever nestle here, and depart in spring, as they arrived in autumn, in numerous assemblages. As they quit us at the epoch of pairing, we are not acquainted with their love-notes. Often, previously to their departure, they are heard chirping all together, but in this loud noisy concert it would be vain to seek for harmony.

In all the species the males and females are of the same size, and their livery is pretty similar. The colours, however, are more lively and better defined in the males. Berries, fruit, and insects constitute the food of all. To these aliments they join earth-worms, in the pursuit of which they are observed to be very eager after rain. They also feed on snails, which, during winter, they seek in those places most exposed to the sun.

Their flesh is excellent for eating, especially that of the thrush, and the redwing when fat. In the vintage time, in the southern countries, it especially acquires that delicacy and exquisite flavour which occasion this small game to be much sought after by gourmands. Among the Romans it was in high esteem. It is said to possess qualities which, if real, should render it still more estimable. It excites, say its eulogizers, the appetite, fortifies the stomach, improves the juices, and is easy of digestion. It is, therefore, considered as peculiarly wholesome for convalescent subjects. It never produces any bad effect, provided it be not eaten to excess. It has been also thought in medicine to be an excellent anti-epileptic; this quality it is said to derive from the bird feeding on mistletoe, to which the same virtue has been attributed.

It may not be unamusing to our readers to notice the manner in which the Romans, with whom thrushes held the first rank among the feathered game, preserved these birds throughout the entire year, and fattened them in their extensive aviaries.

Each of these contained many thousands of thrushes, blackbirds, and other birds good for eating. They were so numerous in the neighbourhood of Rome, that thrush's dung was employed as manure to fertilize the land. It was also employed to fatten oxen and pigs. The thrushes were kept very closely confined, and considerably crowded. But their food was abundant and well chosen, and they grew fat rapidly. These aviaries were vaulted pavilions, furnished within with a great quantity of roosts. The doors were very low, there were but few windows, and always so turned, that the prisoners could see neither the woods nor country, nor even the birds which hovered outside, so that nothing might hinder them from growing fat. They were only left as much light as was necessary to enable them to distinguish what they chiefly wanted. They were fed with millet, which was peeled and pounded and formed into a kind of paste with bruised figs and flour; besides which they received berries of the mastick-tree, of myrtle, and of ivy, and every thing which could render their flesh succulent and high flavoured. A small rivulet of running water traversed the aviary, for them to drink from. Those which were intended to be eaten in succession, received for twenty days before they were taken for that purpose an augmentation of the best nutriment. Particular care was taken to make such as seemed fit for the table pass very quietly into a particular place which communicated with the aviary, and they were not taken until the communication had been closely shut, to prevent the others from being disturbed. To make them support their captivity with greater patience, the aviary was carpeted with green branches, and fresh turf, often renewed, and in fact, the better the proprietor understood his own interests the better the birds were treated. This method succeeded almost invariably in taming birds, however recently they might have been imprisoned. Those, however, which had been newly taken were kept for some time in small separate aviaries; and the better to accustom them to captivity, they were given as companions those who had been already habituated to their prison.

The Roman poets mention these thrushes in many places. Horace declares a thrush to be a very appropriate present from a legacy hunter to a rich old man:

> " —Turdus,
> Sive aliud privum dabitur tibi; devolet illuc
> Res ubi magna nitet domino sene."

Again he puts the praises of a thrush into the mouth of a gormandizing spendthrift.

> " Cum sit obeso
> Nil melius turdo."

And Martial gives it the first rank among esculent birds, as he does to the hare among quadrupeds.

> " Inter aves turdus, si quis me judice certet,
> Inter quadrupedes gloria prima lepus?"

The fieldfare and the redwing are generally supposed to be the Turdi of the Roman writers.

On the approach of vintage time innumerable flocks of thrushes quit the northern regions of Lapland and Siberia, and their abundance is so great on the southern coast of the Baltic, that Klein assures us that the city of Dantzic alone consumes every year eighty thousand pairs of them. The different species do not all arrive at the same time. The thrushes proper, or the song-thrushes, make their appearance first, then come the redwings, and finally the fieldfares and missels. They stop in various places, especially where they find the most abundant food, and the most easily obtained. They thus continue their route southward, arrive in certain countries sooner or later, in greater or less numbers according to the direction of the winds and the changes of temperature. This is universally the case with all the birds which are driven from the north, by the severity of the weather. Of the migratory thrushes, some nestle in the islands of the Mediterranean, and others continue their course even into Africa. They arrive, Sonnini tells us, in Egypt in the month of October, and do not leave that country until March. They remain at no great distance from habitations, and seek the shades of the orange and citron groves which adorn some districts of lower Egypt. They do not all, however, proceed so far south. Many remain during the winter in our more northern climates, where tolerably numerous flocks of redwings and fieldfares are to be seen during this season. They frequent the meadows, and the green borders of woods, of which they quit the interior.

There are more snares laid, perhaps, for thrushes than for any other birds, and the pursuit of them is very profitable. Those which are most easily taken in snares or nooses are the song-thrush and the redwing. These snares are, as every body knows, composed of a few horsehairs twisted together and forming a running knot. They are set around juniper trees, &c., in the neighbourhood of some fountain or pond. If the snares are properly set, in a well-chosen place, many hundreds of thrushes may be caught in a day, while they are on their passage. Snares are also employed baited with different kinds of berries, and placed along the hedges.

Thrushes are also caught in nets in the following ways.

The *Spider-net* is used, and so called because it envelopes the birds in the same way that spiders entangle flies in their web. As these spider-nets are much used in Italy and the South of France, for catching not only thrushes but becaficos and other birds, we shall give a short description of them. The spider-net is seven or eight feet high, by nine or ten wide: it is composed of three nets, the middle one of which is the largest, and is usually made of silk or thread, but silk is the best. The two others are of packthread, and their meshes are square.

This net is sometimes gathered up from one knot to another,

about a foot in height, and sometimes stretched to its entire capacity of tension. Each compartment of this net is about two feet square; it is furnished at the top with rings of horn or iron which slide easily. For hoisting and adjusting the net there are two little cords, called *master-cords*, because they sustain the net by means of the rings. The net is usually set in the middle of a hedge; it is attached to two light poles of about nine or ten feet high, pointed and ironed at the thick end, and to the top of which there is a pulley to hoist and extend the net with greater facility: being once spread, it is fixed, towards the ground below, by the packthreads, which hang down, and which are almost two feet distant from each other. The middle net is then slack, and gathered in a heap; but they draw it with a stick, through the squares of the other, especially towards the centre, that the birds may be entangled more easily: in this part a sort of purse is formed at each square when the net is elevated.

In Switzerland, they use, for thrushes, nets of this description, about fifty feet long by fifteen high. There are several companies of fowlers, and each company has a dozen or fifteen of these nets, which are laid with two poles crossed, and planted perpendicularly in the ground, and by cordages to the edge of some lofty wood. Then the fowlers beat the bushes for about half a league, and force the thrushes to advance gently into the nets.

The net called *rafle* is used during the night. This net is counter-meshed, and usually twelve or fifteen feet wide, by ten high. The poles, which are attached on each side of it, must be very light, and about twelve or thirteen feet long. There is little difference, in general, between the formation of this net and the spider-net. The best nights for operation are the darkest; they are most advantageous when there is least wind: fog is even very favourable.

When the fowlers have discovered hedges which afford a shelter to thrushes and blackbirds during the night, they are certain to catch an abundance of them, provided they act with dexterity. Four persons are necessary to conduct this sport; one carries a lighted torch, two others hold the net, and a fourth, called in French the *traqueur*, incloses the bushes. He, who carries the torch, remains about twenty paces from the end of the hedge; when the net is spread, the traqueur commences at the extremity of the hedge opposite to the net, and the other two hold the net at a proportionate height. The most profound silence must be observed, and the torch must not be lit until they begin to beat the hedge. According to the positions, now described, of the fowlers, it is easy to perceive that the net is between the torch-bearer and the traqueur, and the birds between this last and the net. The birds, awakened by the noise, take wing, and direct their flight towards the torch, and consequently precipitate themselves into the net. It should not be lowered to take out the birds until the traqueur comes up. The net should always be placed as nearly as possible on the side on which the wind blows upon the hedges and bushes; for it is observed, that birds never sleep but with their heads with the wind. Autumn and Spring, when the thrushes and blackbirds are on their passage, are the proper periods for catching them in great quantities, because they then repose in large flocks, in the hedges sheltered from the wind.

Fowlers in France also make use of moveable huts *(huttes ambulantes)*, which are very convenient for killing numbers of thrushes during the vintage time. These birds never repose in the vineyards, but retire into the neighbouring woods and thickets; and generally rest once or twice on the most exposed trees. The hunters have each a hut, which they place near the tree which they judge most advantageous, and there each awaits his game, which he kills easily. It is remarked that the riper the grapes are, the more frequently the birds repose themselves: they appear, as it were, intoxicated; and every kind of snare succeeds in taking them at this time.

The *Song-Thrush* (*Turdus Musicus.*) This bird is well known among us, and is one of the commonest species in the wine-countries in France; its flesh is the most delicate of any. It frequents the vineyards when the grapes are ripe, disappears after the vintage, and makes it appearance again in March or April. All the birds of this species, however, do not migrate; they are sometimes seen in winter in our climates, but few in number. They approach habitations and sojourn in hedges; but as soon as the spring expands its genial influence, they retire into the woods, and announce the return of this delightful season by their varied song. Accordingly, both here and in many other countries, they are called song-thrushes, or some equivalent name. The male usually perches on the summit of some lofty tree, on a thick branch, and remains singing there for entire hours. It continues its notes from the early days of spring to the month of August and sometimes later; it is often heard with us as early as February. At other times these thrushes have only a little whistling note, which may be expressed by the syllables *zipp, zipp*. In flying away, they particularly utter this cry, which may be perfectly imitated by placing the end of the finger in the mouth, pressing it strongly with the lips, and drawing it quickly away. In this manner they are driven into snares, and attracted within reach of gun-shot.

This thrush makes its nest in bushes, and sometimes on a branch of a tree against the trunk, about ten or a dozen feet high: the exterior is composed of dry herbs and moss, and the interior of straws, cemented with clay and rotten wood. The eggs are five or six in number, of a pale blue, with a slight greenish cast, and some reddish and black spots. The male and female share the incubation. After the first brood is hatched, the latter recommences a second, and sometimes even a third, especially when the first has not thriven. Each brood goes separately, and the little ones disperse when they are strong enough to take care of themselves. These thrushes do not fly in flocks; still many are found together, or at no great distance from each other. The species is extended through all Europe, is fonder of woods than other places, especially of such as abound in maple trees. These thrushes possess no great degree of cunning, and suffer themselves easily to be taken with snares and bird-calls. When they cannot find fruits and berries, they subsist on snails, insects, and worms. This is the reason that they are found on the ground so frequently in the woods, and at the foot of hedges and bushes, especially those which border submerged meadows. When they are looked at, they manifest their displeasure by a gnashing of the bill.

To bring up this bird in a cage, it must be taken young, so that it will sing all the better. It is fed with a sort of paste, such as is made for nightingales, or it may be made with crumb of bread, rape-seed, or hemp-seed bruised, and meat cut small. This aliment is varied with grapes or other fruits of which the bird is fond. This thrush is susceptible of education, learns even to speak, and whistles very agreeably many airs of the bird-organ and flageolet. It will live in captivity generally from seven to eight years.

There are many varieties of this thrush, but all of them accidental. Among these may be remarked the white thrush, whose plumage, however, is not in general of a pure white. On some parts of the body spots of a feeble shade and undefined form are observable. In other individuals the plumes of the back are mixed with brown, and some red is observable on the breast. Sometimes the top of the head alone is white, and at others there is only seen a demi-collar of this hue.

The *Chochi*, or thrush of Paraguay, utters a singular sound towards the setting of the sun during the hatching season: it cries in a melancholy tone like the mewling of a cat, yet during the day, at the same epoch, its song is varied, frequent, and agreeable. It preludes with the syllables *chochi-chochi-toropi*, repeated three or four times, from which M. Vieillot has given it its name.

The chochi composes its nest of small and very flexible branches, furnished with slips of roots, and covered with an extremely thick coating of cow dung, mixed with sand.

The *Missel* (*Turdus Viscivorus*) is the largest of all the European thrushes. It is like many other birds that people our woods and orchards, partly migratory, and partly sedentary. In Lorraine, according to Dr. Lottinger, the missels quit the mountains at the approach of winter, always fly in flocks in spring and autumn, return in March, and nestle in the forests with which these mountains are covered. In Brie, according to Hebert, the correspondent of Buffon, they do not unite in flocks at any season of the year. If those two observers speak of the same species of thrush, it would appear that its habits are not the same in all countries. The greater number of the missels quit our northern climates on the approach of winter, but some remain. Those certainly do not live in flocks like the fieldfares, but in families. They pair in the month of January, and once coupled, each pair lives separately.

The missel is one of the first of our sedentary birds which announce the return of spring; for even so early as the fine days of February the male perches on the top of a very lofty tree, and puts forth a varied song, which, though remarkably loud, is not destitute of harmony. The female makes her nest even previously to the setting in of spring, and places it on large trees, but more generally on those of a middling height. She constructs it in the bifurcations of the principal branches, employs moss, leaves, and large weeds outside, cemented with earth, and carpets the nest with fine plants within, horsehair, and wool, and covers the exterior very artfully with moss like that which grows on the tree itself. She seldom lays more than four eggs, of obscure white, spotted with brown, and the male partakes the incubation. They feed the young ones with caterpillars, small worms, slugs, and snails, whose shells they break. A second brood is generally hatched after the first, and when both are ended the families unite, and add to the aliments just mentioned various kinds of berries, cherries, grapes, and other fruits. In winter they feed on flax-seed, hops, ivyberries, buckthorn, and particularly misletoe; from which our name of missel-thrush is given to them. In Burgundy they are called *Draines*, from a peculiar cry which they continually repeat, either as a rallying or a warning signal, and which has some fancied resemblance to this word. Montbeillard tells us that the missel-thrushes are very pacific in their manners; but Le Vaillant, with more appearance of truth, declares that his observation is without foundation. They are, in fact, of a quarrelsome nature, and often fight either for food or the choice of a companion. The males are more numerous than the females, and it is not rare to see two or three of them disputing so bitterly, that they forget their natural distrust, and suffer themselves to be approached very closely. The combat does not cease until the most feeble have abandoned both the object of their quarrel, and the district which she inhabits. Those which establish themselves in orchards prove very vigilant sentinels for our poultry, which they always warn of the approach of birds of prey. They seek to take under their protection all the little birds which nestle in the same quarter with themselves. If a kestril, a hawk, a crow, or a jay should appear in the neighbourhood, the male directly announces its presence by a cry of uneasiness; the female joins him, and on their united cries, repeated with every tone and accent of anger, an entire cohort of little birds, especially finches, join with them in pursuit of the common enemy, and succeed in terrifying him, and obliging him to take to flight before his feeble adversaries.

The missels are very distrustful, much more so than the blackbirds. It is very difficult to surprise them, except at hatching time; then they can be approached more easily: they are so much absorbed in the care of incubation, that they will allow themselves sometimes to be taken on the nest. They generally escape all kinds of snares, and can never be caught with the bird-call. They are sometimes observed to join with the finches in insulting the howlers, which daylight has surprised out of their retreat. The missel may be sometimes taken by the noose, but not so frequently as the song-thrush and the redwing. Their flesh is not so much in estimation as that of other thrushes, at least in our more northern climates, which is attributable to the sort of aliment on which they subsist. When they live on grapes, olives, and other succulent fruits, its flavour must be equal to that of the flesh of the others; but hips, flax-seed, and berries in general, which are deficient in nutritive qualities, impart to it a disagreeable taste, and cannot produce the delicate fat which renders the other thrushes so highly esteemed in some places as an article of game. These birds must be taken in the nest, when they are first covered with feathers, if they are meant to be tamed. Crumbs of bread steeped in water, and the yolks of eggs, constitute a proper food for them at this season; when they will eat of their own accord they may have worms, snails, berries of various kinds, and minced apples.

The *Fieldfare of Canada* (*T. Migratorius*) is a well-tempered and familiar bird. Its song is more varied and melodious than that of the missel, and has equal compass; its throat is more flexible; it is heard to utter the short interrupted cry of our blackbird, which it accompanies with a gnashing of the beak, a vertical motion of the tail, and slight tremor of the wings. It generally places its nest on trees of middling size, and composes it of small roots and dried herbs, bound together with a cement of clay. This nest perfectly resembles that of our song-thrush; the eggs are four or five in number, of a clear blue, varied with obscure spots.

The fieldfares come among us from the north of Europe, in November and December. They delight in fallow-lands, in places where flax-seed is found. Towards the end of winter they prefer humid meadows, and do not frequent woods, cept to pass the night there. During this entire season they live in society, travel together, and remain all the winter without separating, perch all on the same or the most neighbouring trees; it is not rare to see them assembled to the number of two or three thousand, in places where the lotus grows, the fruit of which they eat with avidity. The fieldfares also subsist on slugs and worms, which they are observed to pursue eagerly after rain in humid soils, or grounds newly ploughed. When these aliments are wanting, they eat misletoe, and various berries, among which are those of the whitethorn. They disappear in spring, but a few remain to the end of April. Then they are found in pairs, as this is the coupling time. The male is easily distinguished at this epoch from the female; the gray of his head and neck assumes a bluish tint, tolerably brilliant; the beak is of a fine yellow, and its extremity of a decided black. These couples may be sometimes observed, after a long winter, on the borders of thickets, far remote from habitations, but they are seen no longer when May sets in. Those fieldfares which are late go then to rejoin their companions, and pass the summer in the north, where they hatch the young. We can affirm nothing respecting the song of these birds, as we do not see them during the love season. The male and female with us utter the same cries, whether for warning or rallying. It is said that in Poland and Lower Austria, and Linnæus and Meyer add in Sweden, they nestle on high trees, and lay four or six eggs, of a sea-green, pointed with reddish-brown. M. Vieillot says they never nestle in our climates. This may be true of France, but Dr. Latham mentions an instance or two of the fieldfare's nest being found in this country. Their flesh is not so much esteemed as that of other thrushes; some say it acquires a good flavour when the birds feed on flax-seed, others that it is never better or more succulent than when they live on worms or insects. In general, however, it is insipid enough. The fieldfares may be taken by net, bird-call, or snares of any kind; shooting them is an easy sport.

There are many accidental varieties of this species, in which white predominates more or less.

The *Redwing* has been sometimes confounded with the song-thrush; but besides that its plumage is somewhat different, its habits and mode of life are analogous to those of the fieldfare. Like the latter, it only appears among us twice a year, unites in numerous flocks at certain hours of the day, to chirrup all together. The redwing has some conformity with the song-thrush in the delicacy of its flesh, and fondness for grapes, and they sometimes travel in company, especially in spring.

The redwing generally arrives after the song-thrush, and before the fieldfare, from the north. They are seen in considerable flocks in November, which usually disappear before Christmas. It re-appears towards spring, in the month of March, and is not seen after April. Its cry is *tan, tan, kan, kan.* In constantly repeating this cry it leads the fox, its natural enemy, to a considerable distance after it. It has been remarked that it does not sing in our climate, and has only a chirrup very analogous to that of the linnet; it is said, however, that in its native country its song is very agreeable in the spring season, especially when it perches on the summit of lofty trees. It makes its nest in the woods in the neighbourhood of Dantzic; it nestles also, according to Nozemann, in some parts of Holland, and chooses those which are covered with elder and service-trees, of the berries of which it is very fond. It has two broods every year, in the months of April, May and June: each consists of from four to six eggs, of a greenish-blue, and spotted with blackish. It nestles also in Sweden, and places its nest on the small shrubs and in the hedges. While the female hatches, the male hunts, and brings her her food. From the analogy between this bird and the song-thrush, it would seem probable that the male also partakes the care of incubation. Nozemann says that the male and female of this species swallow the excrement of the young while they remain in the nest. This habit is common

H. Koarsley del.

Mus. Lin. Soc.

THE PUNCTATED THRUSH.

T. PUNCTATUS.

London Published by Whittaker & C.º Ave Maria Lane. 1829.

to them with many other birds, but the excrements remain at the entrance of their œsophagus, and they eject them in some spot away from the nest, so as to remove all suspicion of the place where their young family is concealed. The usual aliment of these birds consists of the small worms, which they procure by scraping up the earth, of berries, of turnips, and caterpillars. When these are wanting, they have recourse to cherries, grapes, and other kinds of tender fruits. Then it is that their flesh acquires the delicacy which renders it in equal estimation with that of the song-thrush. They are not mistrustful, and are more easily ensnared than almost any bird. The fowlers of the continent say, however, that they will avoid any snares that are made only of black or white horsehairs. In Burgundy, therefore, they are made of white and black hairs twisted together. We are almost inclined to regard this as a vulgar prejudice.

Of the *Punctated Thrush*, of which we give a figure, from the Museum of the Linnæan Society, little is known as to habits and manners. It is a native of New Holland, and has been well described by Mr. Vigors and Dr. Horsefield, in the fifteenth volume of the Linnæan Transactions. The general colour of the plumage is brown, inclining to olive; breast ash-colour, and belly rufous-buff; a white streak over the eye, and chin and throat white; tail greatly wedged, and legs pale-yellow.

This species is the type of a new genus proposed by Mr. Vigors and Dr. Horsfield, under the name of *Cinclosoma*, of which these gentlemen observe: "The birds of this genus appear to belong to that subdivision of the thrushes, which, by the weaker conformation of the bill, opens a passage to the slender-billed warblers. They deviate very considerably from the typical form of the *merulidæ*. Besides the more gracile shape of the bill, the *nares* may be observed to be linear and longitudinal, instead of being rounded, as in the true *turdi*: the wings are short and rounded, the first quill-feathers being of moderate length, and the next gradually increasing; they thus differ from the wings of the *Turdus*, where the four quill-feathers succeeding the first are nearly of equal length, and the first almost spurious. The tail is long and graduated, which, in the true thrushes, is even; and the scales on the *acrotarsia* are strongly conspicuous, while the *tarsi* of the thrushes are entire.

The *Ava Thrush*, so named by Mr. Gray, is from Mr. Crawfurd's collection of Indian drawings. It may, probably, when better known, exhibit some deviations from the ordinary type of the genus Turdus, and is therefore referred to it conditionally. The bill is much bent toward the point, the top of the head and nape are bright brown ; the belly, vent, wing-coverts, and spots before and behind the eye, at the base of the lower mandible, and the chin, are yellowish-white.

We shall now speak of the division of the Blackbirds.

This name, *Merula*, is particularly given to the species whose plumage is uniform, or varied only in large masses.

The *Blackbird*, properly so called, is too well known to need description. Some naturalists distinguish the blackbirds generally from the thrushes by the vertical motion of the tail ; but we have already had occasion to see that this is found among some species of the latter.

The blackbird is solitary, living either alone or in company with its female. Though naturally wild, it is more easily tamed than the thrushes. It sojourns and nestles nearer inhabited places ; it is more distrustful and subtle, and is said to have a more piercing sight, which enables it to discover the fowler at a great distance ; it is therefore approached with much more difficulty.

The male has a powerful voice, but hardly supportable except in the woods, or champaign country. It commences its notes from the first fine days in the month of February, and continues to sing until the fine season is pretty well advanced ; it sings one of the longest of any of our birds. The love season begins early with the blackbird, and it is not rare to see young ones at the commencement of May.

This species has two or three broods every year ; it builds its nest in thick bushes, at a moderate height, or in the old trunks of headless trees, covered with ivy ; it is composed of moss, small roots, and dried herbs, bound together with clay, and the interior is furnished with the softest materials. The male and female work together at its construction with so much

Mus. Crawfurd.

By a Native Artist.

AVA THRUSH.

TURDUS AVENSIS, Gray.

London. Published by G. B. Whittaker May, 1828.

assiduity, that we are assured that eight days are sufficient for the finishing of the work. When it is finished, the female deposits in it from four to five eggs, of a bluish-green, with rusty-coloured spots, frequent, and not very distinct. She hatches them with so much ardour, that she sometimes suffers herself to be caught with the hand on the nest. The male provides for her subsistence, and, contrary to the supposition of Montbeillard, is observed to share sometimes the business of incubation. M. Vieillot has seen them on the nest from eleven in the morning, to two or three in the afternoon. Naturally distrustful, these birds often abandon their eggs, or eat them, if they happen to be touched, and they will even serve their young ones so in a similar case, when they are first ejected from the egg. The father and mother find them earth-worms, caterpillars, larvæ, and all kinds of insects. The moment these birds can do without the parent, they follow their natural impulse; each becomes isolated, and unites to its former aliment all kinds of berries and fruits.

These birds are sought after, and brought up in captivity for their song, and more especially for their power of improving it, of retaining the airs which they are taught, and imitating those which they hear. Those who are desirous of bringing them up should take them in the nest, when they are feathered, and feed them at first with a liquid paste, composed of steeped bread, yolk of egg, and bruised hempseed, and afterwards with sheep's-heart, minced meat, crumbs of bread, and different fruits and berries. They must not be shut up with other birds, for, naturally uneasy and petulant, they will pursue and torment them continually, unless in very large aviaries, filled with shrubs and bushes. In this way, indeed, they may have the pleasure of making their own nests, and bringing up their young, if they are provided with a sufficient quantity of the proper aliment. To succeed completely, it is necessary to abstain from approaching the brood while the little ones are not entirely fledged, for otherwise the old ones will either abandon or devour them.

The blackbirds are very fond of bathing themselves; they must, therefore, have plenty of water, which contributes not a little to their gaiety.

Their moulting commences at the end of summer, and is so complete, that some are frequently seen at that period with the head entirely divested of feathers. At this epoch they cease to sing, and, generally, near its termination they proceed to migrate. Some few, however, are observed to remain the winter: they then inhabit hedges and the thickest woods, seeking those where there are warm springs and evergreen trees, as much for a shelter from the cold, as for the purpose of procuring sustenance. They come at this season into gardens, and feed on snails; they even seek them in the holes of walls, and know very well how to break the shell and extract the animal. Their flesh is considered very delicate during the vintage time in wine countries, and is as much in request as that of thrushes; but it grows bitter when they feed only on juniper-berries, ivy-berries, and other such fruits. It is said to have some medicinal properties, and to be good in fluxes and dysenteries. Nevertheless, ulcerated and hemorrhoidal patients should abstain from it; the oil in which blackbirds are cooked is much recommended by foreign physicians, in cases of sciatica; and the dung of these birds, dissolved in vinegar, is said to clear the skin, and disperse redness and blotches, if constantly used.

Though these birds are very distrustful and subtle, they give easily into the snares that are laid for them, provided the fowler be invisible; they are taken in different ways. The methods described for taking the thrushes will succeed equally well with the blackbirds.

A method of taking them, well known to shepherds and the inhabitants of the country, consists in making a little hole in the ground, about five inches broad, eight long, and nine deep. In the bottom are placed various berries, or earth-worms, attached to a little stick with a thread, or transfixed through the body with long thorns. If other birds are wanted to be taken, grains and other aliments are cast into the bottom of the hole, especially those of which they eat in preference. They then take a piece of turf, a tile, or a stone of the size of the hole, and place them on a sort of figure of 4, so arranged on the hole that the bird cannot come to the bait without touching the stick, and making the coverlet fall, which shuts them up in the hole. To draw the blackbirds more effectually, a tame one is sometimes fixed at the side of the snare, either on a stick, or otherwise. This method succeeds well in winter, when the birds are pressed for food, and will go any where in search of it.

They use another mode of catching them in France, towards the close of the vintage season. They choose in the coppices, at no great distance from the vines, a straight and rather high shrub, which they lop down to about five feet; they pierce a hole in it at about four feet and a half of its length. This operation performed, they take another shrub at a distance from the first about four feet. They strip it of all its branches, and attach to the top a small packthread, about half a foot long; they tie to it a collar of horse-hair, formed in a knot. They then take the upper extremity of the last shrub, and bend it so that it advances almost to the other, and they pass the collar into the opening made in the first shrub, drawing it as far as the knot of the packthread, which comes to the level of the hole. They have besides a small stick, about four fingers long, formed on one end into a small hook, and rounded towards the other, which terminates in a point. They insert it a little into the small space which remains from the knot to the edge of the aperture in the shrub, and keep it there rather slack; after which they stretch the collar above, which they open into a circle, and rest flatly on the trap of the little stick. The snare is then laid: they place above, by way of a bait, a cluster of grapes, or some berries, of which the blackbirds are very fond. As soon as they perceive this they come to peck, and perching on the stick it gives way, the bent shrub resumes its former position, and the bird is seized in the noose.

Nothing so opposite as white and black; yet we see the first colour pass abruptly into the second, without going through the intermediate shades. Blackbirds, crows, and other birds of the same hue, present examples of this every day. Among the accidental varieties of this species, we find some completely white, including even the bill and feet. Some have these parts yellow, others have the bill red. Individuals have been observed, whose entire plumage was of a yellowish-rose colour, with the bill and feet yellow. On some specimens the head alone is white, with three oblong black spots placed behind the eyes; the iris, the beak, and the feet are yellow. Others are varied with black and white, in transversal spots on the upper parts, and longitudinal underneath; some have the wings and tail only as white as snow: all the rest of the plumage is a fine black. Finally, young ones are sometimes seen which have the alar and caudal quills white from their origin, and for half their length.

The *Ring-Ouzel* (*T. Torquatus*) is decidedly a different species from the last. To say nothing of the plumage, &c., its habits and manners are different; its usual cry is *cr, cr, cr.* In spring its song is less loud than that of the common blackbird, and varied with sweet and melodious sounds. It is a bird of passage with us, and is never seen but in spring and autumn. It does not always pursue in its migrations a regular route; it usually follows the chains of mountains, and particularly seeks hedges, where ivy is abundant, of the berries of which it is especially fond. It is seen regularly enough in the months of April and October, on the mountains in the neighbourhood of Rouen. It sometimes remains there during the entire summer, but very rarely.

These blackbirds appear to travel in families only, for seldom more than eight or twelve are seen together. They do not quit the hedges, and prefer those which are on the summit of mountains, and on the borders of woods. In both seasons, their passage does not continue for more than from fifteen to twenty days; for all this time they are excessively fat, and their flesh is very delicate eating.

These birds have this peculiarity, that they are as fat in Spring as in Autumn, while the reverse is altogether the case with the other blackbirds and thrushes, and indeed with all other small birds, which are very fat in Autumn, and quite lean in Spring.

Less distrustful than the common blackbirds, the Ouzels suffer themselves to be approached without difficulty. It is said, however, that they are not very easily caught in snares. Still it would appear that they might be taken without much trouble in the spider-nets that we have described; as whenever they are pursued they stick constantly to the hedges, preferring those which are in a right line, and quitting one only, cast themselves into the succeeding.

This species is common in all the high mountains of England and Scotland, of Sweden, Auvergne, Savoy, Switzerland, and Greece. It also inhabits the mountain chain of the Vosges, where it nestles on the fir-trees. It also places its nest at times, at a small distance from the ground, either on a rock covered with bushes and large briars, or at the foot of a very thick bush; branches, roots of heath, and moss heaped together without order form the basis of the nest, the outside of which is furnished with thick weeds, and the inside with clay mixed with filaments of roots and dried leaves: fine and soft plants form the bed, on which the female lays four eggs, of the same size and colour with those of the common blackbird, but very remarkable for the large reddish spots with which they are marked.

Lothinger, who has had occasion to study these birds in Lorraine, assures us, that they nestle very early in the season, and construct and place their nest pretty nearly like the song-thrush; that the young are perfectly capable of providing for themselves by the end of June; that the period of their departure is not fixed; but that they generally commence their migration towards the end of July, and that it continues during the whole month of August, for which time not one of these birds is ever seen in the plain. Lothinger adds, that, though formerly very common, they are now rare in the Vosges.

Montbeillard refers to the ring-ouzel, the white blackbird of which Aristotle and Belon speak. It is certain that this race, which is never found except on the very high mountains of Arcadia, Savoy, Auvergne, Silesia, on the Alps and Apennines, appertains to the species under consideration, both by this peculiar instinct, and by a general mode of life which removes it from the common blackbirds: but still we have seen that, among the latter, accidental varieties occur totally white, and in both species individuals are seen more or less varied with this colour.

The *rock-thrushes* (as their name indicates) are inhabitants of the rocks and mountains, and must be sought for in the wildest and most solitary retreats: continually on their guard, they do not hesitate to stand in exposed places. They are frequently seen at some distance from their haunts, perched on large stones; but they are very difficult to approach, and very rarely stop within range of gun-shot. When they are advanced upon a little too much, they are off to another stone, and always choose one where they can have a full, commanding view of all that surrounds them.

These birds are not a bad eatable, but they are still more in estimation for their voice, which is sweet and varied, approaching the tones of the black-headed warbler. Their throat is so flexible, that they quickly appropriate the song of other birds, and the airs of music. A little before sunrise, and at sunset, they utter the loudest sounds. During the day their song amounts to little more than chirping; but in the middle of the night, if their cage be approached with a light, they begin to sing directly.

The extreme distrustfulness of these birds naturally leads them to choose the most inaccessible places for the security of their young family. They make their nests in the holes of rocks and attach them also to the roofs of caverns. It is not without much risk and labour that their young brood can be got at; and even when the robber arrives at the place, a sure danger awaits him of having his eyes plucked out by these birds, which are not less courageous than distrustful, and will defend their young with desperate obstinacy.

The eggs are four or five in number of a greenish blue.

The young rock-thrushes may be brought up with the same sort of paste used for the nightingale; but they must be taken in the nest, "for," says Montbeillard, "when they have the use of their wings, they will not give in to snares of any kind." He adds, that even if they should be so taken, they will not survive their liberty. M. Vieillot, however, saw one taken on its passage in the neighbourhood of Paris, which swallowed with great avidity all the food presented to it, especially meat, and even took it out of the hand. After three or four days of captivity, it was already as familiar as if it had been always brought up in a cage.

This bird has a very quick motion of the tail, moving it up and down, five or six times successively, especially when it changes place.

The rock-thrushes are found on the Oural Mountains, on the Alps, in the Tyrol, Bugey, Switzerland, Austria, Prussia, and Carniola; but, being migratory birds, they only appear in these places in May, and quit them in September: then extend themselves in Spain, Italy, and the islands of the Archipelago.

The *rose-coloured Blackbird* pleases the eye by the beauty and brilliancy of its plumage, but it also possesses other qualities far more valuable. It is a great destroyer of grasshoppers, locusts, &c. of which it devours an incredible number every day in the various parts of the East. It was regarded by the ancients, who called it *Seleucida*, as a favour of the Gods, when these scourges, more destructive to the productions of the earth than hail and tempest, devastated the country. Even at present, the Arabs, the Indians, and the inhabitants of Aleppo are accustomed by superstitious practices to invoke this bird, which they call the *Samarmar*, to come to the succour of the crops, which are attacked by myriads of locusts. The Turks esteem it a sacred bird, and will not suffer it to be killed in their presence. It would be well if their example was more generally imitated with respect to all birds that render similar services to mankind.

The rose-blackbird has some analogy of habits and disposition with the stare. Like the latter, it is fond of herds and flocks, and will perch upon the animals for the purpose, no doubt, of searching for the insects which lodge in their hair and skin. This species too, like the stare, flies in large flocks, and makes its nest in the holes of rocks; besides locusts, it feeds on various other insects, especially such as live in dunghills. It also eats berries and tender fruits.

It would appear that this blackbird has no song, at least ornithologists and travellers make no mention of it. According to Forskel, its cry is heard at a great distance, and may be expressed thus: *tr, tr, tr.*

This species appears spread through the hottest and coldest parts of the old Continent. Forskel has seen it on the burning sands of Arabia, and in the plains of Aleppo, in July and August. Le Vaillant has met with it in Africa, as high as 24° south latitude. It has been sent into this country from Bengal. Pallas has found it in the north of Siberia, in the mountainous vicinity of the Irtish, where it nestles. Very numerous flocks of these birds traversed Provence and Piedmont, in the autumn of 1817. They are found in the mountains of Lapland, are common on the shores of the Caspian, near Astracan, and along the entire extent of the Volga. They pass every year in large flocks into the southern part of Russia.

The rose-coloured blackbirds, which are seen on the Continent, come only during the passage time of other birds; at this period many are observed in Burgundy. Klein assures us, that they have a name in Spanish, which indicates that they are known in Spain. Aldrovandus, the first naturalist who has mentioned these birds, informs us, that they sometimes appear in the plains of Bologna, where the fowlers call them sea-starlings. They perch on dunghills, grow very fat, and their flesh is good eating. They have been sometimes seen in this country.

The *mocking-thrush*, properly so called, derives its name from the peculiar talent which it possesses of imitating the cries and a part of the song of other birds; but it does not give a caricatured imitation of those foreign sounds its denomination would appear to indicate; on the contrary, if it imitates it is only to embellish. The cries and half-phrases with which it enriches its own naturally varied song, have occasioned the aborigines of Mexico to give it a name far more appropriate and more justly applicable, that of *Cencontlatolli*, which means *four hundred languages*.

This bird not only sings with taste, and without monotony, but also with action and animation. It is, perhaps, one of the first of singing birds; but to place it above the nightingale, with Fernandez, Nieremberg, and others, can only be done by those who have never heard, or who have entirely forgotten the song of that delightful bird. The voice of the mocking-thrush is more loud and powerful, but by no means so agreeable within a certain distance. Its song has little of the softness, delicacy, and plaintive tenderness that so peculiarly characterize the nightingale during the season of love.

As there is no bird among the Americans at all to be compared to the mocking-bird, it is not astonishing that they should have exalted it into so extraordinary a character, and raised it above all other birds. They have, however, exaggerated its talents, in stating that it can imitate completely, and in all their parts, the song of other birds, the cries of different quadrupeds, the crying of infants, the laughter of a young girl, and in being able to repeat entire airs on the same key in which it has heard them. It does not possess the imitative talent to this degree, even in captivity. The mewing of the cat, however, it takes off so completely as to deceive any ear.

This bird is very common in Saint Domingo, where it is called the nightingale; but there it possesses none of those qualities so much vaunted in North America. Its song, however, is the same. It frequents the savannahs, delights to be near habitations, and seems to love the society of man, the sight of whom is alone sufficient to excite it to sing.

This bird moves the tail up and down, and often carries it in a raised position : at such times its wings are pendant.

Bold and courageous, the mocking-thrush is frequently at war with the *pipiris*, and forces the little birds of prey to quit the places which it has adopted for its own abode, especially during the hatching-time.

It places its nest on trees of middle size, or in thick bushes, gives it a similar form to that of the missel, and furnishes the base without with thorny branches. It lays four or five eggs, spotted with red points on a white ground, which points are larger towards the thick end than elsewhere.

It feeds on insects and different berries. It is brought up in cages, but to preserve it, it must be taken in the nest, and its tastes and wants be carefully studied and administered to. When this is done, it will continue to sing many years.

It is about the size of the redwing, and the female is of the same dimensions with the male.

We pass on to the Loriots or Orioles.

The *Oriole, properly so called, (Oriolus galbula,)* and *golden Oriole* of Latham, comes into France about the middle of spring, and quits in autumn to pass the winter in Africa. It migrates at uncertain periods into England and Sweden. On their arrival, the male and female soon couple, and place their nest at the extremity of the branches of very elevated trees.

This nest is constructed with much art and industry : it is attached to the bifurcation of two small branches; the birds enlace around the two branches, which form this bifurcation, long threads of straw, flax, or wool, some of which going right from one branch to another, form the edge of the nest in front, and the others penetrating into the tissue of the nest, or passing underneath and rolling over the opposite branch, give solidity to the work. Between the exterior and interior, there are moss, lichens, and other similar matters. The interior is furnished with wool, spiders'-webs, the silky nests of caterpillars, and feathers, the whole united and tissued most intimately and ingeniously together.

The eggs are four or five in number, of a dirty white, sprinkled with little spots of a blackish-brown, and more numerous towards the thick end. Incubation lasts about one-and-twenty days.

The female has great attachment for the young family, and shows considerable courage in defending it even against man. Montbeillard says, that the father and mother have been seen to dart courageously on those who were carrying off their young; and, what is still more rare, the mother has been known, when taken with the nest, to continue hatching in the cage, and die upon the eggs.

These young birds are a long time before they can provide for themselves; and follow the father and mother a long time before they can eat alone, with the cry of *yo, yo, yo*. Each family assembles together to migrate.

The song of the oriole is tolerably well known, and has given rise to the different names imposed upon the bird, according as the hearers have thought proper to express it, or as they believed that they heard it. Some believe that it always cries *Yo, yo, yo*, syllables which are always preceded or followed by a sort of mewing, like that of a cat. Others that it pronounces *Oriot* or *Loriot*. The absurd fancies of the French have carried them pretty far in this point. Some imagine that the bird cries *compère loriot* (gossip loriot); many that it cries *Louisat bonnes merises* (Louisat, good black cherries); and others have arrived at the very climax of absurdity, in thinking that it articulates "*c'est le compère loriot qui mange les cérises et laisse le noyau.*"

On their first arrival the orioles live on insects, scarabæi, little worms and caterpillars. It is with such food that they bring up their young. They make at this epoch a considerable consumption of these insects, especially of the latter. They bring their young ones as many as the bill will contain. Thus these patient birds clean a multitude of trees of these insects, and return every day upon the same trees until none remain before they proceed to others; still, however, they appear more greedy of berries, figs, red and black cherries, of which they only attack the ripest part. They are not, however, sufficiently numerous to render the mischief which they do in cherry-tree plantations, &c. a counterbalance to the services which they perform in ridding the trees of the quantity of caterpillars which devour the leaves. Their flesh becomes very fat when they subsist on figs, and is then excellent eating; accordingly they are much pursued in the islands of the Archipelago and in Egypt, on their passage at the end of summer. It is quite different, however, on their spring passage. At this epoch their flesh is excessively lean, and they remain in this state until their nutriment grows more abundant.

The oriole is not easily reared in captivity: this, however, may be achieved, and even the old ones taken with the young may be preserved for some time, if they receive plenty of those fruits of which they are particularly fond. As to the young taken from the nest, they are fed at first with the same paste which is given to nightingales, and afterwards with fruits. These birds seldom live more than two years in captivity; they most generally perish, from a species of gout which attacks them in the feet.

The oriole is extremely distrustful, and very difficult of approach. Precaution must be used when it is intended to shoot them, as they fly from tree to tree for a long time, without suffering themselves to remain to be aimed at. They can be attracted by whistling like them, but it must be well done, and exactly like their voice, as, otherwise, they will fly off immediately. In the fruit season they may be caught with various kinds of snares.

All that we have said of the habits of this oriole is applicable to the other species of the genus as far as they are known. We forbear, therefore, to dilate further on them, and proceed to the Ant-eaters.

Sonnini was the first naturalist who made us acquainted with these birds. He has observed them in the interior of the countries of Guiana, in the lofty and sombre forests which cover the soil in this portion of Southern America. They live there, generally speaking, in small flocks, and subsist chiefly on ants, the quantity of which is prodigious in those hot and humid climates. There, where man has been hitherto unable to exercise his destructive imprudence, we may observe the admirable care with which nature has disposed all her works, the harmony of their distribution, the equilibrium which maintains them in a perfect order, the incontestable imprint of a supreme and directing intelligence. In no part of the globe does there exist a greater number of ants than in South America; and in no part, also, do there exist more species of animals destined to subsist on these insects. For some of these species they are not only a preferable article of food, but absolutely a necessary and exclusive aliment. The quadrupeds called ant-eaters have no other, neither have the birds on which we treat at present.

Such a mode of subsistence does not require the frequent exercise of flying. To find it, it is sufficient to flit from one ant-hole to another. Accordingly we find these birds almost continually on the ground. They run there with lightness, and if they ever quit it, it is only to jump upon the bushes or branches of some low tree, where they pass the night. They build their nests there, tissued with dry plants rudely interlaced, and of a hemispherical form. They lay three or four eggs, nearly round. The structure of those parts which serve for the mechanism of flight, in these birds correspond to their mode of life. The wings and tail are extremely short, and consequently little adapted to raise or support them in the air; but their legs are long and well adapted for running, which is all that is necessary for their purposes.

These birds are lively and agile; they are almost always in motion, but invariably at a distance from all inhabited places, where they would not meet with a sufficiency of those insects which constitute their subsistence. Their disposition is social. They not only unite in small troops of the same species, but also join with other birds of different species, but of their own genus. Their plumage, not brilliant, seems, in fact, to indicate this mixture; for, with the exception of the larger species, which are better characterised, it is rare to meet among the small, two individuals which resemble each other perfectly. This is the observation of M. Vieillot, and surely ornithologists would do well to consider the great probability of similar intermixture between birds of other genera, whose size, conformation, &c. is so much alike, before they proceed so rashly, and on such trivial grounds, to the separation of species.

The flesh of the ant-eaters contracts a strong odour of their ordinary food, which renders it disagreeable. They are called in the colony of Guiana *Little Partridges*, and the aborigines of the country term them *Palikours*.

One of the species (*Myrmothera Tinnica*) has a peculiar habit worth remarking here. In the mountainous and wooded deserts of Guiana, where the *Arada* disturbs the traveller by its shrill and repeated whistle, like a bandit calling his companions of plunder, this bird gives the alarm, and appears perpetually on its guard amid the dangers which surround it. It causes the forests and the mountains to re-echo with sounds, grave, yet at the same time sonorous and rapid, like that of a bell repeatedly and quickly rung. M. Vieillot, who resided in Guiana, was some time before he could imagine what animal

C.Hamilton Smith Esq. del.
Paris Mus.

Griffith sc.

RED-BELLIED ANT-EATER.

PITTA ERYTHROGASTER.

London Published by G.B.Whittaker March 1828.

C.Hamilton Smith Esq. del
Paris Mus.

Griffith sc.

VIEILLOT'S GRALLINA, *of New Holland.*

GRALLINA MELANOLEUCA.

London Published by G.B. Whittaker March 1825.

produced this singular noise, which he heard every morning and evening around him. He little thought that this living tocsin was a small bird, which he was in the habit of constantly meeting in these immense solitudes, and which furnished one of the ordinary dishes of his table. He was the first who made known this species to Buffon, who preserved the name given it by M. Vieillot, viz., *Béfroi.*

Of the habits of the *red-bellied ant-eater* we have no information whatever, and its specific characters are sufficiently noticed in the text.

Our figure of the *grallina* is from the Paris Museum, where the specimen has been treated by M. Vieillot as a new genus from Australia. It is entirely black and white, but of its habits and manners there is nothing known.

The Cincle, in consequence of its peculiar habits, has been classed among the grallæ, in the genus tringa, but its conformation proves it to belong to this division. It is a solitary and silent bird, remaining constantly near fountains and limpid streams, whose waters roll over gravel beds in lofty mountains. It is found in Spain, Sardinia, France, and even to the most northern parts of Europe, where it remains all the winter. Sometimes it walks slowly, sometimes it is seen resting on the pebbles, between which the rivulets wind. When it flies, it is in a right line, shaving the ground closely, and uttering a little cry like the king-fisher. Aquatic insects constituting its chief nutriment, it proceeds to seek them even in the bed of the stream, following its declivity, and continuing its progress even when the depth of the water forces it to submerge. It traverses the bottom with the head upright, without appearing to have changed its element. It walks there in all directions with the same facility as on land, only M. Hebert has remarked, that the moment the water passed its knees, it suffered its wings to fall, agitating them a little. The object of this movement may, perhaps, be for the purpose of causing a stratum of air to penetrate the water, and surround it when there. This process has, in all probability, some relation with that of the hydrophilous, and o her aquatic insects, which are always observed to be in the middle of a bubble of air. If this fact can serve to explain the cincle's mode of respiration under water, it cannot explain the cause of its feathers being impermeable by water; but, independently of their thickness, they are provided with a fatty substance, like those of ducks. On plunging one of these birds into a vessel full of water, it was observed that the water fell back in globules, without wetting the feathers.

The cincle is never met with its female but in the season of reproduction, at which season they construct their nest on the ground, often near mill-wheels, with blades of grass, small dry roots, and dead leaves. It is covered with a vaulted dome, and its aperture is furnished with moss. The female lays four or five whitish eggs, an inch long, and six lines in diameter at the thick end.

Of the genus Philedon, the species are very numerous, and appertain to Australia; but nothing sufficiently interesting is known concerning their habits to merit insertion here.

We insert, under the name of *Corniculated Philedon*, a figure from Major Hamilton Smith's collection. The base of the bill, and the greater part of the head, are naked and corneous, of a bright blue colour; the remaining upper parts of the bird are light ashy-brown, and all beneath is white.

The Grakles (*Martins, Fr. gracula*) have been very much mixed-by different authors in various genera. We shall here consider only those which our author has designated Martins.

These birds, all of which appertain to the old Continent, have the manners of the stares, and live like them in large flocks. M. Le Vaillant observes, in a great portion of France, Germany, and Holland, the people are in the habit of applying this name (*Martins*) to the stares which are brought up in cages, as they do that of *Margot* to the pies, and that of *Jacquot* to the parroquets; and he concludes, that if they give the name *Martin* in India to birds which have the habits of the stares, it is most likely to have been introduced by the first Europeans who visited

C. Hamilton Smith Esq.r del.t

Mus. Bonn

CORNICULATED PHILEDON.

PHILEDON CORNICULATA.

London. Published by G.B. Whittaker, Oct. 1828.

these countries. These birds assemble on dunghills, and such other places, where they find either the larvæ of insects, or perfect insects, especially locusts. They also perch on the backs of cattle, to feed on the parasitic insects which infest them. In default of insects, they attack seeds and fruits.

The Common Martin (*Paradise Grakle* of Lath., *Cassyphus Tristis*, Dum., *Paradisea Tristis*, and *Gracula Tristis*, Gm. and Lath.,) is the species whose manners have been most studied. Besides hunting flies, scarabæi, &c., it seeks the vermin from the backs of horses, oxen, and pigs, which willingly submit to the operations of their liberators, until they begin to infringe upon the skin; then these carnivorous birds, which accommodate themselves to all kinds of nutriment, will commence to peck the living flesh.

The discharge of fowling-pieces will scarcely drive away the martins, which assemble at the close of day on the trees which are near habitations, and chatter there in a very troublesome manner, though their song is naturally sufficiently varied and agreeable. In the morning they disperse through the country in groups, or by pairs, according to the season. They have two young broods every year, usually composed of four eggs, in nests of a rude construction, which they attach to the leaves of the palm tree, or other trees, and which they even sometimes place in granaries, when they can find the means. Their attachment for their young is so great, that they will pursue their ravisher, striking with the beak, and uttering piercing cries. If they should discover the place where their young ones are situated, they will enter there for the purpose of feeding them.

The young martins are tamed without difficulty; they are easily taught to speak, and when kept in a barn-yard, learn of themselves to counterfeit the cries of hens, cocks, geese, sheep, and other domestic animals. They even accompany their imitations with accents and motions full of grace and gaiety, and which contrast not a little with the epithet *tristis*, so unaccountably bestowed upon them. It cannot even be derived from their plumage, the varied tints of which have nothing sad or sombre in their appearance.

These birds, very numerous in India, the Philippines, and probably in the intermediate countries, are of a very gluttonous disposition, and great destroyers of locusts. This last circumstance has rendered them celebrated in the island of Bourbon, to which they were for a long time strangers, but where the governor Poivre caused many pairs to be transported to oppose the locusts, which were desolating the island, into which their eggs had been introduced with plants from Madagascar. The views of this excellent statesman were, in the first instance, crowned with complete success, but as the colonists perceived after a few years that the martins tore up with avidity the grounds which had been newly sown, they imagined it was for the purpose of eating the grain, so, after a formal process, they had them all destroyed. The locusts soon reappeared when their enemies were thus put "hors de combat," and causing fresh devastations, the people began to regret the martins, two pairs of which were introduced eight years after, and placed under the protection of the laws. A fresh destruction of these insects was the result of this second introduction of the martins. But this nutriment beginning to fail, these birds attacked an insect, the larvæ of which made continual war with the cotton-tree grubs, so very injurious to the coffee-plants. They also proceeded to devour the fruits and grains. They even killed the young pigeons in the dove-cots, and became in their turn a scourge, which required the adoption of measures to prevent the too great multiplication of their species.

The *Gracula Cristatella* of China, which the Baron scarcely regards as a variety of the last, is said to learn to whistle tunes remarkably well, and articulate words. The Chinese rear them in cages, with rice and insects.

There is another bird of this division, which has been made the type of a separate genus by M. Kuhl, under the name of *Ptilonorhynchus*. It is the *Satin Grakle* of Dr. Latham.